Advanced Information and Knowledge Processing

For other titles published in this series, go to
www.springer.com/series/4738

Advanced Information and Knowledge Processing

For other titles published in this series, go to www.springer.com/series/

Honghai Liu • Dongbing Gu • Robert J. Howlett •
Yonghuai Liu
Editors

Robot Intelligence

An Advanced Knowledge Processing Approach

Editors
Dr. Honghai Liu
University of Portsmouth
Institute of Industrial Research
PO1 3QL Portsmouth
UK
honghai.liu@port.ac.uk

Prof. Dr. Dongbing Gu
University of Essex
Department of Computer Sc
Wivenhoe Park
CO4 3SQ Colchester
UK
dgu@essex.ac.uk

Dr. Robert J. Howlett
University Brighton
School of Engineering
Intelligent Signal Processing
Laboratories (ISP)
Moulsecoomb
BN2 4GJ Brighton
UK
rjhowlett@kesinternational.org

Prof. Dr. Yonghuai Liu
Aberystwyth University
Department of Computer Science
Ceredigion
SY23 3DB Aberystwyth
UK

AI&KP ISSN 1610-3947
ISBN 978-1-4471-2582-2 ISBN 978-1-84996-329-9 (eBook)
DOI 10.1007/978-1-84996-329-9
Springer London Dordrecht Heidelberg New York

British Library Cataloguing in Publication Data
A catalogue record for this book is available from the British Library

Printed on acid-free paper

Springer is part of Springer Science+Business Media (www.springer.com)

Preface

With the growing integration of machine learning techniques into robotics research, there is a need to address this trend in the context of robot intelligence. The multidisciplinary nature of robot intelligence provides a realistic platform for robotics researchers to apply machine learning techniques. One of the principal purposes of this book is to promote idea exchanges and interactions between different communities, which are beneficial and bringing fruitful solutions. Especially when the tasks robots are programmed to achieve become more and more complex, imprecise perception of the environments renders a difficult deliberative control strategy applied for robots for so many years. Understanding the environment where robots operate and then controlling robots gradually rely on machine learning techniques. It is more likely to better off with embedding control problems into the environment perception.

The major challenges for programming autonomous robots stem mainly from firstly the dynamic environment in which it is unable to predict when events will occur and the robots have to perceive their environment repeatedly, secondly uncertain sensory information that is inaccurate, noisy, or faulty, thirdly imperfect actuators that cannot guarantee perfect execution of actions due to mechanical, electrical, and servo problems, and finally limited time that constrains time intervals needed for sensor information processing, actuator control, and goal-oriented planning. As such, the robots cannot rely on their actions to predict motion results. Heavy computation would make the robots move and respond slowly to changes in the environment.

For autonomous mobile robots, early programming approaches followed a sequence: sensing the environment, planning trajectories, and controlling motors to move. With this kind of control strategies, the robot needs to "think" hard, consuming large amounts of time to model the environment and reason about what to do. In addition, modelling and reasoning methods vary with robot tasks and have not reached a widely accepted level of development. Furthermore, this type of control strategies is very fragile, as it can fail to deal with unpredictable events in dynamic environments even if the robot can model and reason precisely. Meanwhile, it is impossible to predict all the potential situations robots may encounter and to specify all the robot behaviors optimally in advance when programming them to achieve complicated tasks in complex environments. Thus, robots have to learn from and adapt to their operating environments.

This volume aims to reflect the latest progresses made on central robotics issues, including robot navigation, human security and surveillance, human-robot interaction, flocking robots, multiple robot cooperation and coordination. The collected chapters not only represent the state-of-the-art research in robot development and investigation, but also demonstrate the application of a wide range of machine learning techniques that vary from artificial neural networks, evolutionary algorithms, fuzzy logic, reinforcement learning, k-means clustering, to multi-agent reinforcing learning. The book can be used as a valuable reference for robotics researchers, engineers, and practitioners for advanced knowledge, and university undergraduates and postgraduates who would like to specialize in robotics research and development.

Thirteen chapters are carefully selected from the extensive body of recent research work, which tackles the challenging issues of robotics development and applications with machine learning techniques. The selection is featured with the breadth of machine learning tools and emphasizes practical robot applications.

Skoglund et al. present a novel approach to robot skill acquisition from human demonstration. Usually the morphology of a robot manipulator is very different from that of the human arm. In this case, a human motion cannot be simply copied. The proposed approach uses a motion planner that operates in an object-related world-frame called hand-state to simplify a skill reconstruction and preserve the essential parts of the skill. In this way, the robot is able to generalize the learned skills to other similar skills without triggering a new learning process.

Palm et al. focus on the robot grasp recognition, which is a major part of the approach for Programming-by-Demonstration. Their work describes three different methods for grasp recognition for a human hand. The finger joint angle trajectories of human grasps are modeled by fuzzy modeling. Three methods for grasp recognition are compared with each other.

Cheng et al. investigate the multiple manipulators which need to achieve the same joint configuration to fulfill certain coordination tasks. Under the multi-agent framework, a robust adaptive control approach is proposed to deal with this consensus problem. Uncertainties and external disturbances in the robot's dynamics are considered, which is more practical in real-world applications. Due to the approximation ability of neural networks, the uncertain dynamics are compensated by the adaptive neural network scheme.

Ji et al. propose an exemplar-based view-invariant human action recognition framework to recognize the human actions from any arbitrary viewpoint image sequence. The proposed framework is evaluated in a public dataset and the results show that it not only reduces computational complexity, but it is also able to accurately recognize human actions using single cameras.

Khoury and Liu introduce the concept of fuzzy Gaussian inference as a novel way to build fuzzy membership functions that map underlying human motions to hidden probability distributions. This method is now combined with a genetic programming fuzzy rule based system in order to classify boxing moves from natural human motion capture data.

Zhou et al. consider the detection of hazards within the ground plane immediately in front of a moving pedestrian. Using epipolar constraints between two views, detected features are matched to compute the camera motion and reconstruct the 3-D geometry. For a less feature based scene a new disparity velocity based obstacle detection scheme is presented.

Tian and Tang explore the feasibility of using monocular vision for robot navigation. The path depth is learned by using the images captured in a single camera. Their work concentrates on finding passable regions from a single still color image and making the robot vision less sensitive to illumination changes.

Liu et al. propose a new model to characterize camera distortion in the process of the camera calibration. This model attempts to blindly characterize the overall camera distortion without taking the specific radial, decentering, or thin prism distortion into account. To estimate the parameters of interest, the well-known Levernburg-Marquardt algorithm is applied. To initialize the Levernburg-Marquardt algorithm, the results from the classical Tsai algorithm are estimated. After both the camera intrinsic and distortion parameters have been estimated, the distorted image points are corrected using again the Levernburg-Marquardt algorithm.

Wang and Gu present an approach to design a flocking algorithm by using fuzzy logic. The design of three basic behaviors in a flocking algorithm is discussed. They are alignment behavior, separation behavior, and cohesion behavior. Navigation control component is used in the design of cohesion behavior. To avoid becoming crowding or collision, an adaptive navigation gain is used. This gain changes with the number of neighbors. The flocking stability is analyzed and stability conditions are acquired from the stability analysis.

Oyekan et al. develop a behavior based control architecture for UAV surveillance mission. This architecture contains two layers: atomic action layer and behavior layer. They have also developed six atomic actions and ten behaviors for these layers. Various techniques have been used in the development, including adaptive PID controller, fuzzy logic controller, SURF algorithm, and Kalman filter.

Guo et al. present a novel anti-disturbance control strategy named hierarchical composite anti-disturbance control for a class of non-linear robotic systems with multiple disturbances. The strategy is established which includes a disturbance observer based controller and an H_∞ controller, stability analysis for two case studies are provided.

Ballantyne et al. present some of the key considerations for human guided navigation in the context of dynamic and complex indoor environments. Solutions and issues related to gesture recognition, multi-cue integration, tracking, target pursuing, scene association and navigation planning are discussed.

Kubota and Nishida discuss the adaptation of perceptual modules of a partner robot based on classification and prediction through actual interactions with a human. They proposed a prediction-based perceptual system consisting of the input layer, clustering layer, prediction layer, and perceptual module selection layer. They apply the proposed method to the actual interaction between a human and a human-like partner robot.

We would like to express our sincere thanks to all the authors who have contributed to the book and support during the book preparation. Without their support, it is impossible to see the advent of this book. Thanks also go to Natasha Harding from Springer UK who kindly and effectively communicated between the publisher and our editors of this book. We feel especially grateful to our publisher, Springer, who kindly supports the research direction of robot intelligence and the publication of the book. Finally, it would be our pleasure that this book would be valuable, for in-depth understanding of robot intelligence from the advanced knowledge processing point of view, to a wide range of audience from multi-disciplinary research communities and industrial practitioners.

Portsmouth, UK Honghai Liu
Colchester, UK Dongbing Gu
Brighton, UK Robert J. Howlett
Aberystwyth, UK Yonghuai Liu

Contents

Contributors

Ala Al-Obaidi Smart Light Devices, Ltd., Aberdeen AB24 2YN, UK, ala@sldltd.com

James Ballantyne Institute of Biomedical Engineering, Imperial College of London, London, UK, james.ballantyne@imperial.ac.uk

Long Cheng Key Laboratory of Complex Sciences, Institute of Automation, Chinese Academy of Sciences, Beijing 100190, China, chenglong@compsys.ia.ac.cn

Patrick R. Green Heriot-Watt University, Edinburgh, UK, P.R.Green@hw.ac.uk

Dongbing Gu School of Computer Science and Electronic Engineering, University of Essex, Wivenhoe Park, Colchester CO3 4SQ, UK, dgu@essex.ac.uk

Lei Guo School of Instrument Science and Opto-Electronics Engineering, Beihang University, Beijing 100191, China; Research Institute of Automation, Southeast University, Nanjing 210096, China, lguo@buaa.edu.cn

Zeng-Guang Hou Key Laboratory of Complex Sciences, Institute of Automation, Chinese Academy of Sciences, Beijing 100190, China, hou@compsys.ia.ac.cn

Huosheng Hu School of Computer Science and Electronic Engineering, University of Essex, Wivenhoe Park, Colchester CO3 4SQ, UK, hhu@essex.ac.uk

Boyko Iliev Department of Technology, Orebro University, 70182 Orebro, Sweden, boyko.iliev@aass.oru.se

Anthony Jakas Department of Computer Science, Aberystwyth University, Ceredigion SY23 3DB, UK, ajj08@aber.ac.uk

Xiaofei Ji Intelligent Systems and Biomedical Robotics Group, School of Creative Technologies, The University of Portsmouth, Eldon Building, Portsmouth PO1 2DJ, UK; School of Automation, Shenyang Institute of Aeronautical Engineering, No. 37 Daoyi South Avenue, Shenyang 110136, China, xiaofei.ji@port.ac.uk

Edward Johns Institute of Biomedical Engineering, Imperial College of London, London, UK, edward.johns09@imperial.ac.uk

Bourhane Kadmiry Department of Technology, Orebro University, 70182 Orebro, Sweden, bourhane.kadmiry@tech.oru.se

Mehdi Khoury Intelligent Systems and Biomedical Robotics Group, School of Creative Technologies, The University of Portsmouth, Eldon Building, Portsmouth PO1 2DJ, UK, mehdi.khoury@port.ac.uk

Naoyuki Kubota Department of System Design, Tokyo Metropolitan University, 6-6 Asahigaoka, Hino, Tokyo 191-0065, Japan, kubota@tmu.ac.jp

Bo Li School of Computer Science and Electronic Engineering, University of Essex, Wivenhoe Park, Colchester CO3 4SQ, UK

Yibo Li School of Automation, Shenyang Institute of Aeronautical Engineering, No. 37 Daoyi South Avenue, Shenyang 110136, China, lyb20040612@yahoo.com.cn

Honghai Liu Intelligent Systems and Biomedical Robotics Group, School of Creative Technologies, The University of Portsmouth, Eldon Building, Portsmouth PO1 2DJ, UK, honghai.liu@port.ac.uk

Junjie Liu Department of Computer Science, Aberystwyth University, Ceredigion SY23 3DB, UK, jul06@aber.ac.uk

Yonghuai Liu Department of Computer Science, Aberystwyth University, Ceredigion SY23 3DB, UK, yyl@aber.ac.uk

Bowen Lu School of Computer Science and Electronic Engineering, University of Essex, Wivenhoe Park, Colchester CO3 4SQ, UK, blv@essex.ac.uk

Kenichiro Nishida Multimedia Soc Development Dept., Toshiba Corporation Semiconductor Company, 580-1, Horikawa-Cho, Saiwai-Ku, Kawasaki 212-8520, Japan, Kenichiro.nishida@toshiba.co.jp

John Oyekan School of Computer Science and Electronic Engineering, University of Essex, Wivenhoe Park, Colchester CO3 4SQ, UK, jooyek@essex.ac.uk

Rainer Palm Department of Technology, Orebro University, 70182 Orebro, Sweden, rub.palm@t-online.de

Alexander Skoglund Department of Technology, Orebro University, 70182 Orebro, Sweden, alexander.skoglund@aass.oru.se

Min Tan Key Laboratory of Complex Sciences, Institute of Automation, Chinese Academy of Sciences, Beijing 100190, China, tan@compsys.ia.ac.cn

Yandong Tang State Key Laboratory of Robotics, Shenyang Institute of Automation, Chinese Academy of Sciences, Shenyang, China, ytang@sia.cn

Jiandong Tian State Key Laboratory of Robotics, Shenyang Institute of Automation, Chinese Academy of Sciences, Shenyang, China; Graduate School of the Chinese Academy of Sciences, Beijing, China, tianjd@sia.cn

Salman Valibeik Institute of Biomedical Engineering, Imperial College of London, London, UK, salman.valibeik05@imperial.ac.uk

Andrew M. Wallace Heriot-Watt University, Edinburgh, UK, A.M.Wallace@hw.ac.uk

Xu Wang Key Laboratory of Complex Sciences, Institute of Automation, Chinese Academy of Sciences, Beijing 100190, China, wangxu.zju@163.com

Zongyao Wang School of Computer Science and Electronic Engineering, University of Essex, Essex, UK, zwangf@essex.ac.uk

Xin-Yu Wen Research Institute of Automation, Southeast University, Nanjing 210096, China,

Charence Wong Institute of Biomedical Engineering, Imperial College of London, London, UK, charence.wong05@imperial.ac.uk

Xin Xin Faculty of Computer Science and Systems, Okayama Prefectural University, 111 Kuboki, Japan,

Guang-Zhong Yang Institute of Biomedical Engineering, Imperial College of London, London, UK, gzy@doc.ic.ac.uk

Huiyu Zhou Queen's University Belfast, Belfast, UK, H.Zhou@ecit.qub.ac.uk

Chapter 1
Programming-by-Demonstration of Robot Motions

Alexander Skoglund, Boyko Iliev, and Rainer Palm

Abstract In this chapter a novel approach to skill acquisition from human demonstration is presented. Usually the morphology of a robot manipulator is very different from the human arm and cannot simply copy a human motion. Instead the robot has to execute its own version of the skill demonstrated by the operator. Once a skill has been acquired by the robot it must also be able to generalize to other similar skills without starting a new learning process. By using a motion planner that operates in an object-related world-frame called hand-state, we show that this representation simplifies a skill reconstruction and preserves the essential parts of the skill.

1.1 Introduction

This article presents a method for imitation learning based on fuzzy modeling and a next-state-planner in a Programming-by-Demonstration (PbD) framework. For a recent comprehensive overview of PbD, (also called Learning from Demonstration) see [1]. PbD refers to a variety of methods where the robot learns how to perform a task by observing a human teacher, which greatly simplifies the programming process [2–5]. One major scientific challenge in PbD is how to make the robot *capable* of imitating a human demonstration. Although the idea of copying human motion trajectories using a simple teaching-playback method seems straightforward, it is not realistic for several reasons. Firstly, there is a significant difference in morphology between the human and the robot, known as the correspondence problem in imitation [6]. The difference in the location of the human demonstrator and the robot might force the robot into unreachable parts of the workspace or singular arm configurations even if the demonstration is perfectly feasible from human viewpoint. Secondly, in grasping tasks the reproduction of human hand motions is not possible

A. Skoglund (✉), B. Iliev, and R. Palm
Department of Technology, Orebro University, 70182 Orebro, Sweden
e-mail: alexander.skoglund@aass.oru.se; boyko.iliev@aass.oru.se; rub.palm@t-online.de

H. Liu et al. (eds.), *Robot Intelligence*,
Advanced Information and Knowledge Processing,
DOI 10.1007/978-1-84996-329-9_1,

since even the most advanced robot hands cannot match neither the functionality of the human hand nor its sensing capabilities. However, robot hands capable of autonomous grasping can be used in PbD provided that the robot can generate an appropriate reaching motion towards the target object, as we will demonstrate in this article.

In this article, we present an approach to learning of reaching motions where the robot uses human demonstrations in order to collect essential knowledge about the task. This knowledge, i.e., grasp-related object properties, hand-object relational trajectories, and coordination of reach and grasp motions is encoded and generalized in terms of *hand-state space* trajectories. The hand-state components are defined such that they are perception-invariant and define the correspondence between the human and robot hand. The hand-state representation of the task is then embedded into a next-state-planner (NSP) which enables the robot to perform reaching motions from an arbitrary robot configuration to the target object. The resulting reaching motion ensures that the robot hand will approach the object in such way that the probability for a successful grasp is maximized.

An NSP plans one step ahead from its current state. This contrasts to traditional robotic approaches which plan the entire trajectory in advance. One of the first researchers to use a NSP approach in imitation learning were Ijspeert et al. [7], where they encode the trajectory in an autonomous dynamical system with internal dynamic variables that shapes a "landscape" used for both point attractors and limit cycle attractors. For controlling a humanoid's reaching motion, Hersch and Billad [8] considered a combined controller with two controllers running in parallel; one controller acts in joint space, while the other one acts in Cartesian space. To generate reaching motions and avoiding obstacles simultaneously Iossifidis and Schöner [9] used attractor dynamics, where the target object acts as a point attractor on the end effector. The end-effector as well as a redundant elbow joint avoids an obstacle as the arm reaches for an object.

In our approach, a human demonstration guides the robot to grasp an object. Our use of an NSP differs from previous work [7–9] in the way it combines the demonstrated path with the robot's own plan. The use of hand-state trajectories distinguishes our work from most previous work on imitation. According to [7], most approaches in the literature use the joint space for motion planning while some other approaches use the Cartesian space.

To illustrate the approach we describe three scenarios where human demonstrations of goal-directed reach-to-grasp motions are reproduced by a robot. Specifically, the generation of reaching and grasping motions in pick-and-place tasks is addressed. In the experiments we test how well the skills perform the demonstrated task and how well they generalize over the workspace. The contributions of the work are as follows:

1. We introduce a novel next-state-planner based on a *fuzzy modeling* approach to encode human and robot trajectories.
2. We apply the *hand-state concept* [10] to encode motions in hand-state trajectories and apply this in PbD.

3. The combination of the NSP and the hand-state approach provides a tool to address the *correspondence problem* resulting from the different morphology of the human and the robot. The experiments show how the robot can generalize and use the demonstration despite the fundamental difference in morphology.

1.2 Learning from Human Demonstration

In PbD the idea is that the robot programmer (here called demonstrator) shows the robot what to do and from this demonstration an executable robot program is created. In our case, the demonstrator shows the task by performing it in a way that seems to be feasible for the robot. This means that we assume the demonstrator to be aware of the particular restrictions of the robot. In this work we consider only the body language of the demonstrator, i.e., the approach is entirely based on proprioceptive information. Interpretation of human demonstrations is done under two assumptions: the type of tasks and grasps that can be demonstrated are *a priori* known by the robot; we consider only demonstrations of power grasps (e.g., cylindrical and spherical grasps) which can be mapped to–and executed by–the robotic hand.

1.2.1 Interpretation of Demonstrations in Hand-State Space

To create the associations between human and robot reaching/grasping we employ the hand-state hypothesis from the Mirror Neuron System (MNS) model of [10]. The aim is to mimic the functionality of the MNS to enable a robot to interpret human goal-directed motions in the same way as its own motions. Following the ideas behind the MNS-model, both human and robot motions are represented in hand-state space. A hand-state trajectory encodes a goal-directed motion of the hand during reaching and grasping. Thus, the hand-state space is common for the demonstrator and the robot and preserves the necessary execution information. Hence, a particular demonstration can be converted into executable robot code and experience from multiple demonstrations is used to control/improve the execution of new skills. Thus, when the robot tries to imitate an observed reach and grasp motion, it has to move its own hand so that it follows a hand-state trajectory similar to the demonstrated one. If such a motion is successfully executed by the robot, a new robot skill is acquired. Seen from a robot perspective, human demonstrations are interpreted as follows.

If hand motions with respect to a potential target object are associated with a particular grasp type G_i, it is assumed that there must be a target object that matches the observed grasp type. In other words, the object has certain grasp-related features, also called *affordances* [10], which makes this particular grasp type appropriate. The position of the object can be retrieved by a vision system, or it can be estimated from the grasp type and the hand pose, given some other motion capturing device. For each grasp type G_i, a subset of suitable object affordances is identified a priori and learned from a set of training data. In this way, the robot is able to associate observed grasp types G_i with their respective affordances A_i.

According to [10], the hand-state must contain components describing both the hand configuration and its spatial relation with respect to the affordances of the target object. Thus, the hand-state is defined in the form:

$$H = \{h_1, h_2, \dots, h_{k-1}, h_k, \dots, h_p\} \tag{1.1}$$

where $h_1 \dots h_{k-1}$ are *hand-specific components* which describe the motion of the fingers during grasping. The remaining components $h_k \dots h_p$ describe the motion of the hand in relation to the object. Thus, a hand-state trajectory contains a record of both the reaching and the grasping motions as well as their synchronization in time and space.

The hand-state representation equation (1.1) is invariant with respect to the actual location and orientation of the target object. Thus, demonstrations of object-reaching motions at different locations and initial conditions can be represented in a common domain. This is both the strength and weakness of the hand-state approach. Since the hand-state space has its origin in the goal object, a displacement of the object will not affect the hand-state trajectory. However, when an object is firmly grasped then the hand-state is fixed and will not capture a change in the object position relative to the base coordinate system. This implies that for object handling and manipulation the use of hand-state trajectories is limited.

1.2.2 Skill Encoding Using Fuzzy Modeling

Once the hand-state trajectory of the demonstrator is determined, it has to be modeled for several reasons. In [11] Ijspeert enumerates five important desirable properties for encoding movements have been identified. These are:

1. The representation and learning of a goal trajectory should be simple.
2. The representation should be compact (preferably parameterized).
3. The representation should be reusable for similar settings without a new time consuming learning process.
4. For recognition purpose, it should be easy to categorize the movement.
5. The representation should be able to act in a dynamic environment and be robust to perturbations.

Several methods for encoding human motions include Splines [12]; Hidden Markov Models (HMM) [13]; HMM combined with Non-Uniform Rational B-Splines [14]; Gaussian Mixture Models [2]; dynamical systems with a set of Gaussian kernel functions [11]. We developed a method based on fuzzy logic which deals with the above properties in a sufficient manner [15].

Let us examine the properties of fuzzy modeling with respect to the above enumerated desired properties. Fuzzy modeling is simple to use for trajectory learning and is a compact representation in form of a set of weights, gains and offsets (i.e., they fulfill property 1 and 2) [16]. To change a learned trajectory into a new one for a similar task with preserved characteristics of a motion, we proposed modification to the fuzzy time modeling algorithm [17], thus addressing property 3. Furthermore, the method satisfies property 4, as it was successfully used for grasp recognition by [15].

The algorithm for fuzzy time modeling of motion trajectories is briefly described as follows. Takagi and Sugeno proposed a structure for fuzzy modeling of input-output data of dynamical systems [18]. Let $\mathbf{X}$ be the input data set and $\mathbf{Y}$ be the output data set of the system with their elements $x \in \mathbf{X}$ and $y \in \mathbf{Y}$. The fuzzy model is composed of a set of c rules R from which rule R_i reads:

$$\text{Rule } i\text{: IF } x \text{ IS } \mathbf{X}_i \text{ THEN } y = A_i x + B_i \tag{1.2}$$

$\mathbf{X}_i$ denotes the ith fuzzy region in the fuzzy state space. Each fuzzy region $\mathbf{X}_i$ is defuzzified by a fuzzy set $\int w_{x_i}(x)|x$ of a standard triangular, trapezoidal, or bell shaped type. $W_i \in \mathbf{X}_i$ denotes the fuzzy value that x takes in the ith fuzzy region $\mathbf{X}_i$. A_i and B_i are fixed parameters of the local linear equation on the right hand side of (1.2).

The variable $w_i(x)$ is also called degree of membership of x in $\mathbf{X}_i$. The output from rule i is then computed by:

$$y = w_i(x)(A_i x + B_i). \tag{1.3}$$

A composition of all rules $R_1 \ldots R_c$ results in a summation over all outputs from (1.3):

$$y = \sum_{i=1}^{c} w_i(x)(A_i x + B_i) \tag{1.4}$$

where $w_i(x) \in [0, 1]$ and $\sum_{i=1}^{c} w_i(x) = 1$.

The fuzzy region $\mathbf{X}_i$ and the membership function w_i can be determined in advance by design or by an appropriate clustering method for the input-output data. In our case we used a clustering method to cope with the different non linear characteristics of input-output data-sets (see [19] and [20]). For more details about fuzzy systems see [21].

In order to model time dependent trajectories $x(t)$ using fuzzy modeling, the time instants t take the place of the input variable and the corresponding points $\mathbf{x(t)}$ in the state space becomes the outputs of the model.

The Takagi-Sugeno fuzzy model is constructed from captured data from the end-effector trajectory described by the nonlinear function:

$$\mathbf{x}(t) = \mathbf{f}(t) \tag{1.5}$$

where $\mathbf{x}(t) \in R^3$, $\mathbf{f} \in R^3$, and $t \in R^+$.

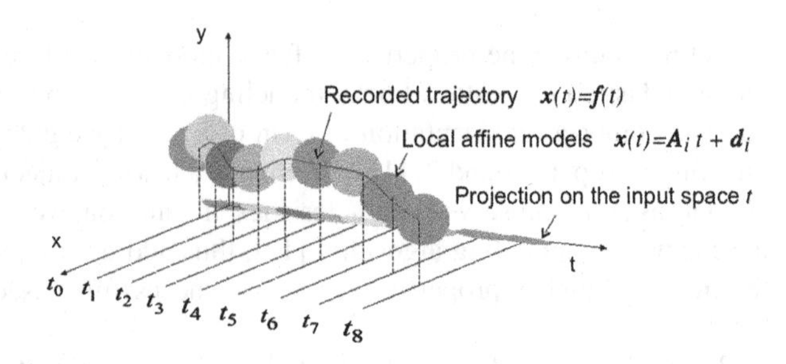

Fig. 1.1 Time-clustering principle

Equation (1.5) is linearized at selected time points t_i with

$$\mathbf{x}(t) = \mathbf{x}(t_i) + \left.\frac{\Delta\mathbf{f}(t)}{\Delta t}\right|_{t_i} \cdot (t - t_i) \tag{1.6}$$

resulting in a locally linear equation in t.

$$\mathbf{x}(t) = \mathbf{A}_i \cdot t + \mathbf{d}_i \tag{1.7}$$

where $\mathbf{A}_i = \frac{\Delta\mathbf{f}(t)}{\Delta t}|_{t_i} \in R^3$ and $\mathbf{d}_i = \mathbf{x}(t_i) - \frac{\Delta\mathbf{f}(t)}{\Delta t}|_{t_i} \cdot t_i \in R^3$. Using (1.7) as a local linear model one can express (1.5) in terms of an interpolation between several local linear models by applying Takagi-Sugeno fuzzy modeling [18] (see Fig. 1.1)

$$\mathbf{x}(t) = \sum_{i=1}^{c} w_i(t) \cdot (\mathbf{A}_i \cdot t + \mathbf{d}_i) \tag{1.8}$$

$w_i(t) \in [0, 1]$ is the degree of membership of the time point t to a cluster with the cluster center t_i, c is number of clusters, and $\sum_{i=1}^{c} w_i(t) = 1$.

The degree of membership $w_i(t)$ of an input data point t to an input cluster C_i is determined by

$$w_i(t) = \frac{1}{\sum_{j=1}^{c}\left(\frac{(t-t_i)^T M_{i\,pro}(t-t_i)}{(t-t_j)^T M_{j\,pro}(t-t_j}\right)^{\frac{1}{\tilde{m}_{proj}-1}}}. \tag{1.9}$$

The projected cluster centers t_i and the induced matrices $M_{i\,pro}$ define the input clusters C_i $(i = 1 \dots c)$. The parameter $\tilde{m}_{pro} > 1$ determines the fuzziness of an individual cluster [19].

1.3 Generation and Execution of Robotic Trajectories Based on Human Demonstration

This section covers generation and execution of trajectories on the actual robot manipulator. We start with a description of how the mapping from human to robot hand is achieved and how the hand-state components are defined. Then follows a description of the next-state-planner, which produces the actual robot trajectories.

1.3.1 Mapping Between Human and Robot Hand States

In the PbD framework, the hand-state components $h_1, \ldots h_p$ must be such that they can be recovered from both human demonstrations and the perception system of the robot. That is, the definition of H is *perception invariant* and can be updated from arbitrary types of sensory information. Figure 1.2 shows the definition of the hand-state in this article.

Let the human hand be at some initial state H_1. Then the hand moves along a certain path and reaches the final state H_f where the target object is held by the hand [17]. That is, the recorded motion trajectory can be seen as a sequence of states, i.e.,

$$H(t) : H_1(t_1) \rightarrow H_2(t_2) \rightarrow \cdots \rightarrow H_f(t_f). \tag{1.10}$$

To determine the hand-state representation of a demonstration the robot needs to have access to the complete motion trajectories of the teacher's hand since the motion must be in relation to the target object. This means that the hand-state trajectories can only be computed *during* a motion if the target object is known in advance.

Let $H_{des}(t)$ be the desired hand-state trajectory recorded from a demonstration. Since $H_{des}(t)$ cannot be executed by the robot without modification in the general case, we have to construct the *robotic* version of $H_{des}(t)$, denoted by $H^r(t)$, see Fig. 1.3 for an illustration.

To find $H^r(t)$ a mapping from the human grasp to the robot grasp a transformation is needed, denoted by T_h^r. This mapping is created as follows. We can measure the pose of the demonstrator hand and the robot hand holding the same object at fixed position and obtain T_h^r as a static mapping between the two poses. Thus, the target state H_f^r will be derived from the demonstration by mapping the goal configuration of the human hand H_f into a goal configuration for the robot hand H_f^r, using the transformation T_h^r:

$$H_f^r = T_h^r H_f. \tag{1.11}$$

The pose of the robot hand at the start of a motion defines the initial state H_1^r. Since H_f^r represents the robot hand holding the object, it has to correspond to a stable

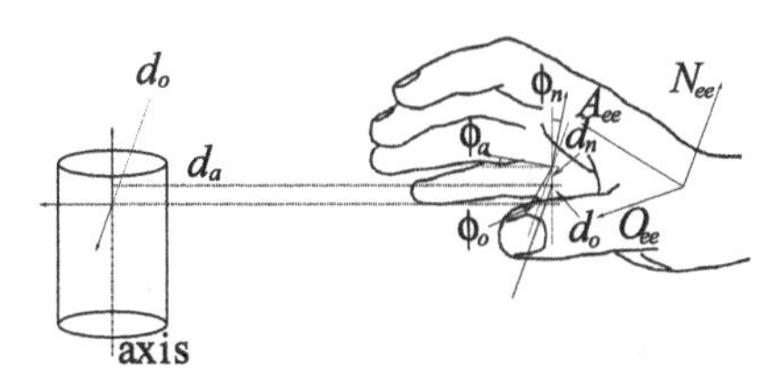

Fig. 1.2 The hand-state describes the relation between the hand pose and the object affordances. N_{ee} is the normal vector, O_{ee} the side (orthogonal) vector and A_{ee} is the approach vector. The vector Q_{ee} is the position of the point. The same definition is also valid for boxes, but with the restriction that the hand-state frame is completely fixed, it cannot be rotated around the symmetry axis

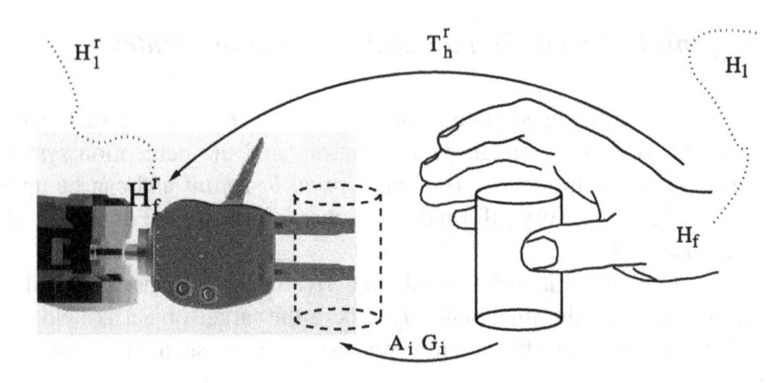

Fig. 1.3 Mapping from human hand to robotic gripper

grasp. For a known object, suitable H_f^r can either be obtained by simulation [22], grasp planning or by learning from experimental data. Thus, having a human hand-state H_f and their corresponding robot hand-state H_f^r, T_h^r is obtained as:

$$T_h^r = H_f^r H_f^{-1}. \tag{1.12}$$

It should be noted that this method is only suitable for power grasps. In the general case it might produce ambiguous results or rather inaccuarate mappings.

One advantage of using one demonstrated trajectory as the desired trajectory over trajectory averaging (e.g., [2] or [23]) is that the average might contain two essentially different trajectories [14]. By capturing a human demonstration of the task, the synchronization between reach and grasp is also captured, demonstrated in [24]. Other ways of capturing the human demonstrating, such as kinesthetics [2] or by a teach pendant (a joystick), cannot capture this synchronization easily.

1.3.2 Definition of Hand-States for Specific Robot Hands

Having the initial state H_1^r and the target state H_f^r defined, we have to generate the trajectory between the two states. In principle, we could transform $H_{des}(t)$ using (1.11) in such way that it has its final state in H_f^r. Then, the robot starts at H_1^r, approaches the displaced demonstrated trajectory and tracks it until the target state is reached. However, such an approach would not take trajectory constraints into account. Thus, it is also necessary to specify exactly how to approach $H_{des}(t)$ and what segments must be tracked accurately. driving the hand.

A hand-state trajectory must be constructed from the demonstrated trajectory. From the recorded demonstration we reconstruct the end-effector trajectory, represented by a time dependent homogeneous matrix $T_{ee}(t)$. Each element is represented by the matrix

$$T_{ee} = \begin{pmatrix} N_{ee} & O_{ee} & A_{ee} & Q_{ee} \\ 0 & 0 & 0 & 1 \end{pmatrix} \tag{1.13}$$

where N_{ee}, O_{ee} and A_{ee} are the normal vector, the side vector, and the approach vector respectively. The last vector Q_{ee} is the position. The matrix T_{ee} is defined differently for different end-effectors, for example, the human hand is defined as in Fig. 1.2.

There is evidence that the internal models of arm dynamics found in biological systems are state-dependent rather than time-dependent [25]. Therefore, when we transform human demonstrations into robot motions we define *distance to object* d, as an additional scheduling variable for hand-state trajectories. To preserve the velocity profile from the human demonstration the distance to the target is modeled as a function of time using fuzzy time-modeling, see Sect. 1.2.2. The inputs to the fuzzy modeling is the Euclidean distance at each instance t of time:

$$d(t) = \sqrt{(Q_{ee}(t) - P)^2} \tag{1.14}$$

where Q_{ee} and P are the end-effector position and object position respectively.

The same procedure is applied to the hand-state trajectories. Two types of models are needed: one modeling of the hand-state as a function of time, and one as a function of distance. In this article a general formulation of the hand-state is adopted to fit the two states (open and close) for the anthropomorphic hand. We formulate the hand-state as:

$$H(t) = [d_n(t)\ d_o(t)\ d_a(t)\ \phi_n(t)\ \phi_o(t)\ \phi_a(t)]. \tag{1.15}$$

The individual components denote the position and orientation of the end-effector. The first three components, $d_n(t)$, $d_o(t)$ and $d_a(t)$, describe the distance from the object to the hand along the three axes n, o and a with the object as the base frame. The next three components, $\phi_n(t)$, $\phi_o(t)$ and $\phi_a(t)$, describe the rotation of the hand in relation to the object around the three axes n, o and a. The notion of the hand-state used in this section is illustrated in Fig. 1.2.

The components of the hand-state, as a function of distance, are given by:

$$H(d) = [d_n(d)\ d_o(d)\ d_a(d)\ \phi_n(d)\ \phi_o(d)\ \phi_a(d)] \tag{1.16}$$

where the hand-state components are the same as in (1.15), but with $d \in R^1$ instead of t. The role of the scheduling variable d is important since it expresses *when* the robot should move to the next state, while the hand-state variables reflect *where* the hand should move. Thus, d synchronizes the motions' *when* (dynamics and synchronization) and *where* (desired path) of the reach and grasp.

Note that with this simplified definition of H we cannot determine the human grasp type, since we have omitted the finger specific components of the hand-state. In [24] we give an account of how these components can be used to synchronize reaching with grasping. Grasp classification is out of scope of this article, because only power grasps are used in our experiments. Thus, the grasp type is assumed to be known $G = \{cylindrical, spherical, plane\}$; the affordances are: position, size, and cylinder axis $A = \{width, axis\}$ or box $A = \{width, length, N\text{-}axis, O\text{-}axis, A\text{-}axis\}$. See [26] for grasp taxonomy.

1.3.3 Next-State-Planners for Trajectory Generation

In this section we present the next-state-planner (NSP) that balances its actions between *following a demonstrated trajectory* and *approaching the target*, first presented in [24]. The NSP is inspired by the Vector Integration To Endpoint (VITE) planner suggested by Bullock and Grossberg [27]. The VITE planner is a biologically inspired planner for human control of reaching motions. The NSP-approach requires a control policy, i.e., a set of equations describing the next action from the current state and some desired behavior.

The proposed NSP generates a hand-state trajectory for the robot using the TS fuzzy-model of a demonstration. As the resulting hand-state trajectory $H^r(t)$ can easily be converted to Cartesian space, we can use the inverse kinematics provided by the controller for the robot arm. The TS fuzzy-model serves as a motion primitive for the arm's reaching motion. The initial hand-state of the robot is determined from its current configuration and the position and orientation of the target object, since these are known at the end of the demonstration. Then, the desired hand-state H_d^r is computed from the TS fuzzy time-model (see (1.8)). The desired hand-state H_d is fed to the NSP. Instead of using only one goal attractor as in VITE [27], and additional attractor—the desired hand-state trajectory—is used at each state. The system has the following dynamics:

$$\ddot{H} = \alpha(-\dot{H} + \beta(H_g - H) + \gamma(H_d - H)) \tag{1.17}$$

where H_g is the hand-state goal, H_d the desired state, H is the current hand-state, $\dot{H}$ and $\ddot{H}$ are the velocity and acceleration respectively. α is a positive constant and β, γ are positive weights for the goal and tracking point, respectively.

If the last term $\gamma(H_d - H)$ in (1.17) is omitted, i.e., $\gamma = 0$, then the dynamics is *exactly* as the VITE planner [27]. Indeed, if no demonstration is available the planner can still produce a motion if the target is known. Similarly, if the term $\beta(H_g - H)$ is omitted, the planner becomes a trajectory following controller. If the final position from the demonstration can be used for gasping, as in [28], it is possible to set $\beta = 0$, which we do in our experiments in Sects. 1.5.1 and 1.5.2. The reason for setting β to zero is that the demonstration towards the goal will end at the goal, making the term $\beta(H_g - H)$ redundant if the goal cannot be estimated more accurately using some other sensor system as in [29]. Then the variance across multiple demonstrations is used to determine γ, which controls the behavior of the NSP.

Analytically, the poles in (1.17) are:

$$p_1, p_2 = -\frac{\alpha}{2} \pm \sqrt{\frac{\alpha^2}{4} - \alpha\gamma}. \tag{1.18}$$

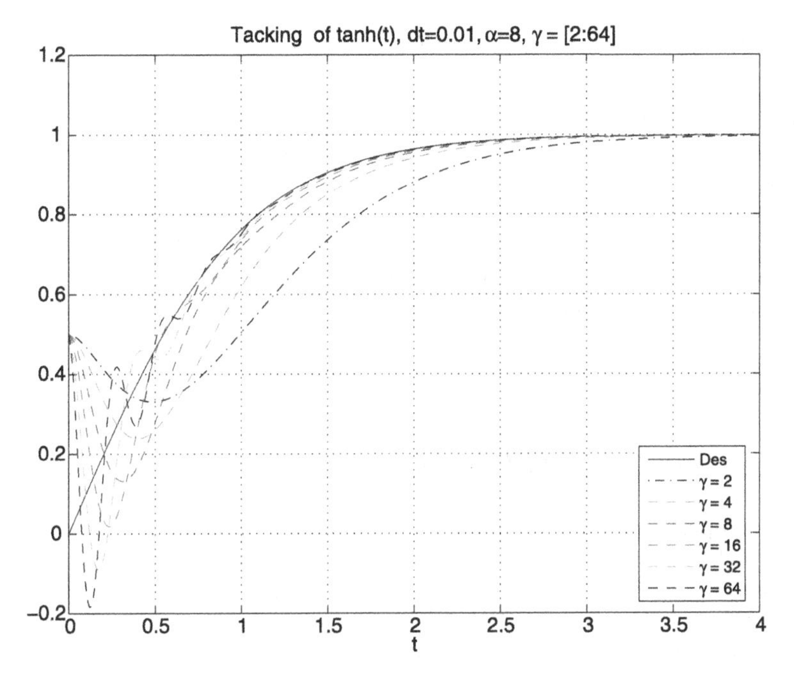

Fig. 1.4 The dynamics of the planner for six different values of γ. The tracking point is $\tanh(t)$, with $dt = 0.01$ and α is fixed at 8. A low value on $\gamma = 2$ produces slow dynamics (*black dot-dashed line*), while a high value $\gamma = 64$ is fast but overshoots the tracking point (*black dashed line*)

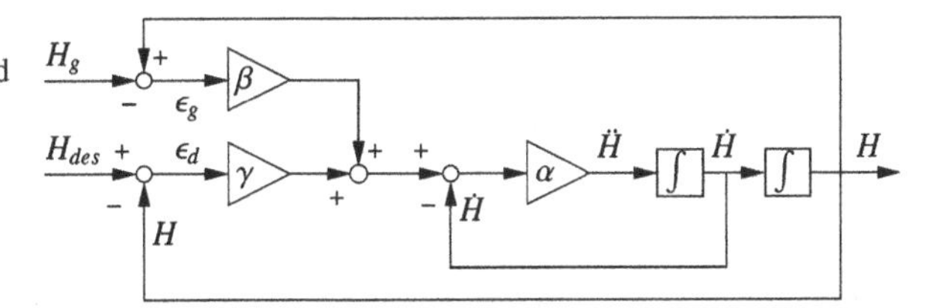

Fig. 1.5 Hand-state planner architecture. H_g is the desired hand-state goal, H_{des} is the desired hand-state at the current distance to target

The real part of p_1 and p_2 will be ≤ 0, which will result in a stable system [30]. Moreover, $\alpha \not\leq 4\gamma$ and $\alpha \geq 0$, $\gamma \geq 0$ will contribute to a critically damped system, which is fast and has small overshoot. Figure 1.4 shows how different values γ affect the dynamics of the planner.

The controller has a feedforward structure as in Fig. 1.5. The reason for this structure is that a commercial manipulator usually has a closed architecture, where the controller is embedded in the system. For this type of manipulators, a trajectory is usually pre-loaded and then executed. Therefore, we generate the trajectories in batch mode for the ABB140 manipulator. Since our approach is general, for a given different robot platform with hetroceptive sensors (e.g., vision) our method can be implemented in a feedback mode, but this requires that the hand-state $H(t)$ can be measured during execution.

1.3.4 Demonstrations of Pick-and-Place Tasks

The demonstration of the task is performed with the demonstrator (teacher) standing in front of the robot. Then the *task*, i.e., a pick-and-place of an object, is shown. The target object is determined from the task demonstration, where the center point of the grasp can be estimated from the grasp type. For example, grasp recognition (see [31]) can improve the estimate of the object position and orientation. In our experiments we assume a power grasp to determine the orientation of the object, while the motion capturing system is used to record the position of the object. The task demonstration contains the trajectories which the robot should execute to perform the task.

1.3.4.1 Variance from Multiple Demonstrations

When multiple demonstrations of a skill are available to the robot we can obtain a generalized version of that skill. We exploit the fact that when humans grasp the same object several times they seem to repeat the same grasp type which leads to similar approach motions. Based on that, multiple demonstrations of a skill become more and more similar to each other the closer one gets to the target state. This implies that successful grasping requires an accurate positioning of the hand in a region near the object while the path towards this area is subject to less restrictions. Therefore, by looking at the variance of several demonstrations the importance of each hand-state component can be determined. The variance of the hand-state as a function of the distance to target d is given by:

$$var({}^k h(d)) = \frac{1}{n-1} \sum_{i=1}^{n} ({}^k h_i(d) - mean({}^k h(d)))^2 \tag{1.19}$$

where d is the Euclidean distance to the target, ${}^k h_i$ is the kth hand-state parameter of ith demonstration and n is the number of demonstrations. Figure 1.6 shows how the variance decreases as the distance to the object decreases. This means that the position and orientation of the hand are less relevant when the distance to the target increases.

1.4 Experimental Platform

For these experiments human demonstrations of a pick-and-place task are recorded with two different subjects, using the PhaseSpace Impulse motion capturing system described below. The Impulse motion capturing system consists of four cameras mounted around the operator to register the position of the LEDs. Each LED has a unique ID by which it is identified. Each camera can process data at 480 Hz and has 12 Mega pixel resolution resulting in sub-millimeter precision. The Impulse systems

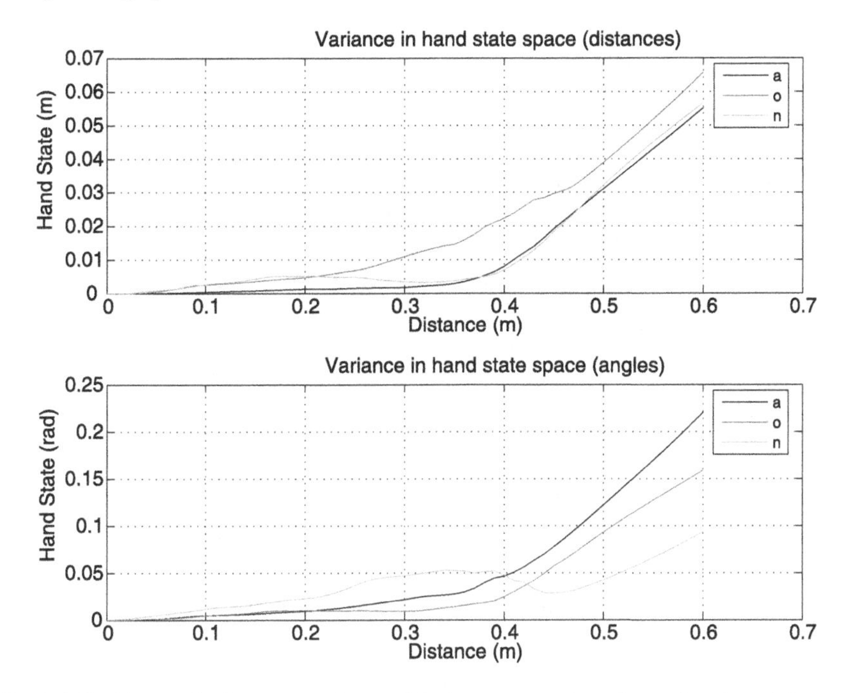

Fig. 1.6 Position- and orientation-variance of the hand-state trajectories as function of distance, across 22 demonstrations of a reaching to grasp motion. Note that distances over 0.47 are extrapolations made by the clustering method

Fig. 1.7 *Left*: The glove used in the Impulse motion capturing system from PhaseSpace. The glove from the top showing the LEDs. *Right*: The system in use showing one of the cameras and the LED on the glove

can be seen in the upper right picture in Fig. 1.7. The operator wears a glove with LEDs attached to it, see upper left picture in Fig. 1.7. Thus, each point on the glove can be associated with a finger, the back of the hand or the wrist. To compute the orientation of the wrist, three LEDs must be visible during the motion. The back of the hand is the best choice since three LEDs are mounted there and they are most

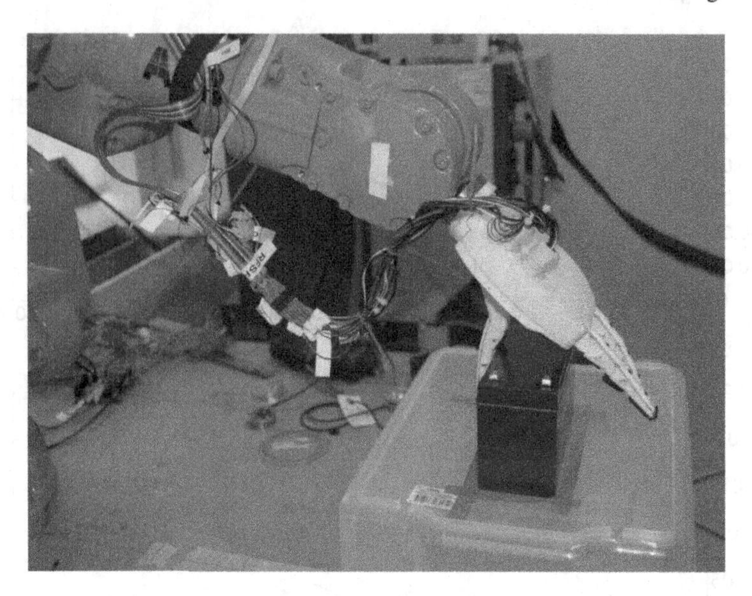

Fig. 1.8 The anthropomorphic gripper KTHand used in the experiments

of the time visible for at least three cameras. One LED is mounted on each finger tip, and the thumb has one additional LED in the proximal joint. One LED is also mounted on the target object.

The motions are automatically segmented into reach and retract motions using the velocity profile and distance to the object. The robot used in the experiments is the industrial manipulator ABB IRB140. In this experiment we use the anthropomorphic gripper KTHand (Fig. 1.8), which can perform power grasps (i.e., cylindrical and spherical grasps) using a hybrid position/force controller. For details on the KTHand, see [32].

1.5 Experimental Evaluation

In this section we provide an experimental evaluation of the methods presented. The 1st experiment deals with the task to learn a robot skill from human demonstration. The 2nd experiment shows how well the learned trajectories can be generalized w.r.t. different workspaces especially the workspace of the human operator and the robot's workspace. In the 3rd experiment a complete pick-and-place task is executed.

1.5.1 Experiment 1: Learning from Demonstration

For this experiment 26 task demonstrations of a pick-and-place task were performed using a soda can for performing a spherical grasp. To make the scenario more realistic the object is placed with respect to what is convenient for the human operator and what *seems* to be feasible for the robot.

Five of the 26 demonstrations were discarded in the segmentation (see [4]) and modeling process for reasons such as failure to segment the demonstrations into three distinct motions (approach, transport and retract) or the amount of data were not enough for modeling because of occlusions. Only the reach-to-grasp phase is considered in this experiment. All 21 demonstrations were used for trajectory generation and to compute the variance, shown in Fig. 1.6. Moreover, the variance is used to compute the γ-gain, which determines how much the robot can deviate from the followed trajectory. The trajectory generator produced 21 reaching motions, one from each demonstration, which are loaded to the robot controller and executed. By using each demonstrated trajectory as the desired trajectory H_d instead of building an average of them we avoid fusing of essentially different trajectories into a possibly incoherent trajectory. Large differences will instead affect the variance, resulting in a small γ-gain. In eight attempts, the execution succeeded while 13 attempts failed because of unreachable configurations in joint space. This could be prevented by placing the robot at a different location with better reachability. Moreover, providing the robot with more demonstrations, with higher variations in the path, will lead to fewer constraints. Two sample hand-state trajectories of the successfully generated ones are shown in Fig. 1.9. In the top graphs it is shown how for different initial locations the generated trajectory converges towards the desired trajectory. The bottom graphs shows how γ varies over time, to make the generated trajectory H^r follow the desired H_d.

In the eight successfully executed reaching motions we measured the variation in position of the gripper, shown in Fig. 1.10, which is within the millimeter range. This means that the positioning is accurate enough to enable successful grasping using an autonomous gripper, such as the Barrett hand [33] or the KTHand.

1.5.1.1 Importance of the Demonstration

The weight γ reflects the importance of the path, acquired from variance, see Sect. 1.3.4.1. For experiment 1 and 2, we have empirically found γ to produce satisfying results at:

$$\gamma_{pos} = 0.3\frac{1}{\sqrt{Var(H_{xyz}(d))}},$$

$$\gamma_{ori} = 5\frac{1}{\sqrt{Var(H_{rpy}(d))}}$$

where γ_{pos} and γ_{ori} are the weights for position and orientation, respectively. $Var(H_{xyz}(d))$ and $Var(H_{rpy}(d))$ are the variance for the position and orientation respectively, from (1.19), of the respective hand state component. α_{pos} and α_{ori} are fixed during our experiments at 8 and 10, respectively, with a time step $dt = 0.01$.

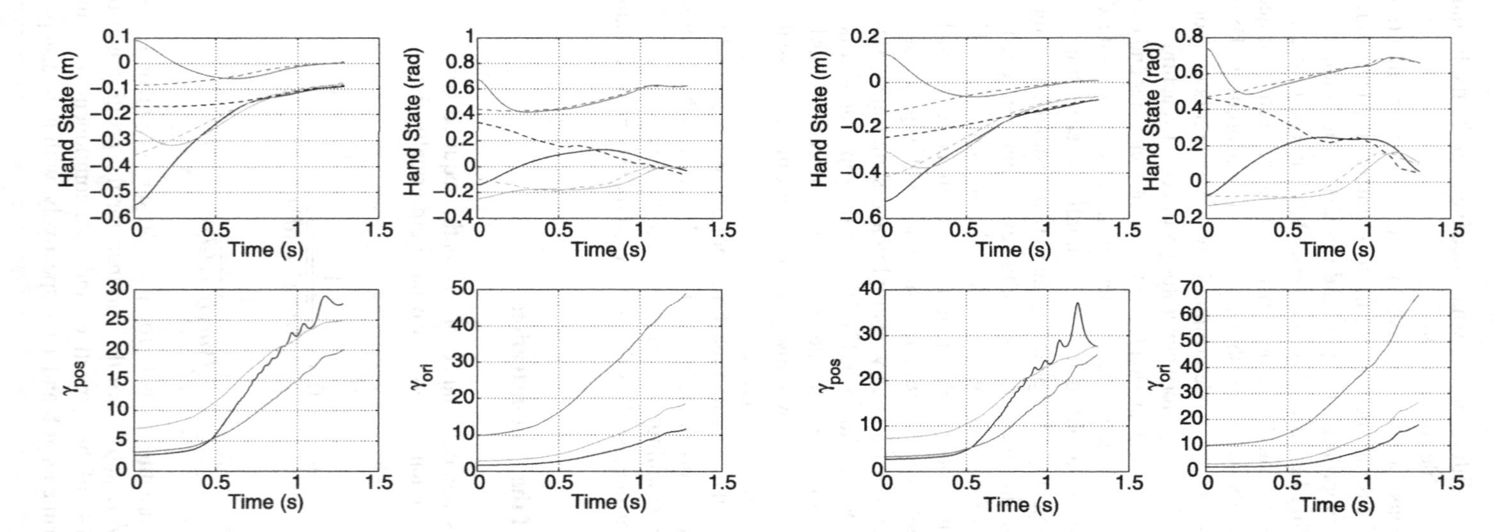

Fig. 1.9 Two samples of demonstrations, and the corresponding imitations. The *top graphs* shows the hand-state trajectories, position and orientation for each of the demonstrations/imitation. The *solid lines* are the trajectories produced by the planner, *dashed lines* are the model constructed from the recorded demonstration

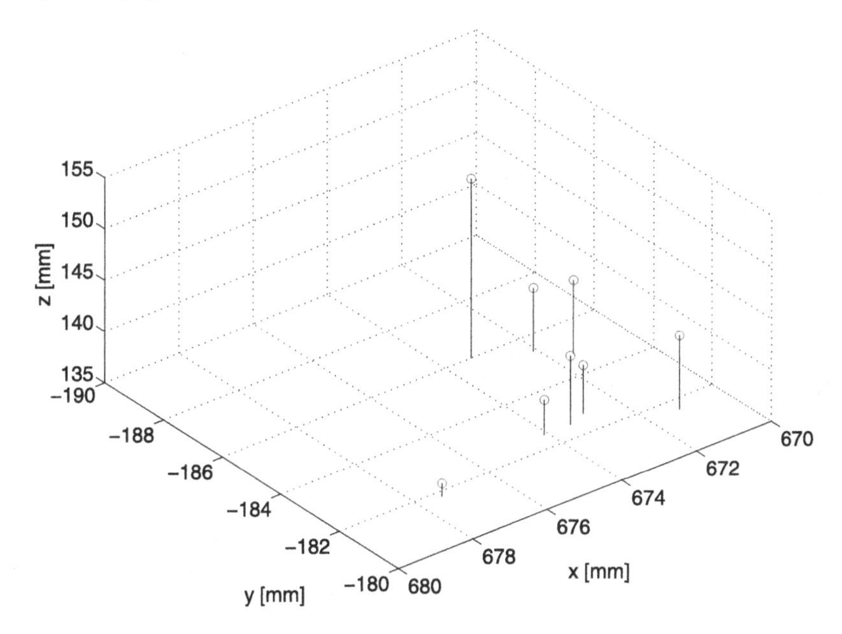

Fig. 1.10 The end effector position at the end of the motion for the 8 successfully executed trajectories. The positioning accuracy is within the millimeter range; 6 mm along x, 4 mm along y and 12 mm along z

These gains were chosen to provide dynamic behavior similar to the demonstrated motions, but other criteria can also be used.

The next-state planner uses the demonstration to generate a similar hand-state trajectory, using the distance as a scheduling variable. Hence, the closer to the object the robot is the more important it becomes to follow the demonstrated trajectory. This property is reflected by adding a higher weight to the trajectory-following dynamics when we get closer to the target; in reverse a long distance to the target leads to a lower weight to the trajectory following dynamics.

1.5.2 Experiment 2: Generalization in Workspace

In this experiment, the generalization of the method is tested. This is done by examining whether feasible trajectories are generated when the object is placed at arbitrary locations and when the initial configuration of the manipulator is very different from the demonstration. This determines how the trajectory planner handles the correspondence problem in terms of morphological differences. In experiment 2 the same data set was used as in experiment 1. Three tests were performed to evaluate the trajectory generator in different parts of the workspace.

1. Trajectories are generated when the manipulator's end-effector starts directly above the object at the desired final position with the desired orientation, that

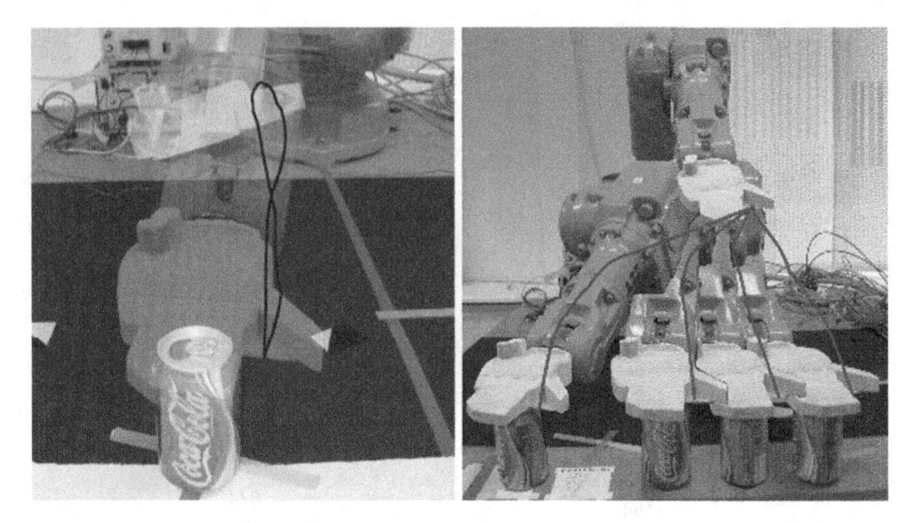

Fig. 1.11 *Left*: A trajectory generated when the initial position is the same as the desired final position, showing that the method generate trajectories as similar to the demonstration as possible based on the distance. *Right*: The object is placed at four new locations within the workspace

is $H_1^r = H_f^r$. The resulting trajectory is shown to the left in Fig. 1.11. Four additional cases are also tested displacing the end-effector by 50 mm in $+x$, $-y$, $+y$, and $+z$ direction from H_f, all with very similar results (from the robot's view: x is forward, y left and z up).

2. The object is placed at four different locations within the robot's workspace; displaced 100 mm along the x-axis, and -100 mm, $+100$ mm, $+200$ mm, and $+300$ mm along the y-axis, seen to the right in Fig. 1.11. The initial pose of the manipulator is the same in all reaching tasks. The planner successfully produces four executable trajectories to the respective object position.
3. We tested the reaching of the object at a fixed position from a random initial configuration. Figure 1.12 shows the result from two random initial positions where one trajectory is successfully tracked but the other one fails. The failure is a result of operation in hand-state space instead of joint space, and it might therefore have a tendency to go onto unreachable joint space configurations, as seen in the right column of Fig. 1.12. To prevent this it is possible to combine two controllers: one operating in joint space and the other in hand-state space, similar to the approach suggested by [34], but at the price of violating the demonstration constraints.

The conclusion from this experiment is that the method generalizes well in the tested scenarios, thus adequately addressing the correspondence problem. However, the unreachability problem has to be addressed in future research to investigate how the robot should balance the two contradiction goals: reaching an object in its own way, with the risk of collision, or reaching an object as the demonstrator showed. Indeed, if the robot has more freedom to choose the path it is more likely to avoid unreachable configurations. However, such freedom increases the risk for collision.

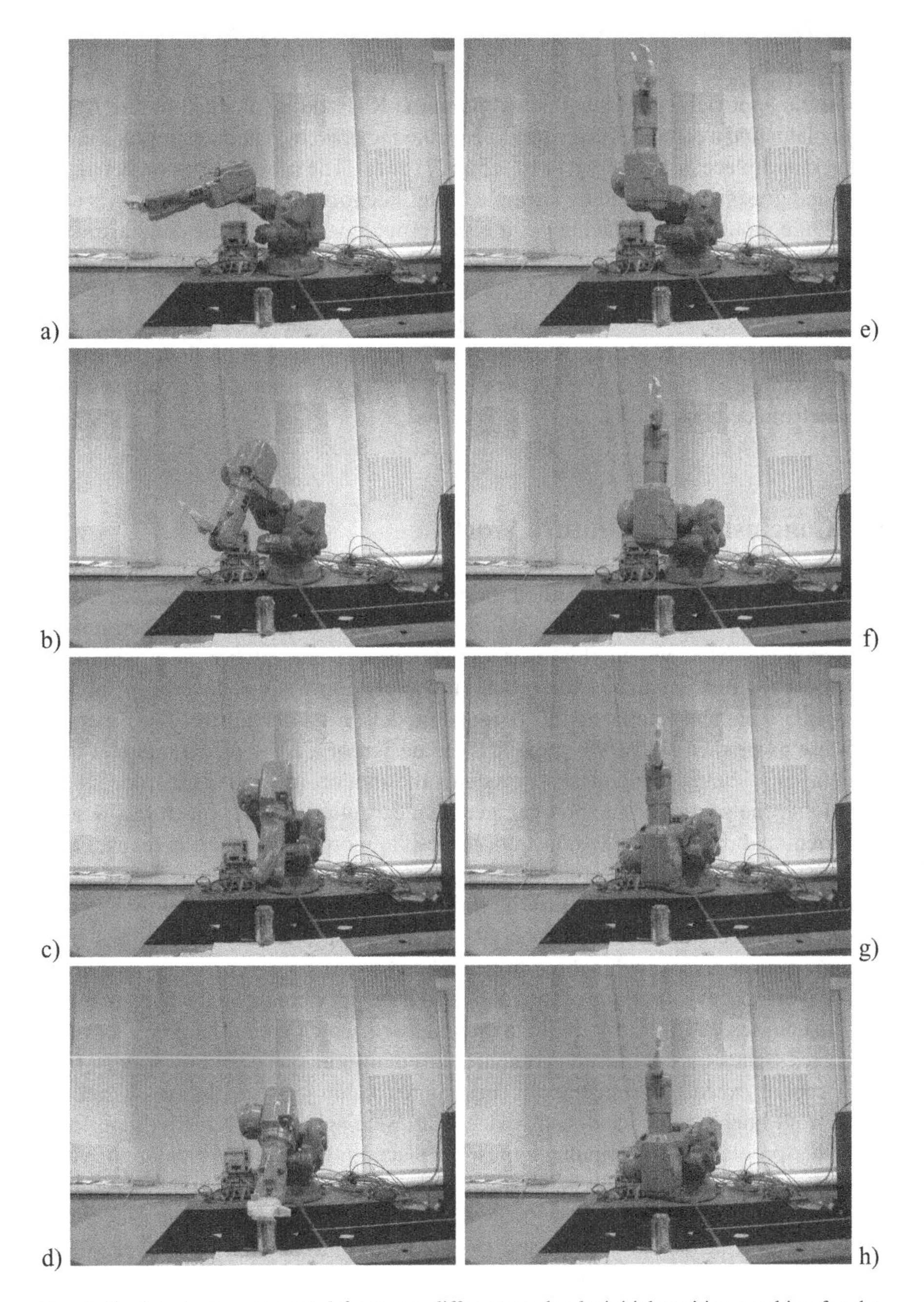

Fig. 1.12 A trajectory generated from two different randomly initial position reaching for the same object. *In the left column (a)–(d)*, a successful reaching motion is generated where the final position is on top of the can. *The right column (e)–(h)* shows a case where the robot reaches an unreachable joint configuration and cannot move along the trajectory

1.5.3 Experiment 3: a Complete Pick-and-Place Task

To test the approach on an integrated system the KTHand is mounted on the ABB manipulator and a pick-and-place task is executed, guided by a demonstration showing pick-and-place task of a box (110 × 56 × 72 mm). The synchronization between reach and grasp is performed by a simple finite state machine. After the grasp is executed, the motion to the placing point is performed by following the demonstrated trajectory (see Sect. 1.2.2). Since the robot grasp pose corresponds approximately to the human grasp pose it is possible for the planner to reproduce the human trajectory almost exactly. This does not mean that the robot actually can always execute the trajectory, due to workspace constraints. The retraction phase follows the same strategy as the reaching motion, but in reverse. Figure 1.13 shows the complete task learned from demonstration.

1.6 Conclusions and Future Work

In this article, we presented a method for programming-by-demonstration of reaching motions for robotic grasping tasks. To allow the robot to interpret the human motions as its own, we employ a hand-state space representation as a common basis between the human and the robot. Human demonstrations have been shown to provide sufficient knowledge to produce models good enough for the robot to them use as its own skills. We presented the design of a NSP, which includes the advantages of fuzzy modeling and executes the motion in hand-state space. It is shown that the suggested method can generate executable robot trajectories based on current and past human demonstrations despite morphological differences. Furthermore we have shown that he robot gains experience from human demonstration. The generalization abilities of the trajectory planner are illustrated by several experiments where an industrial robot arm executes various reaching motions and performs power grasps with a three-fingered hand.

The workspace restrictions of the robot also have to be considered when creating new trajectories. A trajectory might contain regions which are out of reach, or two connected points on the trajectory require different joint space solutions, thus, the robot cannot execute the trajectory. These unreachable joint configurations are a result from operating in hand-state space (Cartesian space). To remedy the effect from this problem the manipulator must be placed at a position/orientation with good reachability. Although it is possible to avoid unreachable joint configurations [34], this will lead to a trajectory which will *not* follow the demonstration. Either the robot performs the task in a way that is possible but not as demonstrated because of joint constraints, or it can ask for more information from the teacher. Other solutions include a mobile platform, a larger robot, or more degrees of freedom (DOF) to mimic the redundancy of the human arm.

In our future work we plan to extend the theoretical and experimental work to include all feasible grasp types of the KTHand. To remedy the effect of the small

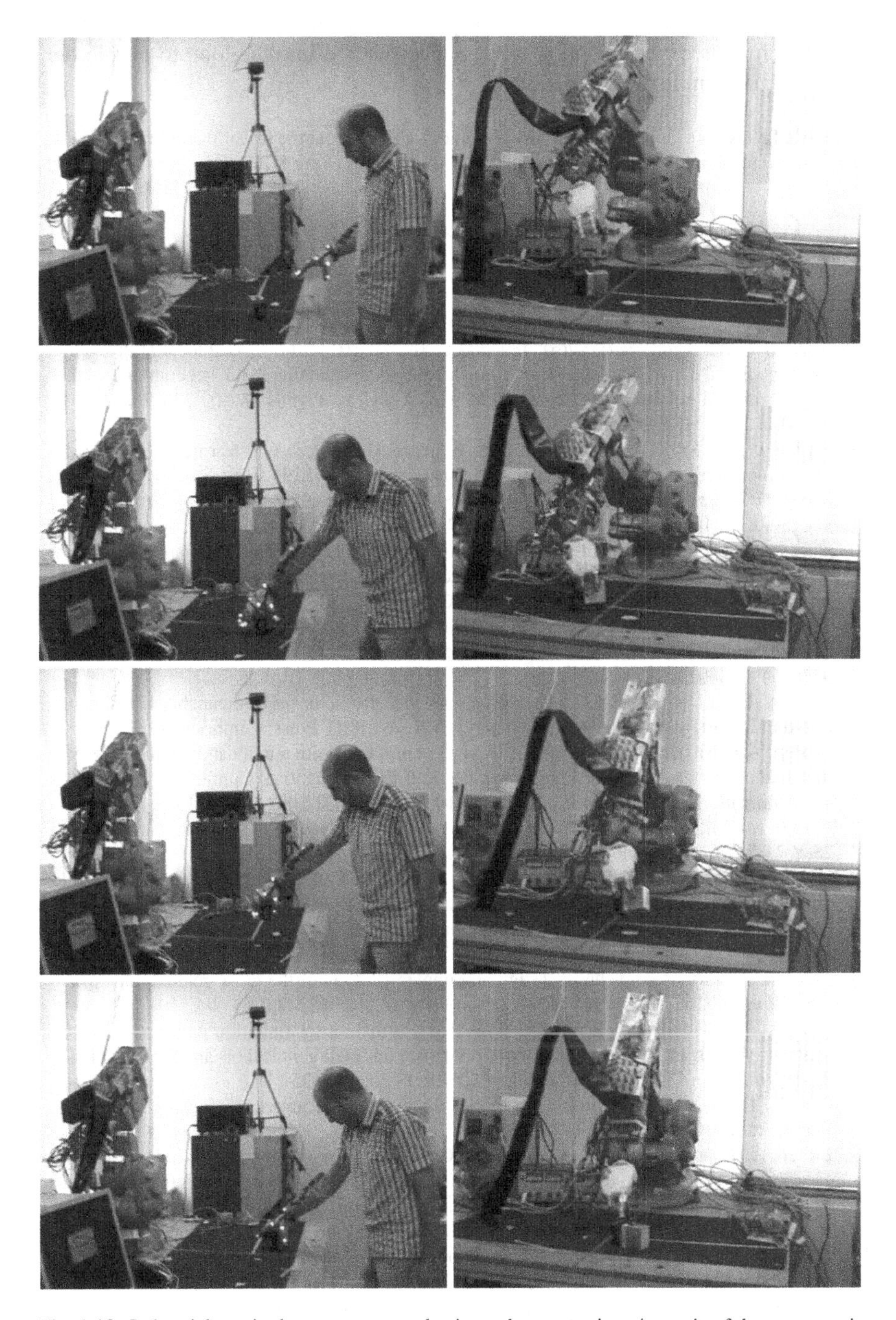

Fig. 1.13 Industrial manipulator programmed using a demonstration. A movie of the sequence is available at: http://www.aass.oru.se/Research/Learning/arsd.html

workspace of the robot a different workspace configuration will be used. Furthermore, the robot's own perception will be incorporated into the loop to enable the robot to learn from its own experience.

Acknowledgements Johan Tegin at Mechatronics Laboratory, at the Royal Institute of Technology, Stockholm, should be acknowledged for providing access to the KTHand.

References

1. Argall, B.D., Chernova, S., Veloso, M., Browning, B.: A survey of robot learning from demonstration. Robot. Auton. Syst. **57**(5), 469–483 (2009). doi:10.1016/j.robot.2008.10.024
2. Calinon, S., Guenter, F., Billard, A.: On learning, representing, and generalizing a task in a humanoid robot. IEEE Trans. Syst. Man Cybern., Part B **37**(2), 286–298 (2007). doi:10.1109/TSMCB.2006.886952
3. Pardowitz, M., Knoop, S., Dillmann, R., Zöllner, R.D.: Incremental learning of tasks from user demonstrations, past experiences, and vocal comments. IEEE Trans. Syst. Man Cybern., Part B **37**(2), 322–332 (2007). doi:10.1109/TSMCB.2006.886951
4. Skoglund, A., Iliev, B., Kadmiry, B., Palm, R.: Programming by demonstration of pick-and-place tasks for industrial manipulators using task primitives. In: IEEE International Symposium on Computational Intelligence in Robotics and Automation, Jacksonville, Florida, 20–23 June 2007, pp. 368–373 (2007). doi:10.1109/CIRA.2007.382863
5. Takamatsu, J., Ogawara, K., Kimura, H., Ikeuchi, K.: Recognizing assembly tasks through human demonstration. Int. J. Robot. Res. **26**(7), 641–659 (2007). doi:10.1177/0278364907080736
6. Nehaniv, C.L., Dautenhahn, K.: The correspondence problem. In: Dautenhahn, K., Nehaniv, C. (eds.) Imitation in Animals and Artifacts, pp. 41–61. MIT Press, Cambridge (2002)
7. Ijspeert, A.J., Nakanishi, J., Schaal, S.: Movement imitation with nonlinear dynamical systems in humanoid robots. In: Proceedings of the 2002 IEEE International Conference on Robotics and Automation, pp. 1398–1403 (2002). doi:10.1109/ROBOT.2002.1014739
8. Hersch, M., Billard, A.G.: Reaching with multi-referential dynamical systems. Auton. Robots **25**(1–2), 71–83 (2008). doi:10.1007/s10514-007-9070-7
9. Iossifidis, I., Schöner, G.: Dynamical systems approach for the autonomous avoidance of obstacles and joint-limits for an redundant robot arm. In: Proceedings of 2006 IEEE International Conference on Robotics and Automation, pp. 580–585 (2006). doi:10.1109/IROS.2006.282468
10. Oztop, E., Arbib, M.A.: Schema design and implementation of the grasp-related mirror neurons. Biol. Cybern. **87**(2), 116–140 (2002). doi:10.1007/s00422-002-0318-1
11. Ijspeert, A., Nakanishi, J., Schaal, S.: Trajectory formation for imitation with nonlinear dynamical systems. In: IEEE International Conference on Intelligent Robots and Systems (IROS 2001), vol. 2, pp. 752–757 (2001). doi:0.1109/IROS.2001.976259
12. Ude, A.: Trajectory generation from noisy positions of object features for teaching robot paths. Robot. Auton. Syst. **11**(2), 113–127 (1993)
13. Billard, A., Epars, Y., Calinon, S., Schaal, S., Cheng, G.: Discovering optimal imitation strategies. Robot. Auton. Syst. **47**(2–3), 69–77 (2004). doi:10.1016/j.robot.2004.03.002
14. Aleotti, J., Caselli, S.: Robust trajectory learning and approximation for robot programming. Robot. Auton. Syst. **54**(5), 409–413 (2006). doi:10.1016/j.robot.2006.01.003
15. Palm, R., Iliev, B., Kadmiry, B.: Recognition of human grasps by time-clustering and fuzzy modeling. Robot. Auton. Syst. **57**(5), 484–495 (2009). doi:10.1016/j.robot.2008.10.012
16. Palm, R., Iliev, B.: Learning of grasp behaviors for an artificial hand by time clustering and Takagi-Sugeno modeling. In: Proceedings of the IEEE International Conference on Fuzzy Systems, Vancouver, BC, Canada, 16–21 July 2006, pp. 291–298 (2006). doi:10.1109/fuzzy.2006.1681728

17. Iliev, B., Kadmiry, B., Palm, R.: Interpretation of human demonstrations using mirror neuron system principles. In: Proceedings of the 6th IEEE International Conference on Development and Learning, Imperial College London, 11–13 July 2007, pp. 128–133 (2007). doi:10.1109/DEVLRN.2007.4354036
18. Takagi, T., Sugeno, M.: Fuzzy identification of systems and its applications to modeling and control. IEEE Trans. Syst. Man Cybern. **SMC-15**(1), 116–132 (1985)
19. Gustafson, D.E., Kessel, W.C.: Fuzzy clustering with a fuzzy covariance matrix. In: Proceedings of the 1979 IEEE CDC, pp. 761–766 (1979)
20. Palm, R., Stutz, C.: Generation of control sequences for a fuzzy gain scheduler. Int. J. Fuzzy Syst. **5**(1), 1–10 (2003)
21. Palm, R., Driankov, D., Hellendoorn, H.: Model Based Fuzzy Control. Springer, Berlin (1997)
22. Tegin, J., Ekvall, S., Kragic, D., Wikander, J., Iliev, B.: Demonstration based learning and control for automatic grasping. J. Intell. Serv. Robot. **2**(1), 23–30 (2009). doi:10.1007/s11370-008-0026-3
23. Delson, N., West, H.: Robot programming by human demonstration: adaptation and inconsistency in constrained motion. In: IEEE International Conference on Robotics and Automation, pp. 30–36 (1996). doi:10.1109/ROBOT.1996.503569
24. Skoglund, A., Iliev, B., Palm, R.: A hand state approach to imitation with a next-state-planner for industrial manipulators. In: Proceedings of the 2008 International Conference on Cognitive Systems, University of Karlsruhe, Karlsruhe, Germany, 2–4 April 2008, pp. 130–137 (2008)
25. Conditt, M.A., Mussa-Ivaldi, F.A.: Central representation of time during motor learning. Proc. Natl. Acad. Sci. USA **96**, 11625–11630 (1999)
26. Palm, R., Iliev, B.: Segmentation and recognition of human grasps for programming-by-demonstration using time-clustering and fuzzy modeling. In: Proceedings of the IEEE International Conference on Fuzzy Systems, London, UK, 23–26 July 2007
27. Bullock, D., Grossberg, S.: VITE and FLETE: neural modules for trajectory formation and postural control. In: Hershberger, W.A. (ed.) Volitonal Action, pp. 253–297. Elsevier Science, Amsterdam (1989)
28. Skoglund, A., Tegin, J., Iliev, B., Palm, R.: Programming-by-demonstration of reach to grasp tasks in hand-state space. In: Proceedings of the 14th International Conference on Advanced Robotics, Munich, Germany, 22–26 June 2009
29. Skoglund, A.: Programming by demonstration of robot manipulators. PhD thesis, Örebro University, June 2009
30. Levine, W.S.: The root locus plot. In: Levine, W.S. (ed.) The Control Handbook, pp. 192–198. CRC Press, Boca Raton (1996)
31. Palm, R., Iliev, B., Kadmiry, B.: Grasp recognition by fuzzy modeling and hidden Markov models (hmm). In: Robot Intelligence. Springer, Berlin (2010)
32. Tegin, J., Wikander, J., Iliev, B.: A sub €1000 robot hand for grasping—design, simulation and evaluation. In: International Conference on Climbing and Walking Robots and the Support Technologies for Mobile Machines, Coimbra, Portugal, September 2008
33. Tegin, J., Ekvall, S., Kragic, D., Iliev, B., Wikander, J.: Demonstration based learning and control for automatic grasping. In: International Conference on Advanced Robotics, Jeju, Korea, August 2007
34. Hersch, M., Billard, A.G.: A biologically-inspired controller for reaching movements. In: Proceedings of the IEEE/RAS-EMBS International Conference on Biomedical Robotics and Biomechatronics, Pisa, pp. 1067–1071 (2006). doi:10.1109/BIOROB.2006.1639233

Chapter 2
Grasp Recognition by Fuzzy Modeling and Hidden Markov Models

Rainer Palm, Boyko Iliev, and Bourhane Kadmiry

Abstract Grasp recognition is a major part of the approach for Programming-by-Demonstration (PbD) for five-fingered robotic hands. This chapter describes three different methods for grasp recognition for a human hand. A human operator wearing a data glove instructs the robot to perform different grasps. For a number of human grasps the finger joint angle trajectories are recorded and modeled by fuzzy clustering and Takagi-Sugeno modeling. This leads to grasp models using time as input parameter and joint angles as outputs. Given a test grasp by the human operator the robot classifies and recognizes the grasp and generates the corresponding robot grasp. Three methods for grasp recognition are compared with each other. In the first method, the test grasp is compared with model grasps using the difference between the model outputs. The second method deals with qualitative fuzzy models which used for recognition and classification. The third method is based on Hidden-Markov-Models (HMM) which are commonly used in robot learning.

2.1 Introduction

The field of human-like robotic hands has attracted significant research efforts in the last two decades aiming at applications like service robots, prosthetic hands and also industrial applications. However, due to the lack of appropriate sensor systems and some unsolved problems with the human-robot interaction such applications are relatively few so far. One particular reason is the difficult programming procedure due to the high dimensionality of grasping and manipulation tasks. An approach to solve this problem is *Programming-by-Demonstration (PbD)* which is used in complex robotic applications such as grasping and dexterous manipulation. That is, the

R. Palm (✉), B. Iliev, and B. Kadmiry
Department of Technology, Orebro University, 70182 Orebro, Sweden
e-mail: rub.palm@t-online.de; boyko.iliev@tech.oru.se; bourhane.kadmiry@tech.oru.se

H. Liu et al. (eds.), *Robot Intelligence*,
Advanced Information and Knowledge Processing,
DOI 10.1007/978-1-84996-329-9_2,

operator performs a task while the robot captures the data by a motion capture device or a video camera and analyzes the demonstrated actions. Then the robot has to recognize these actions and replicate them in a framework of a complex application. One of the most complicated tasks is the recognition procedure because of the ambiguous nature of a human grasp. Different techniques for grasp recognition have been applied in PbD. Kang et al. [1] describe a system which observes, recognizes and maps human grasps to a robot manipulator using a stereo vision system and a data glove. Zoellner et al. [2] use a data glove with integrated tactile sensors where the recognition is based on support vector machines (SVM). Ikeuchi et al. [3] apply Hidden Markov Models (HMMs) to segment and recognize grasp sequences. Ekvall and Kragic [4] use also HMM methods and address the PbD-problem using the arm trajectory as an additional feature for grasp classification. Li et al. [5] use the singular value decomposition (SVD) for the generation of feature vectors of human grasps and support vector machines (SVM) which are applied to the classification problem. Aleotti and Caselli [6] describe a virtual reality-based PbD-system for grasp recognition where only final grasp postures are modeled based on the finger joint angles. Palm and Iliev presented two methods based on fuzzy models [7] and [8]. Having a look at the rich variety of cited methods it is evident that they do not provide equally successful results. Moreover, the experimental setups often include different sensor suits which makes the comparison of results very difficult.

Therefore, in this article we compare three methods. The first two methods are described in detail in [7] and [8], while the third approach is a hybrid method of fuzzy clustering and HMM-methods. We choose to compare our methods with a HMM-approach since the latter is widely used in robot learning and considered as state-of-the-art. All three methods start with fuzzy time clustering. The 1st method, which is the simplest one, classifies a given test grasp using the distances between the time clusters of the test grasp and the time clusters of a set of model grasps [7]. The 2nd method, which is more complex, is based on qualitative fuzzy recognition rules and solves the segmentation problem and the recognition problem at once [8, 9]. The 3rd method deals with fuzzy time clustering and grasp recognition using HMM's [10]. All three methods are tested on the same set of grasp data in order to provide a fair comparison of the methods. This chapter is organized as follows: Sects. 2.2 and 2.3 describe the experimental platform consisting of a data glove and a hand simulation tool. Section 2.4 discusses the learning of grasps by time-clustering and the training of model grasps. Section 2.5 describes the three recognition methods. Section 2.6 presents the experimental results and gives a comparison of the three methods. Finally Sect. 2.7 draws some conclusions and directions for future work.

2.2 An Experimental Platform for PBD

Robotic grasping involves two main tasks: *segmentation* of human demonstrations and *grasp recognition*. The first task is to partition the data record into a sequence

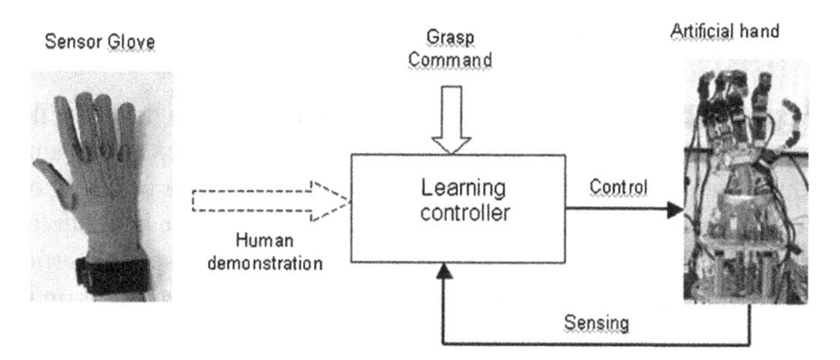

Fig. 2.1 Learning grasp primitives from human demonstrations

of episodes, where each one contains a single grasp. The second task is to recognize the grasp performed in each episode. Then the demonstrated task is (automatically) converted into a program code that can be executed on a particular robotic platform (see Fig. 2.1). If the system is able to recognize the corresponding human grasps in a demonstration, the robot will also be able to perform the demonstrated task by activating the respective grasp primitives.

The experimental platform consists of a hand motion capturing device and a hand simulation environment. The motions of the human operator are recorded by a data glove (CyberGlove) which measures 18 joint angles in the hand and the wrist (see [11]). Since humans mostly use a limited number of grasp types, the recognition process can be restricted to a certain grasp taxonomy, such as those developed by Cutkosky [12] and Iberall [13].

To test the grasp primitives, we developed a simulation model of a five-fingered hand with 3 links and 3 joints in each finger. The simulation environment allows us to perform a kinematic simulation of the artificial hand and its interaction with modeled objects (see Fig. 2.3).

Moreover, we can simulate recorded demonstrations of human operators and compare them with the result from the execution of corresponding grasping primitives. Inspired by the grasp taxonomy of Iberall 15 different grasps have been tested (see Fig. 2.2). These grasps are special cases of the following general classes [13]:

1. cylindrical grasp (grasps 1, 2, 3, 14)
2. power grasp (grasps 4, 5)
3. spherical grasp (grasps 6, 7, 12, 13)
4. extension grasp (grasps 10, 11)
5. precision grasp, nippers pinch (grasps 8, 9)
6. penholder grasp (writing grip) (grasp 15)

The advantage of this selection is that the quality of a grasp recognition both between classes and within a class can be analyzed (see Sect. 2.6: Experiments and Simulations).

We tested the grasping of 15 different objects some of them belonging to same class in terms of an applied type of grasps. For example, *cylinder* and *small bottle* correspond to cylindrical grasp, *sphere* and *cube* to precision grasp, etc.

2.3 Simulation of Grasp Primitives

For the purpose of PbD we need a model of the human hand which allows the simulation of demonstrated grasps. In order to test the grasp primitives a hand simulation was developed with the help of which one can mimic the hand poses recorded by the data glove. This hand model allows to compute the trajectories of both the finger angles and the fingertips. Since the size of the object to be grasped determines the starting and end points of the fingertips to a great extend we used fingertip trajectories instead of finger angles for modeling.

2.3.1 Geometrical Modeling

In order to study grasp primitives and to develop specific grasp models a geometrical simulation of the hand is required (see Fig. 2.3). The hand model consists of 5 fingers which are linked to the wrist in such a way that the poses of a human operator can be displayed in a realistic way. The kinematic relations can be studied by means of the example of a single finger (see Fig. 2.4). Each finger is modeled with 3 links and 3 joints moving like a small planar robot. This turned out to be sufficient for the simulation of the grasp primitives in Fig. 2.2. The calculation of fingertip trajectories requires the formulation of transformations between the fingertips and

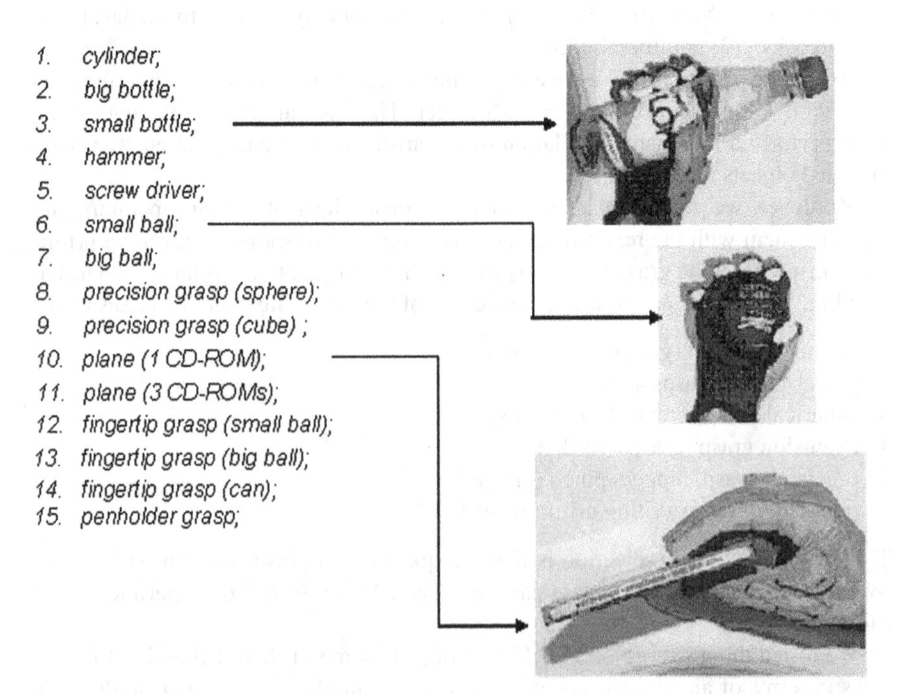

1. cylinder;
2. big bottle;
3. small bottle;
4. hammer;
5. screw driver;
6. small ball;
7. big ball;
8. precision grasp (sphere);
9. precision grasp (cube) ;
10. plane (1 CD-ROM);
11. plane (3 CD-ROMs);
12. fingertip grasp (small ball);
13. fingertip grasp (big ball);
14. fingertip grasp (can);
15. penholder grasp;

Fig. 2.2 Grasp primitives

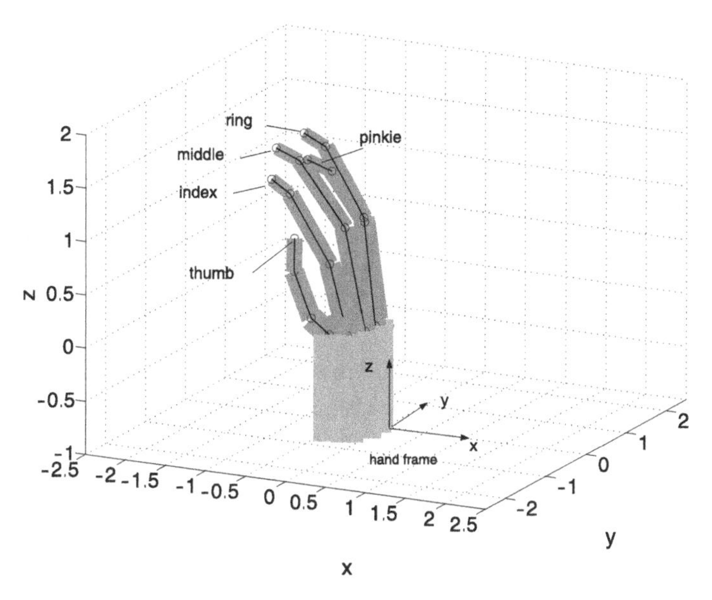

Fig. 2.3 Simulation of the hand

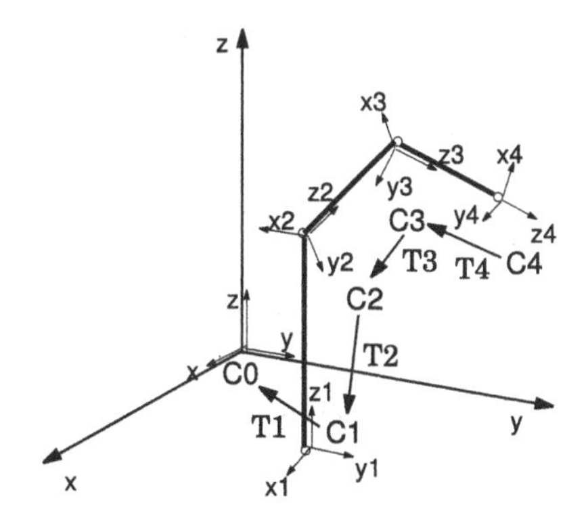

Fig. 2.4 Configuration of a single finger

the base frame of the hand. Translations and rotations between coordinate frames are calculated by homogeneous transformations with the help of which a point $P_{C4} = (x_4, y_4, z_4, 1)^T$ in local homogeneous fingertip coordinates can be transformed into the base frame C_0 by $P_{C0} = T_1 \cdot T_2 \cdot T_3 \cdot T_4 \cdot P_{C4}$. The transformation matrix T_i defines the transformation between the coordinate systems C_i and C_{i-1}.

2.3.2 Modeling of Inverse Kinematics

An important modeling aspect is the *inverse problem* which is crucial both for the simulation of grasps and the control of robotic hands as a feedforward component of the control law [7]. Given the fingertip position vector $\mathbf{x}(t)$, compute the corresponding joint angle vector $\mathbf{q}(t)$. Let

$$\mathbf{x}(t) = \mathbf{f}(\mathbf{q}); \qquad \mathbf{q}(t) = \mathbf{f}^{-1}(\mathbf{x}) \tag{2.1}$$

be the nonlinear direct and inverse transformation for a single finger where the inverse transformation is not necessarily unique for the existing finger kinematics. Therefore we deal with the differential kinematics which makes the computation of the *inverse* much easier. From (2.1) one obtains the differential transformations

$$\dot{\mathbf{x}}(t) = J(\mathbf{q})\dot{\mathbf{q}}; \qquad \dot{\mathbf{q}}(t) = J^{+}(\mathbf{q})\dot{\mathbf{x}} \tag{2.2}$$

where $J(\mathbf{q}) = \frac{\partial \mathbf{q}}{\partial \mathbf{x}}$ is the Jacobian and $J^{+}(\mathbf{q})$ is the pseudo inverse Jacobian. Assuming $\mathbf{x}(t)$ or $\dot{\mathbf{x}}(t)$ to be given from the task, i.e. from captured human demonstrations, the inverse kinematics in (2.2) remains to be computed. In order to avoid the time-consuming calculation of the inverse Jacobian at every time step the inverse differential kinematics is approximated by a TS fuzzy model

$$\dot{\mathbf{q}}(t) = \sum_{i=1}^{c_x} w_i(\mathbf{x}) J_{inv,i}(\mathbf{x}_i) \cdot \dot{\mathbf{x}} \tag{2.3}$$

where $w_i(\mathbf{x}) \in [0, 1]$ is the degree of membership of the vector $\mathbf{x}$ to a cluster C_{xi} with the cluster center $\mathbf{x}_i$. $J_{inv,i}(\mathbf{x}_i)$ are the inverse Jacobians in the cluster centers $\mathbf{x}_i$. c_x is the number of clusters. Due to the errors $\Delta\mathbf{x} = \mathbf{x}(t) - \mathbf{x}_m(t)$ between the desired position $\mathbf{x}(t)$ and the position $\mathbf{x}_m$ computed by the forward kinematics a correction of the angles is calculated via the analytical forward kinematics $\mathbf{x}_m(t) = \mathbf{f}(\mathbf{q}(t))$ of the finger. This changes (2.3) into

$$\dot{\mathbf{q}}(t) = \sum_{i=1}^{c_x} w_i(\mathbf{x}) J_{inv,i}(\mathbf{x}_i) \cdot (\dot{\mathbf{x}} + K \cdot (\mathbf{x}(t) - \mathbf{x}_m(t))). \tag{2.4}$$

It has to be emphasized that the correction or optimization loop using the forward kinematics $\mathbf{f}(\mathbf{q}(t))$ is started at every new time instant and stops until either a lower bound $\|\Delta\mathbf{x}\| < \varepsilon$ is reached or a given number of optimization steps is executed. The gain K has to be determined so that the optimization loop is stable. This TS-modeling is based on a clustering algorithm whose steps are described in the next section in more detail. The degree of membership $w_i(\mathbf{x})$ of an input vector $\mathbf{x}$ belonging to a cluster C_{xi} is defined by a bell-shape-like function

$$w_i(\mathbf{x}) = \frac{1}{\sum_{j=1}^{c_x} \left(\frac{(\mathbf{x}-\mathbf{x}_i)^T M_{xi} (\mathbf{x}-\mathbf{x}_i)}{(\mathbf{x}-\mathbf{x}_j)^T M_{xj} (\mathbf{x}-\mathbf{x}_j}\right)^{\frac{1}{\tilde{m}_x - 1}}} \tag{2.5}$$

M_{xi} define the induced matrices of the input clusters C_{xi}, $(i = 1 \ldots c_x)$, $\tilde{m}_x > 1$ determines the fuzziness of an individual cluster. The complexity of the on-line calculation of (2.4) is much lower than the complexity of (2.2) because (2.4) avoids the on-line calculation of numerous trigonometric functions. The time consuming clustering algorithm leading to the inverse Jacobians $J_{inv,i}$ is computed off-line.

2.4 Modeling of Grasp Primitives

2.4.1 Modeling by Time-Clustering

The recognition of a grasp type is achieved by a model that reflects the *behavior of the hand in time*.

In the following an approach to learning of human grasps from demonstrations by *time-clustering* [7] is shortly described. The result is a set of grasp models for a selected number of human grasp motions. According to Sect. 2.2 experiments were performed in which time sequences for 15 different grasps were collected using a data glove with 18 sensors (see [11]).

Each demonstration has been 10 times repeated by the same test person to collect enough samples of every particular grasp. The time period for a single grasp is about 3 seconds. From those data models for each individual grasp have been developed using fuzzy clustering and Takagi-Sugeno fuzzy modeling [14]. We consider the time instants as model inputs and the 3 finger joint angles as model outputs. Let the angle trajectory of a finger be described by

$$\mathbf{q}(t) = \mathbf{f}(t) \tag{2.6}$$

where $\mathbf{q}(t) \in R^3$, $\mathbf{f} \in R^3$, and $t \in R^+$. Linearization of (2.6) at selected time points t_i yields

$$\mathbf{q}(t) = \mathbf{A}_i \cdot t + \mathbf{d}_i \tag{2.7}$$

where $\mathbf{A}_i = \frac{\Delta \mathbf{f}(t)}{\Delta t}|_{t_i} \in R^3$ and $\mathbf{d}_i = \mathbf{q}(t_i) - \frac{\Delta \mathbf{f}(t)}{\Delta t}|_{t_i} \cdot t_i \in R^3$. Using (2.7) as a local linear model one can express (2.6) in terms of a Takagi-Sugeno fuzzy model [15]

$$\mathbf{q}(t) = \sum_{i=1}^{c} w_i(t) \cdot (\mathbf{A}_i \cdot t + \mathbf{d}_i) \tag{2.8}$$

where $w_i(t) \in [0, 1]$ is the degree of membership of the time point t to a cluster with the cluster center t_i, c is the number of clusters, and $\sum_{i=1}^{c} w_i(t) = 1$.

Let t be the time and $\mathbf{x} = [q_1, q_2, q_3]^T$ the finger angle coordinates. Then the general clustering and modeling steps are

- Choose an appropriate number c_t of local linear models (data clusters)
- Find c_t cluster centers $(t_i, q_{1i}, q_{2i}, q_{3i})$, $i = 1 \ldots c_t$, in the product space of the data quadruples (t, q_1, q_2, q_3) by Fuzzy-c-elliptotype clustering
- Find the corresponding fuzzy regions in the space of input data (t) by projection of the clusters of the product space first into so-called Gustafson-Kessel clusters (GK) and then onto the input space [16]
- Calculate c_t local linear (affine) models (2.8) using the GK clusters from step 2.

The degree of membership $w_i(t)$ of an input data point t to an input cluster C_{ti} is determined by

$$w_i(t) = \frac{1}{\sum_{j=1}^{c_t} \left(\frac{(t-t_i)^T M_{ti} (t-t_i)}{(t-t_j)^T M_{tj} (t-t_j)} \right)^{\frac{1}{\tilde{m}_t - 1}}}. \tag{2.9}$$

The projected cluster centers t_i and the induced matrices M_{ti} define the input clusters C_{ti} $(i = 1 \dots c_t)$. The parameter $\tilde{m}_t > 1$ determines the fuzziness of an individual cluster. A detailed description of this very effective clustering method can be found in [14]. In this way for each of the 15 grasp primitives in Fig. 2.2 a TS-fuzzy model is generated. These so-called *model grasps* are used to identify demonstrated grasps from a test sequence of a given combination of grasps.

2.4.2 Training of Time Cluster Models Using New Data

A grasp model can be built in several ways

- A single user trains the grasp model by repeating the same grasp n times
- m users train the grasp model by repeating the same grasp n times

The 1st model is generated by the time sequences

$$[(t_1, t_2, \dots, t_N)_1 \dots (t_1, t_2, \dots, t_M)_n]$$

and the finger angle sequences

$$[(\mathbf{q}_1, \mathbf{q}_2, \dots, \mathbf{q}_N)_1 \dots (\mathbf{q}_1, \mathbf{q}_2, \dots, \mathbf{q}_M)_n].$$

The 2nd model is generated by the time sequences

$$[((t_1, t_2, \dots, t_N)_1^1 \dots (t_1, t_2, \dots, t_M)_n^1) \dots ((t_1, t_2, \dots, t_N)_1^m \dots (t_1, t_2, \dots, t_M)_n^m)]$$

and the finger angle sequences

$$[((\mathbf{q}_1, \mathbf{q}_2, \dots, \mathbf{q}_N)_1^1 \dots (\mathbf{q}_1, \mathbf{q}_2, \dots, \mathbf{q}_M)_n^1) \dots$$
$$((\mathbf{q}_1, \mathbf{q}_2, \dots, \mathbf{q}_N)_1^m \dots (\mathbf{q}_1, \mathbf{q}_2, \dots, \mathbf{q}_M)_n^m)]$$

where m is the number of users in the training process, N, M are lengths of time sequences where $N \approx M$.

Once a particular grasp model has been generated it might be necessary to take new data into account. These data may originate from different human operators to cover several ways of performing the same grasp type. Let for simplicity the old model be built by a time sequence $[t_1, t_2, \dots, t_N]$ and a respective finger angle sequence

$$[\mathbf{q}_1, \mathbf{q}_2, \dots, \mathbf{q}_N].$$

The old model is then represented by the input cluster centers t_i and the output cluster centers $\mathbf{q}_i$ $(i = 1 \dots c)$. It is also described by the parameters $\mathbf{A}_i$ and $\mathbf{d}_i$ of the local linear models. Let

$$[\tilde{t}_1, \tilde{t}_2, \dots, \tilde{t}_M], [\tilde{\mathbf{q}}_1, \tilde{\mathbf{q}}_2, \dots, \tilde{\mathbf{q}}_M]$$

be the new training data. A new model can be built by "chaining" the old and the new training data leading for the time sequences to

$$[t_1, t_2, \dots, t_N, \tilde{t}_1, \tilde{t}_2, \dots, \tilde{t}_M]$$

and for the finger angle sequences to

$$[\mathbf{q}_1, \mathbf{q}_2, \ldots, \mathbf{q}_N, \tilde{\mathbf{q}}_1, \tilde{\mathbf{q}}_2, \ldots, \tilde{\mathbf{q}}_M].$$

The result is a model that involves properties of the old model and the new data. If the old sequence of data is not available, a corresponding sequence can be generated by running the old model with the time instants

$$[t_1, t_2, \ldots, t_N]$$

as inputs and the finger angles

$$[\mathbf{q}_1, \mathbf{q}_2, \ldots, \mathbf{q}_N]$$

as outputs.

2.5 Recognition of Grasps—Three Methods

In the previous section we showed that TS fuzzy models can be successfully used for modeling and imitation of human grasp behaviors. Now, we will show that they can also be used for classification of grasps in data from recorded human demonstrations. If we just observe captured motions of a human arm while executing several grasp actions it is difficult to identify the exact moment when a grasp sequence starts and ends. Related research shows that this task can be solved efficiently only by fusion of additional information sources such as tactile sensing and vision (see [3] and [4]). Since the scope of this chapter is only the recognition we assume the segmentation already to be finished. In the following we present three different recognition methods all of them being based on the time clustering of human grasps [10]. The first method classifies a test grasp by comparing the time clusters of the test grasp and a set of model grasps. The second method uses fuzzy recognition rules for segmentation and recognition. The third method classifies a test grasp using HMM which are applied to the output cluster centers of the grasp models. It should be stressed that methods 1 and 2 are related with each other both of them using distances between fuzzy clusters for recognition. Method 3 is a completely different approach using a probabilistic approach for recognition and classification.

2.5.1 Recognition of Grasps Using the Distance Between Fuzzy Clusters

Let the model of each grasp have the same number of clusters $i = 1 \ldots c$ so that each duration T_l $(l = 1 \ldots L)$ of the l-th grasp is divided into $c - 1$ time intervals Δt_i, $i = 2 \ldots c$ of the same length. Let the grasps be executed in an environment comparable with the modeled grasp in order to avoid calibration and re-scaling procedures. Furthermore let

$$\begin{aligned} V_{model l} &= [V_{index}, V_{middle}, V_{ring}, V_{pinkie}, V_{thumb}]_l, \\ V_{index l} &= [\mathbf{q}_1, \dots, \mathbf{q}_i, \dots, \mathbf{q}_c]_{index,l}, \\ &\vdots \\ V_{thumb l} &= [\mathbf{q}_1, \dots, \mathbf{q}_i, \dots, \mathbf{q}_c]_{thumb,l}, \\ \mathbf{q}_i &= [q_1, q_2, q_3]^T \end{aligned} \tag{2.10}$$

where matrix $V_{model l}$ includes the output cluster centers $\mathbf{q}_i$ of every finger for the l-th grasp model. $\mathbf{q}_i$ is the vector of joint angles of each finger.

A model of the grasp to be classified (the test grasp) is built by the matrix

$$V_{grasp} = [V_{index}, V_{middle}, V_{ring}, V_{pinkie}, V_{thumb}]_{grasp}. \tag{2.11}$$

A decision on the grasp is made by applying the Euclidean matrix norm

$$N_l = \| V_{model l} - V_{grasp} \|. \tag{2.12}$$

The unknown grasp is classified to the grasp model with the smallest norm $\min(N_l), l = 1 \dots L$ and the recognition of the grasp is finished.

2.5.2 Recognition Based on Qualitative Fuzzy Recognition Rules

The goal of this method is to recognize and classify all individual grasps types in a data sequence containing a combination of several grasps. That is, it also performs the segmentation of the data sequence. The identification of a grasp from a combination of grasps is based on a recognition model. This model is represented by a set of recognition rules using the model grasps mentioned in the last section. The generation of the recognition model is based on the following steps:

1. Computation of distance norms between a test grasp combination and the model grasps involved.
2. Computation of extrema along the sequence of distance norms.
3. Formulation of a set of fuzzy rules reflecting the relationship between the extrema of the distance norms and the model grasps.
4. Computation of a vector of similarity degrees between the model grasps and the grasp combination.

2.5.2.1 Distance Norms

Let, for example, $grasp_2$, $grasp_5$, $grasp_7$, $grasp_{10}$, $grasp_{14}$ be a combination of grasps taken from the list of grasps shown in Fig. 2.2. In the training phase a time series of these grasps is generated using the existing time series of the corresponding

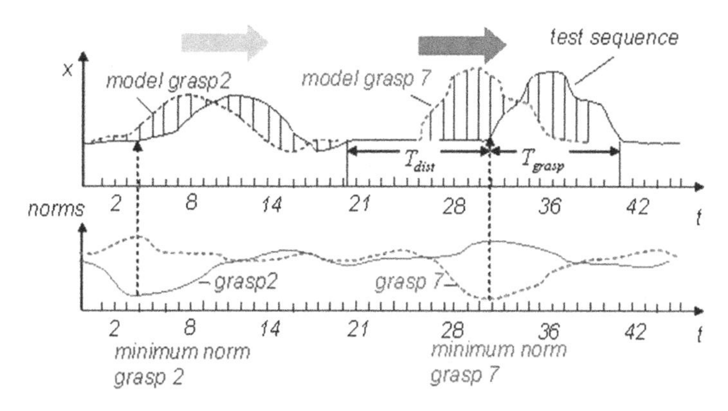

Fig. 2.5 Overlap principle

grasp models. Then each of the model grasps $i = 2, 5, 7, 10, 14$ is shifted along the time sequence of the grasp combination and compared with parts of it while taking the norm2 $\|\mathbf{Q}_{ci} - \mathbf{Q}_{mi}\|$ between the difference of the finger angles

$$\mathbf{Q}_{mi} = (\mathbf{q}_m(t_1), \dots \mathbf{q}_m(t_{n_c}))^T{}_i$$

of a $grasp_i$ and the finger angles of the grasp combination

$$\mathbf{Q}_{ci} = (\mathbf{q}_{ci}(\tilde{t}_1), \dots, \mathbf{q}_c(\tilde{t}_{n_c}))^T.$$

The vectors $\mathbf{q}_m$ and $\mathbf{q}_c$ include the 3 finger angles for each of the 5 fingers. Because of scaling reasons the norm of the difference is divided by the norm $\|\mathbf{Q}_{mi}\|$ of the model grasp. Then we obtain for the scaled norm

$$n_i = \frac{\|\mathbf{Q}_{ci} - \mathbf{Q}_{mi}\|}{\|\mathbf{Q}_{mi}\|} \tag{2.13}$$

where n_i are functions of time. With this for each grasp $i = 2, 5, 7, 10, 14$ a time sequence $n_i(t_1)$ is generated. Once the model grasp starts to overlap a grasp in the grasp combination, the norms n_i reach an extremum at the highest overlap which is either a minimum or a maximum (see Fig. 2.5).

2.5.2.2 Extrema in the Distance Norms and Segmentation

Using a model $grasp_i$ for comparison with a sequence of M test grasps the norm n_i forms individual patterns at M distinct time intervals of the norm sequence. Within each of these time intervals the norm sequence n_i reaches an extremum, i.e. either a local minimum or a maximum. In order to find the local extrema in n_i the total time interval T_{n_i} of n_i is partitioned into l time slices within which the search takes place (see Fig. 2.6). To be able to identify all relevant extrema, the lengths T_{slice} of the time slices have to be bounded by

$$T_{grasp,min} < T_{slice} < T_{dist,min}/2 \tag{2.14}$$

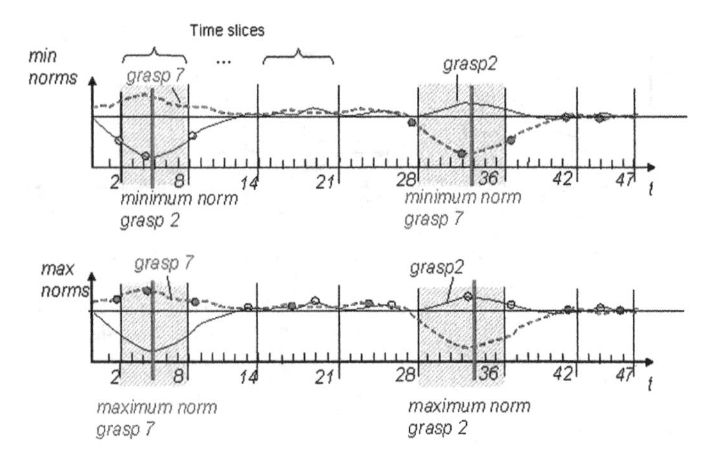

Fig. 2.6 Time slices

where $T_{grasp,min}$ is the minimum time length of a grasp. $T_{dist,min}$ is the minimum time distance between the end of a grasp and the starting point of a new grasp which is equal to the length of the pause. $T_{grasp,min}$ and $T_{dist,min}$ are supposed to be known. This search yields two pairs of vectors

$$\begin{aligned} \mathbf{z}_{min\,i} &= (z_{min,1i}, \ldots, z_{min,li})^T, \\ \mathbf{t}_{min\,i} &= (t_{min,1i}, \ldots, t_{min,li})^T \end{aligned} \tag{2.15}$$

and

$$\begin{aligned} \mathbf{z}_{max\,i} &= (z_{max,1i}, \ldots, z_{max,li})^T, \\ \mathbf{t}_{max\,i} &= (t_{max,1i}, \ldots, t_{max,li})^T \end{aligned} \tag{2.16}$$

where $l = \lceil T_{ni}/T_{slice} \rceil$. The elements of $\mathbf{z}_{min\,i}$ and $\mathbf{z}_{max\,i}$ contain l absolute values of local minima and maxima of n_i, respectively. $\mathbf{t}_{min\,i}$ and $\mathbf{t}_{max\,i}$ contain the corresponding l time stamps of the local minima and maxima. Usually there are more elements (extrema) included in (2.15) and (2.16) than grasps exist in the grasp sequence $l \geq M$. The segmentation task is to find the time slices that include the beginnings of grasps. To deal with an unknown number of grasps solutions different strategies are possible. A 'soft' solution requires a variable number of time clusters and a repetitive search for the most likely starting points of grasps. A mixed hardware-software solution is to utilize sensor information about the established contact between the fingers and the object to be grasped. In the following we assume the number M of grasps in a grasp sequence to be known. A segmentation procedure finds those extrema that indicate the starting points of the grasps. The segmentation is done by time clustering where the time vectors $\mathbf{t}_{max\,i}$ and $\mathbf{t}_{min\,i}$ are the model inputs, $\mathbf{z}_{min\,i}$ and $\mathbf{z}_{max\,i}$ are the model outputs. We expect the elements of $\mathbf{z}_{max\,i}$ and $\mathbf{z}_{min\,i}$ to form M clusters $\mathbf{t}_{seg} = (t_{seg,1} \ldots t_{seg,M})^T$. The result of the clustering procedure is a vector of M time cluster centers pointing to the starting points of the grasps. For each time point $t_{seg,r}$ there is a pair $(z_{min,ij}, z_{max,ij})$. Index j denotes a $grasp_j$ in the grasp sequence executed at time $t_{seg,r}$. Index $i = 2, 5, 7, 10, 14$ denote the model $grasps_i$. This finalizes the segmentation procedure.

2.5.2.3 Set of Fuzzy Rules

The two sets of vectors $\mathbf{z}_{mini}$ and $\mathbf{z}_{maxi}$ build '*fingerprint patterns*' for each grasp in a specific grasp combination. On the basis of these patterns a set of rules decides whether a special combination of minima and maxima by consideration of their absolute values belong to a certain grasp or to another one. Obviously, for a selected grasp these patterns change with the change of a grasp combination. For example, the pattern for $grasp_2$ in the grasp combination 2, 5, 7, 10, 14 differs significantly from the pattern in grasp combination 1, 2, 3, 4, 5 etc. This is taken into account by the formulation of an individual set of rules for each grasp combination. In order to recognize a model $grasp_i$ from a specific grasp combination a set of 5 rules is formulated, one rule for each grasp in the combination.

A general recognition rule for $grasp_i$ to be identified from the combination reads:

$$\begin{aligned}
&\text{IF } (n_j \text{ is } ex_{ji}) \text{ AND} \ldots \\
&\text{AND } (n_k \text{ is } ex_{ki}) \\
&\text{THEN } grasp \text{ is } grasp_i
\end{aligned} \tag{2.17}$$

Rule (2.17), for example, can be read

"IF (norm n_2 of model $grasp_2$ is $max_{2,5}$)

...

AND (norm n_{14} of model $grasp_{14}$ is $max_{14,5}$)

THEN grasp is $grasp_5$".

The full rule to identify $grasp_5$ reads

$$\begin{aligned}
&\text{IF } (n_2 \text{ is } max_{2,5}) \\
&\text{AND } (n_5 \text{ is } min_{5,5}) \\
&\text{AND } (n_7 \text{ is } max_{7,5}) \\
&\text{AND } (n_{10} \text{ is } min_{10,5}) \\
&\text{AND } (n_{14} \text{ is } max_{14,5}) \\
&\text{THEN } grasp \text{ is } grasp_5
\end{aligned} \tag{2.18}$$

$j = 2 \ldots k = 14$ are the indexes of the grasps in the grasp combination. i is the index of $grasp_i$ to be identified. ex_{ji} indicate fuzzy sets of local extrema which can be either minima min_{ji} or maxima max_{ji}. Extrema appear at the time points $\tilde{t} = t_j$ at which *model* $grasp_i$ meets $grasp_j$ in the grasp combination with a maximum overlap.

Let the total extremum $z_{ex_{tot}}$ either be a total minimum $z_{min_{tot}}$ or a maximum $z_{max_{tot}}$ over all 5 rules and all time slices (see Fig. 2.7)

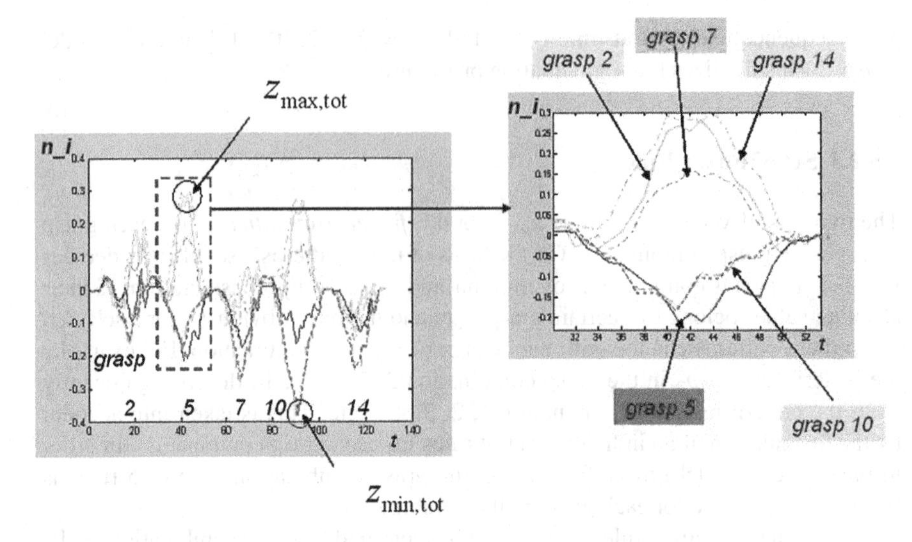

Fig. 2.7 Norms of a grasp sequence

$$z_{min_{tot}} = \min(z_{j_{min,i}}), \qquad z_{max_{tot}} = \max(z_{j_{max,i}}),$$
$$j = 1, \ldots, l; \; i = 2, 5, 7, 10, 14. \tag{2.19}$$

Then a local extremum $z_{j_{ex,i}}$ can be expressed by the total extremum $z_{ex_{tot}}$ and a weight $w_{ji} \in [0, 1]$

$$z_{j_{ex,i}} = w_{ji} \cdot z_{ex_{tot}}, \qquad z_{j_{min,i}} = w_{ji} \cdot z_{min_{tot}},$$
$$z_{j_{max,i}} = w_{ji} \cdot z_{max_{tot}}, \qquad j, i = 2, 5, 7, 10, 14. \tag{2.20}$$

2.5.2.4 Similarity Degrees

The special form of data requires the design of a specific similarity degree and a regarding membership function. From the time plots in Fig. 2.7 of the norms n_i for the training sets the analytical form of (2.18) for the identification of $grasp_i$ is chosen as follows

$$\mathbf{a}_i = \prod_j \mathbf{m}_{ji}, \quad j = 2, 5, 7, 10, 14,$$
$$\mathbf{m}_{ji} = \mathrm{Exp}(-|w_{ji} \cdot \mathbf{z}_{ex_{tot}} - \mathbf{z}_{ex\,i}|), \; \mathbf{z}_{ex\,i} = (z_{1_{ex,i}}, \ldots, z_{l_{ex,i}})^T, \tag{2.21}$$

$\mathbf{a}_i = (a_{1,i}, \ldots, a_{l,i})^T$; $a_{m,i} \in [0, 1]$, $m = 1 \ldots l$ is a vector of *similarity degrees* between the *model* $grasp_i$ and the individual grasps 2, 5, 7, 10, 14 in the grasp combination at the time point t_m. The vector $\mathbf{z}_{ex\,i}$ represents either the vector of minima $\mathbf{z}_{min\,i}$ or maxima $\mathbf{z}_{max\,i}$ of the norms n_i, respectively.

The product operation in (2.21) represents the AND-operation in the rules (2.18). The exponential function in (2.21) is a *membership function* specially designed for

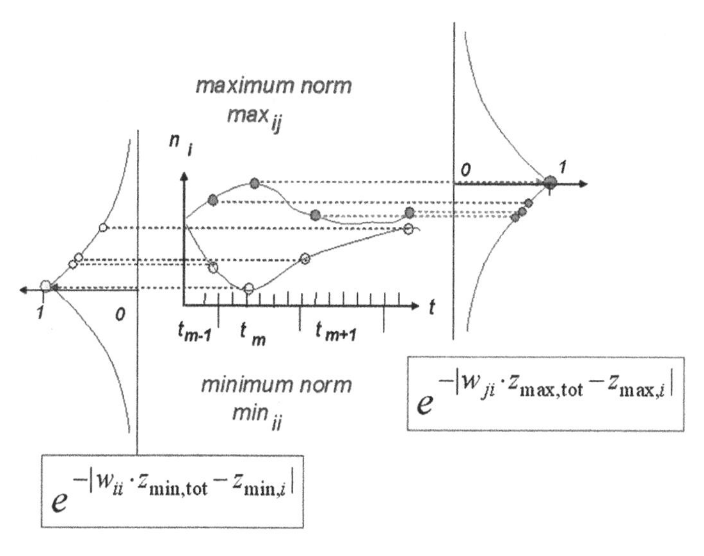

Fig. 2.8 Grasp membership functions

the indication of the distance of a norm n_i to a local extremum $w_{ji} \cdot z_{ex_{tot}}$. With this the exponential function reaches its maximum at exactly that time point t_m when $grasp_i$ in the grasp combination has its local extremum (see, e.g., Fig. 2.8).

If, for example, $grasp_5$ occurs at the time point t_m in the grasp combination then we obtain for $a_{m,5} = 1$. All the other grasps lead to smaller values of $a_{k,i}$, $k = 2, 7, 10, 14$. With this the type of grasp is identified and the grasp recognition is finished.

2.5.3 Recognition Based on Time-Cluster Models and HMM

The task is to classify an observation sequence of a test grasp given a set of observation sequences of model grasps using HMM. The HMM used here are of discrete nature which requires the formulation of a of number discrete states and discrete observations. One condition for the use of HMM in our approach is that all model grasps and the test grasp to be recognized are modeled by time-clustering described before.

The elements of a discrete HMM can be described in the compact notation [17]

$$\lambda = (A, B, \pi, N, M) \tag{2.22}$$

where N is the number of states S, M is the number of observations O, $A = \{a_{ij}\}$ is the matrix of state transition probabilities, $B = \{b_{jk}\}$ is the observation symbol probability of symbol O_k in state j, π is the initial state distribution vector. As an example, Figs. 2.9 and 2.10 show for grasp 1 (cylinder, only *close operation*) the graphs of the initially chosen state transitions $\{a_{ij}\}$ and the state transitions after

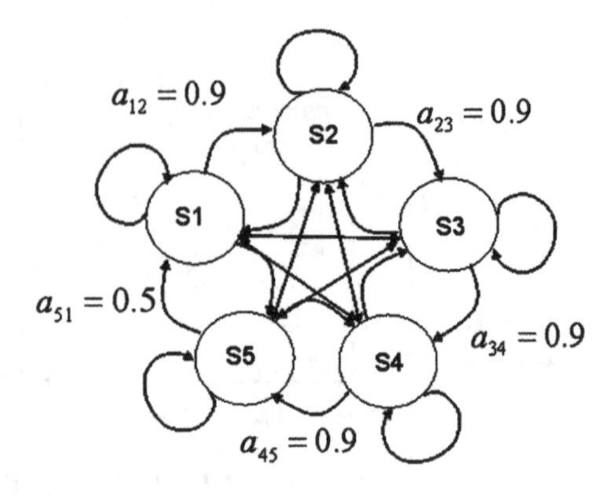

Fig. 2.9 Initial state transitions

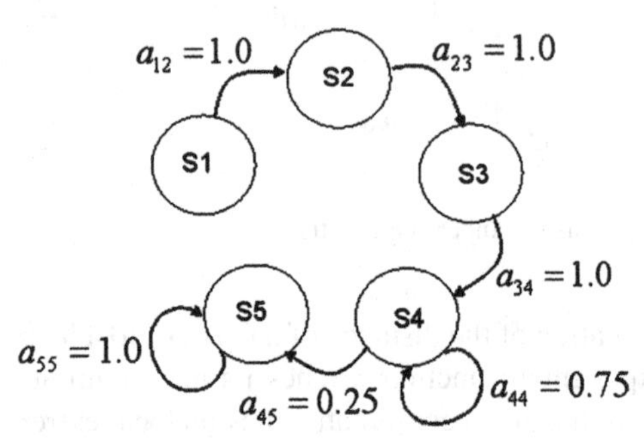

Fig. 2.10 Computed state transitions

the computation via HMM, respectively. Connections in Fig. 2.9 without explicit transition probabilities are denoted as $a_{ij} = 0.1$. Observe that after the computation most of the connections in the initial graph can be cut because of $a_{ij} = 0$.

To prepare the HMM for the recognition a number of steps has to be done:

Step 1: Determine a number N of states S. The states need not necessarily to be directly connected with a physical meaning, but it is of high advantage to do so. Therefore, $M = 5$ states are chosen getting the following labels:

state $S1$: open hand
state $S2$: half open hand
state $S3$: middle position
state $S4$: half closed hand
state $S5$: closed hand

Step 2: Generate a number M of discrete observations O. To generate discrete observations one has to deal first with the continuous observations, meaning the output cluster centers in V_{grasp} and the corresponding joint angles $\mathbf{q}_i$. It should be mentioned that the clustering process leads to vectors of cluster centers whose elements are, although being 'labeled' by a time stamp, not sorted in an increasing order of time. Since clustering of several grasps is done independently of each other the orders of time stamps of the cluster centers are in general different. This makes a comparison of test clusters V_{grasp} and model clusters V_{model} impossible. Therefore after time-clustering has been performed the output clusters have to be sorted in an increasing order of time. In the following, cluster centers are assumed to be sorted in that way. Next, one has to transform the continuous output cluster centers $V_{model}(i)$, $i = 1 \dots 10$ of the model into discrete numbers or 'labels'. If one would attach each cluster center an individual label one would obtain $M = 10 \times 15 = 150$ observation labels, 10—number of clusters, 15—number of grasps. This number of observations is unnecessarily high because some of the cluster centers form almost the same hand poses. Therefore two observations are reserved for the starting pose and end pose of all grasps since it can be assumed that every grasp starts and ends with nearly the same pose. Then, three poses for each grasp are chosen at the cluster numbers $(3, 5, 6)$ which makes $M = 3 \times 15 + 2 = 47$ observations. The result is obviously a set of possible observations labeled by the numbers $1 \dots 47$ representing 47 poses of 15 time-clustering models of grasps. In order to label a specific pose of a given grasp one finds the minimal norms

$$I_j(i) = \min(\| V_{grasp_j}(i) - out_1 \|, \dots, \| V_{grasp_j}(i) - out_{47} \|), \quad i = 1 \dots 10 \tag{2.23}$$

where $I_j(i) \in [1 \dots 47]$ is the observation label, $i \in [1 \dots 10]$ is the number of a time cluster for test grasp $j \in [1 \dots 15]$, $O(k) = V_{model m}(l)$, $k \in [1 \dots 47]$ is the k-th observation, $m \in [1 \dots 15]$ is a corresponding model grasp, $l \in [2, 3, 5, 6, 9]$ is a corresponding number of a time cluster in model grasp m. This procedure is done for all model grasps V_{model} with the result of 15 sequences $I_j(i)$ of 10 observations each, and for the test grasp V_{grasp} to be recognized.

Step 3: Determine the initial matrices $A \in R^{M \times M}$, $B \in R^{N \times M}$ and the initial state distribution vector $\pi \in R^{1 \times N}$. Since in the experiments the hand always starts to move with almost the same pose and keeps on moving through the states defined above we can both estimate the initial matrices A, and B, and the initial state distribution vector π easily.

Step 4: Generate 15 observation sequences $O_{train} \in R^{10 \times 15}$ for the 15 model grasps according to *step 2*.

Step 5: Generate 1 observation sequence $O_{test} \in R^{10 \times 1}$ for the test grasp according to *step 2*.

Step 6: Train the HMM with every model sequence O_{train} separately using the iterative expectation-modification procedure (EM), also known as Baum-Welsch method. The training process evaluates a log-likelihood LL of the trained model during iteration and stops as the change of LL undergoes a certain threshold. Observe here that $LL \leq 0$.

Step 7: Classify the observation sequence O_{test} by evaluating the log-likelihood LL of the m-th trained HMM for a model grasp m given the test data O_{test}. In addition, the most probable sequence of states using the Viterbi algorithm is computed.

Step 8: Compute the most probable model grasp number m to be closest to the test model by computing $\max(LL_i)$, $i = 1 \dots 15$. With step 8 the grasp recognition is completed.

2.6 Experiments and Simulations

In this section time clustering and fuzzy modeling results are presented first. Then an experimental evaluation of the three methods for grasp recognition follows together with a comparison of the three methods.

2.6.1 Time Clustering and Modeling

The choice of the numbers of clusters both for the fingertip models and for the inverse kinematics depend on the quality of the resulting TS fuzzy models. On the basis of a performance analysis for each grasp and finger, 10 fingertip position models with 10 cluster centers have been generated from collected data.

Furthermore, 3 inverse Jacobian models for each grasp primitive and finger with 3 cluster centers have been built which are 15 Jacobians to be computed off-line. Since there are 3 angles (q_1, q_2, q_3) and 3 fingertip coordinates (x, y, z) for a single finger the Jacobians and their inverses are 3×3 square matrices. The 3rd link of each finger (next to the fingertip) does not have a sensor in the data glove. Therefore the angle of this link gets a fixed value greater than zero so that neither ill-conditioned Jacobians nor their inverses can computationally occur. For the identification of inverse Jacobians small random noise excitation is added to the angles to prevent ill-conditioned Jacobians while modeling from data. The motion of a grasp lasts 3.3 s in the average which adds up to 33 timesteps $\Delta t = 0.1$ s. The time clustering procedure results in the cluster centers $t_i = 2.04, 5.43, 8.87, 12.30, 15.75, 19.19, 22.65, 26.09, 29.53, 32.94$ where the time labels are measured in steps of $\Delta t = 0.1$ s. The time cluster centers are then complemented by the corresponding cluster centers for the x, y, z coordinates of the fingertips. This equidistant spacing can be found for every individual grasp primitive as a result of the time clustering. Figures 2.11, 2.12, 2.13, 2.14 and 2.15 shows modeling results for grasp 10 (plane (1 CD-ROM)) for the x,

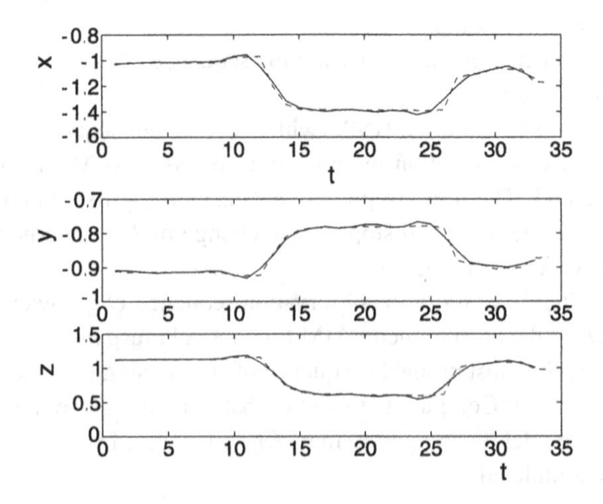

Fig. 2.11 Index finger, original:*solid*, model:*dashed*

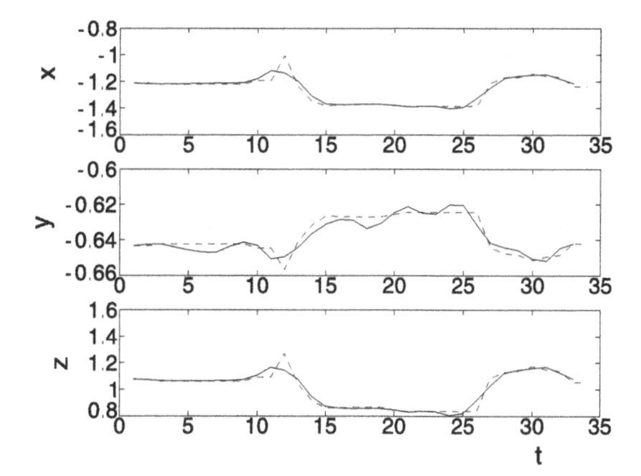

Fig. 2.12 Middle finger, original:*solid*, model:*dashed*

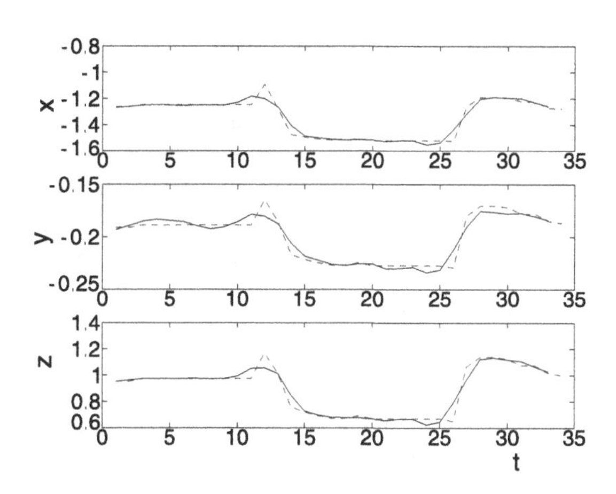

Fig. 2.13 Ring finger, original:*solid*, model:*dashed*

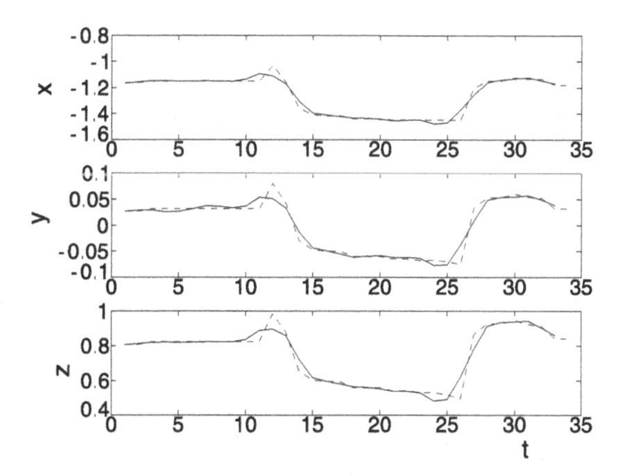

Fig. 2.14 Pinkie finger, original:*solid*, model:*dashed*

y, and z coordinates for the index, middle, ring, and pinkie finger plus the thumb. These results show a good or even excellent modeling quality.

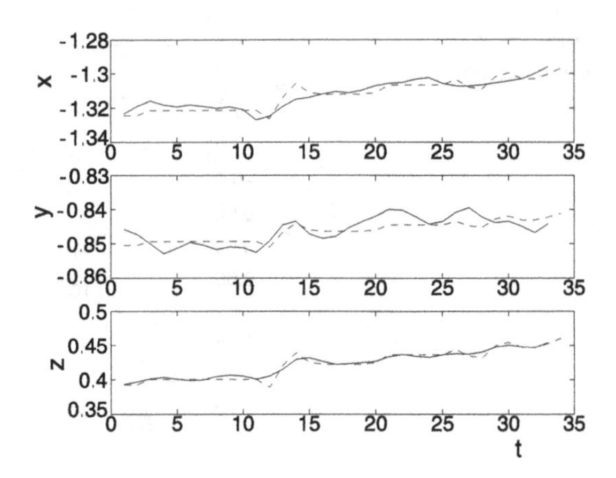

Fig. 2.15 Thumb finger, original:*solid*, model:*dashed*

2.6.2 Grasp Segmentation and Recognition

In this section an experimental evaluation of the three methods for grasp recognition is presented and a comparison of the methods is made. 10 test grasps for each of the 15 different grasp primitives have been tested according to Sect. 2.2. A grasp starts with an open hand and is completed when the fingers establish contact with the object. The experimental results are divided into 3 groups of recognition rates:

1. grasps with a recognition rate $\geq 75\%$
2. grasps with a recognition rate $< 75\% - \geq 50\%$
3. grasps with a recognition rate $< 50\%$.

In the following, the recognition rates of the three discussed methods are listed in Tables 2.1, 2.2, and 2.3. The experimental results confirm the assumption that distinct grasps can be discriminated quite well from each other while the discrimination between similar grasps is difficult. Therefore, merging of similar grasps and building of larger classes can improve the recognition process significantly. Examples of such classes are grasps (4, 5, 15), grasps (10, 11), and grasp (8, 9).

Table 2.1 shows the recognition rates for method 1.

The 1st group with a recognition rate $\geq 75\%$ is the largest one where 4 of 7 grasps show a recognition rate 100%. It follow the equally large groups 2 and 3. Table 2.2 shows the recognition rates for method 2. In this experiment 12 grasp combinations of 5 grasps each have been tested. It could be shown that grasps with distinct maxima and minima in their n_i patterns can be recognized better than grasps without this feature. Reliable grasps are also robust against variations in the time span of an unknown test grasp compared to the time span of the respective model grasp. Our results show that this method can handle a temporal difference up to 20%. By temporal difference we mean the difference in the length of the test grasp and the respective model grasp. The 1st group with a recognition rate $\geq 75\%$ is again the largest one where 3 of 8 grasps show a recognition rate 100%.

Table 2.3 shows the recognition rates for method 3. The 2nd group is here the largest one with a recognition rate $< 75\%$, $\geq 50\%$ followed by the 1st group where

Table 2.1 Recognition rates, method 1

Class	Grasp	Percentage
$\geq 75\%$	4. Hammer	100%
	8. Precision. grasp sphere	87%
	10. Small plane	100%
	11. Big plane	85%
	12. Fingertip small ball	100%
	14. Fingertip can	100%
	15. Penholder grip	85%
$< 75\%, \geq 50\%$	1. Cylinder	71%
	2. Big bottle	57%
	3. Small bottle	57%
	13. Fingertip big ball	71%
$< 50\%$	5. Screwdriver	0%
	6. Small ball	14%
	7. Big ball	28%
	9. Precision grasp cube	42%

Table 2.2 Recognition rates, method 2

Class	Grasp	Percentage
$\geq 75\%$	4. Hammer	100%
	5. Screwdriver	93%
	8. Precision. grasp sphere	80%
	9. Precision grasp cube	100%
	10. Small plane	100%
	11. Big plane	88%
	13. Fingertip big ball	75%
	15. Penholder grip	75%
$< 75\%, \geq 50\%$	1. Cylinder	55%
	2. Big bottle	60%
	3. Small bottle	66%
	6. Small ball	55%
$< 50\%$	7. Big ball	16%
	12. Fingertip small ball	33%
	14. Fingertip can	33%

2 of 5 grasps show a recognition rate 100%, and by the 3rd one. For more than half of the grasp primitives all three methods provide similar results. This is true for the grasps 1, 2, 3, 4, 7, 8, 10, and 15. However similarities between grasps may give space for misinterpretations which explains the low percentages for some grasps e.g. grasps 5 and 9 in method 1 or grasps 5 and 8 in method 3. Looking at groups 1,

Table 2.3 Recognition rates, method 3

Class	Grasp	Percentage
$\geq 75\%$	4. Hammer	100%
	9. Precision grasp cube	85%
	10. Small plane	85%
	14. Fingertip can	85%
	15. Penholder grip	100%
$< 75\%, \geq 50\%$	1. Cylinder	71%
	2. Big bottle	57%
	3. Small bottle	71%
	5. Screwdriver	71%
	6. Small ball	57%
	11. Big plane	57%
	12. fingertip small ball	57%
	13. Fingertip big ball	71%
$< 50\%$	7. Big ball	0%
	8. Precision. grasp sphere	28%

method 1 is the most successful one which is also a solution with the easiest implementation. Then it follows method 2 with a quite high implementation effort and finally method 3 based on HMM. It should be stated that the HMM principle may allow some improvement of the results especially in the case of an extended sensory suit in the experimental setup.

2.7 Conclusions

The goal of grasp recognition is to develop an easy way of 'programming by demonstration' of grasps for a humanoid robotic arm. In this chapter, three different methods of grasp recognition are presented. Grasp primitives are captured by a data glove and modeled by TS-fuzzy models. Fuzzy clustering and modeling of time and space data are applied to the modeling of the finger joint angle trajectories of grasp primitives. The 1st method being the simplest one classifies a human grasp by computing the minimum distance between the time-clusters of the test grasp and a set of model grasps. In the 2nd method a qualitative fuzzy model is developed with the help of which both the segmentation and grasp recognition can be achieved. The 3rd method uses Hidden Markov Models (HMM) for grasp recognition. A comparison of the three methods showed that the 1st method is the most effective one, followed by the 2nd and the 3rd method. In order to achieve a further increase of the recognition rates methods 1 and 2 could be combined because of their close relationship whereas method 3 is only connected with methods 1 and 2 via the time cluster modeling of the grasps. Therefore, the HMM principle may lead to better results using more haptic sensors in the experimental setup. To improve the PbD-process in general, all 3 methods will be further developed for the recognition and classification of operator motions in a robotic environment using more sensor information about the robot workspace and the objects to be handled.

References

1. Kang, S.B., Ikeuchi, K.: Towards automatic robot instruction form perception—mapping human grasps to manipulator grasps. IEEE Trans. Robot. Autom. **13**(1), 81–95 (1997)
2. Zoellner, R., Rogalla, O., Dillmann, R., Zoellner, J.: Dynamic grasp recognition within the framework of programming by demonstration. In: Proceedings Robot and Human Interactive Communication, 10th IEEE International Workshop, Bordeaux, Paris, France, 18–21 September 2001. IEEE, New York (2001)
3. Bernardin, K., Ogawara, K., Ikeuchi, K., Dillman, R.: A sensor fusion approach for recognizing continuous human grasp sequences using hidden Markov models. IEEE Trans. Robot. **21**(1), 47–57 (2005)
4. Ekvall, S., Kragic, D.: Grasp recognition for programming by demonstration. In: Proceedings of International Conference on Robotics and Automation, ICRA 2005, Barcelona, Spain (2005)
5. Li, C., Khan, L., Prabhaharan, B.: Real-time classification of variable length multi-attribute motions. Knowl. Inf. Syst. (2005). doi:10.1007/s10115-005-0223-8
6. Aleotti, J., Caselli, S.: Grasp recognition in virtual reality for robot pregrasp planning by demonstration. In: International Conference on Robotics and Automation, ICRA 2006, Orlando, FL, USA, May 2006. IEEE, New York (2006)
7. Palm, R., Iliev, B.: Learning of grasp behaviors for an artificial hand by time clustering and Takagi-Sugeno modeling. In: Proceedings FUZZ-IEEE 2006—IEEE International Conference on Fuzzy Systems, Vancouver, BC, Canada, 16–21 July 2006. IEEE, New York (2006)
8. Palm, R., Iliev, B.: Segmentation and recognition of human grasps for programming-by-demonstration using time clustering and Takagi-Sugeno modeling. In: Proceedings FUZZ-IEEE 2007—IEEE International Conference on Fuzzy Systems, London, UK, 23–26 July 2007. IEEE, New York (2007)
9. Palm, R., Iliev, B., Kadmiry, B.: Recognition of human grasps by time-clustering and fuzzy modeling. Robot. Auton. Syst. **57**(5), 484–495 (2009)
10. Palm, R., Iliev, B.: Grasp recognition by time clustering, fuzzy modeling, and hidden Markov models (hmm)—a comparative study. In: Proceedings FUZZ-IEEE 2008—IEEE International Conference on Fuzzy Systems, Hong Kong, 1–5 July 2008. IEEE, New York (2008)
11. Asada, H.H., Fortier, J.: Task recognition and human-machine coordination through the use of an instrument-glove. Progress report No. 2-5, pp. 1–39, March 2000
12. Cutkosky, M.: On grasp choice, grasp models, and the design of hands for manufacturing tasks. IEEE Trans. Robot. Autom. **5**(3), 269–279 (1989)
13. Iberall, T.: Human prehension and dexterous robot hands. Int. J. Robot. Res. **16**, 285–299 (1997)
14. Palm, R., Stutz, C.: Open loop dynamic trajectory generator for a fuzzy gain scheduler. Eng. Appl. Artif. Intell. **16**, 213–225 (2003)
15. Takagi, T., Sugeno, M.: Identification of systems and its applications to modeling and control. IEEE Trans. Syst. Man Cybern. **SMC-15**(1), 116–132 (1985)
16. Gustafson, D., Kessel, W.C.: Fuzzy clustering with a fuzzy covariance matrix. In: Proceedings of the 1979 IEEE CDC, pp. 761–766 (1979)
17. Rabiner, L.R.: A tutorial on hidden Markov models and selected applications in speech recognition. Proc. IEEE **77**(2), 257–286 (1989)

Chapter 3
Distributed Adaptive Coordinated Control of Multi-Manipulator Systems Using Neural Networks

Zeng-Guang Hou, Long Cheng, Min Tan, and Xu Wang

Abstract On many occasions, all the manipulators in the multi-manipulator system need to achieve the same joint configuration to fulfill certain coordination tasks. In this chapter, a distributed adaptive approach is proposed for solving this coordination problem based on the leader-follower strategy. The proposed algorithm is distributed because the controller for each follower manipulator is solely based on the information of connected neighbor manipulators, and the joint value of leader manipulator is only accessible to partial follower manipulators. The uncertain term in the manipulator's dynamics is considered in the controller design, and it is approximated by the adaptive neural network scheme. The neural network weight matrix is adjusted on-line by the projection method, and the pre-training phase is no longer required. Effects of approximation error and external disturbances are counteracted by employing the robustness signal. According to the theoretical analysis, all the joints of follower manipulators can be regulated into an arbitrary small neighborhood of the value of leader's joint. Finally, simulation results are given to demonstrate the satisfactory performance of the proposed method.

3.1 Introduction

Recently, the coordinated control of multi-manipulator system has become an attractive research area owing to its important role in the assembly automation and flexible manufacturing industries. On many occasions, all the manipulators in the multi-manipulator system need to achieve the same joint configuration to fulfill certain coordination tasks, such as loading a workpiece, etc. A great deal of existing control strategies for the coordinated control can be employed to solve this problem.

Z.-G. Hou (✉), L. Cheng, M. Tan, and X. Wang
Key Laboratory of Complex Sciences, Institute of Automation, Chinese Academy of Sciences, Beijing 100190, China
e-mail: hou@compsys.ia.ac.cn; chenglong@compsys.ia.ac.cn; tan@compsys.ia.ac.cn; wangxu.zju@163.com

H. Liu et al. (eds.), *Robot Intelligence*,
Advanced Information and Knowledge Processing,
DOI 10.1007/978-1-84996-329-9_3,

Among them, the leader-follower approach is more appealing due to the advantage of simplicity in theory analysis [1]. However, the leader-follower strategy requires that each follower manipulator has the information access to the leader manipulator. When the number of coordinated manipulators increases, the implementation cost of this control architecture becomes unacceptable. And the coordination performance degrades badly if there is the communication failure between the leader and followers. A reasonable alternative is to design the distributed leader-follower controller for the multi-manipulator system. By the distributed approach, each manipulator only needs the information of its connected neighbor manipulators, which is more suitable for real-world applications.

It is noted that there is a growing research interest on the coordination of multi-agent systems in the control community [2–7]. These coordinated methods are characterized by the feature that each agent's controller is designed by only using the information of its neighbor agents (named the "nearest neighbor rule"), and the system consensus behavior can be achieved. Therefore, the multi-agent theory can provide a general framework for the distributed leader-follower control of interconnected multi-manipulator system. Moreover, considerable work has been done on the multi-agent system with an active leader [2, 8–12]. In [2], Jadbabaie *et al.* first investigated the synchronization problem of networks of agents. They rigorously proved that all the states of follower agents could be synchronized with the leader agent's state by the "nearest neighbor rule". In [8], Hong *et al.* considered the case where partial velocity information of leader agent is not available to the follower agents. By constructing a distributed velocity observer, each follower agent could track the leader with a bounded tracking error. In [9], the virtual leader approach was proposed. The virtual leader could be considered as a common reference signal for each agent. In [10], Ren showed that all the follower agents could follow the time-varying state of leader agent by local interactions. The coupling time delays among the agents' communication were taken into account in [11], and the channel noise was considered in [12]. For the state of art of multi-agent coordination research, the readers are referred to [13, 14].

According to the survey [13], most of existing works on the multi-agent coordination focus on the agent with simple first-order or second-order integral dynamics [2–4, 8–12]. There are few results on the agent with the linear dynamics [5] and non-linear dynamics [6, 7]. The manipulator dynamic characteristics are highly non-linear because of the coupling between joints. Furthermore, in practical applications, the exact dynamics model of manipulator is hard to obtain due to the imprecision measurement of manipulator parameters and interactions between manipulator and different environments. Therefore, designing the distributed leader-follower approach for manipulators with uncertain non-linear dynamics is worth studying from both the theory and the application. There are several attempts made on this subject, such as the consensus control of manipulator with uncertain kinematics [15] or uncertain dynamics [16], and the leader-follower control of manipulation with uncertain dynamics [17]. However, the traditional adaptive approach was employed to deal with the manipulator uncertainty, which suffered from the "linearity-in-parameters" assumption and the tedious analysis and computation of the regression matrix.

In the past two decades, neural networks have been successfully used for the system identification and control owing to their "universal approximation" property [18]. The "linearity-in-parameters" assumption in traditional adaptive control approaches can be overcome by replacing the standard adaptive unit with the neural network structure [19, 20]. The stability of this neural-network-based adaptive controller is guaranteed by the Lyapunov synthesis method, and synaptic weights of neural networks are tuned on-line without any off-line learning phases. In addition, several successfully applications on the tracking control of the single manipulator with uncertainties have verified its effectiveness [21–23]. Hence, the neural-network-based approach can be regarded as a good alternative for the traditional adaptive algorithm. For the general framework of this neural-network-based method and the state of art of intelligent adaptive control, the readers are referred to [24].

Motivated by the aforementioned discussion, a distributed leader-follower approach is proposed for the coordinated control of the multi-manipulator system with uncertainties. All the manipulators in the system can achieve the same joint configuration. The neural network is employed to approximate the uncertainties in the manipulator's dynamics. The robustness signal is utilized to counteract the approximation error and external disturbances. The controller of each follower manipulator is designed based on the "nearest neighbor rule", and the joint value of leader manipulator is not available for all the follower manipulators. By the theoretical analysis, all the joints of follower manipulators can be regulated at the value of leader's joint with the regulation error as small as desired. At last, the theoretical analysis and controller performance are validated by simulation examples.

The remainder of this chapter is organized as follows. Section 3.2 introduces the problem formulation and some preliminary results. Section 3.3 provides the proposed leader-follower coordinated controller. Section 3.4 analyses the controller performance. The illustrative example is shown and discussed in Sect. 3.5. Section 3.6 concludes this chapter with final remarks.

3.2 Preliminaries

The following notations will be used throughout this chapter: $1_n = (1, 1, \dots, 1) \in \mathbb{R}^n$; $0_n = (0, 0, \dots, 0) \in \mathbb{R}^n$; I_n denotes the $n \times n$ dimensional identity matrix; $\otimes$ denotes the Kronecker operator; For a given matrix X, $\|X\|$ denotes its Euclidean norm; $\|X\|_F$ denotes its Frobenius norm; $\lambda_{\min}(X)$ denotes its smallest eigenvalue; $\lambda_{\max}(X)$ denotes its largest eigenvalue. For a given vector x_i, x_{ij} denotes its jth element.

3.2.1 Multi-Manipulator System Description

Consider a multi-manipulator system composed of $N + 1$, rigid n-link, manipulators. The leader manipulator is indexed by 0, and the follower manipulators are indexed by $1, 2, \ldots, N$, respectively. According to [21], the dynamics of the ith manipulator can be expressed

$$M_i(q_i)\ddot{q}_i + V_i(q_i, \dot{q}_i)\dot{q}_i + G_i(q_i) + \tau_{di} = \tau_i, \quad i = 0, 1, \ldots, N, \tag{3.1}$$

where $q_i, \dot{q}_i, \ddot{q}_i \in \mathbb{R}^n$ denote the joint position, velocity, and acceleration vectors of the ith manipulator, respectively; $M_i(q_i) \in \mathbb{R}^{n\times n}$ is the inertia matrix; $V_i(q_i, \dot{q}_i) \in \mathbb{R}^{n\times n}$ is the centripetal-Coriolis matrix; $G_i(q_i) \in \mathbb{R}^n$ is the gravitational vector; $\tau_{di} \in \mathbb{R}^n$ denotes the bounded unknown disturbance vector including unstructured unmodeled dynamics and mechanism noise; $\tau_i \in \mathbb{R}^n$ represents the torque input of the ith manipulator.

Two important properties of the dynamics described by (3.1) are given as follows [21]

Property 3.1 The inertia matrix $M_i(q_i)$ is symmetric and positive definite, and satisfies the following inequalities:

$$m_{i1}\|y\|^2 \le y^T M_i(q_i) y \le m_{i2}\|y\|^2, \quad \forall y \in \mathbb{R}^n,$$

where m_{i1} and m_{i2} are known positive constants.

Property 3.2 The time derivative of the inertia matrix and the centripetal-Coriolis matrix satisfy the skew symmetric relation; that is,

$$y^T \left(\dot{M}_i(q_i) - 2V_i(q_i, \dot{q}_i)\right) y = 0, \quad \forall y \in \mathbb{R}^n.$$

The communication topology of the follower manipulators can be described by the weight undirected graph $\mathcal{G} = (\mathcal{V}_\mathcal{G}, \mathcal{E}_\mathcal{G}, \mathcal{A}_\mathcal{G})$. $\mathcal{V}_\mathcal{G} = \{v_1, \ldots, v_N\}$ denotes the set of nodes; $\mathcal{E}_\mathcal{G} \subseteq \mathcal{V}_\mathcal{G} \times \mathcal{V}_\mathcal{G}$ denotes the set of undirected edges; and $\mathcal{A}_\mathcal{G} = [a_{ij}] \in \mathbb{R}^{N\times N}$ denotes the weighted adjacency matrix, where $a_{ii} = 0$ and $a_{ij} = a_{ji}$. Node v_i represents the ith follower manipulator. An undirected edge in $\mathcal{E}_\mathcal{G}$ is denoted by the pair $e_{ij} = (v_i, v_j)$. $e_{ij} \in \mathcal{E}_\mathcal{G}$ if and only if there is the information exchange between the manipulator i and manipulator j, and $e_{ij} \in \mathcal{E}_\mathcal{G} \Leftrightarrow e_{ji} \in \mathcal{E}_\mathcal{G}$; the adjacency element a_{ij} represents the quality of communication channel, and $e_{ij} \in \mathcal{E}_\mathcal{G} \Leftrightarrow a_{ij} > 0$.

The Laplacian matrix $\mathcal{L}_\mathcal{G}$ of graph $\mathcal{G}$ is defined by

$$\mathcal{L}_\mathcal{G} = \mathcal{D}_\mathcal{G} - \mathcal{A}_\mathcal{G}, \tag{3.2}$$

where $\mathcal{D}_\mathcal{G} = \mathrm{diag}(d_1, d_2, \ldots, d_n)$ and $d_i = \sum_{j=1}^{n} a_{ij}$. It is easy to see that $\mathcal{L}_\mathcal{G}$ is a symmetric matrix.

The communication between the leader manipulator and follower manipulators is given by the adjacency matrix $\mathcal{B}_\mathcal{G} = \mathrm{diag}(b_1, b_2, \ldots, b_N)$, where $b_i \ge 0$ ($i = 1, 2, \ldots, N$). The inequality holds if and only if the ith manipulator could receive the information of the leader manipulator. And it is noted that the leader manipulator does not receive the information from any other manipulators.

Now, consider an extended directed graph $\bar{\mathcal{G}} = (\bar{V}_{\bar{\mathcal{G}}}, \bar{E}_{\bar{\mathcal{G}}}, \bar{A}_{\bar{\mathcal{G}}})$ associated with the entire multi-manipulator system composed of N follower manipulators and one leader manipulator. $\bar{V}_{\bar{\mathcal{G}}} = \{v_0, v_1, \dots, v_N\}$ contains $N+1$ nodes and v_0 denotes the leader manipulator. $\bar{E}_{\bar{\mathcal{G}}}$ is extended by adding the directed edges from the leader manipulator to the follower manipulators. $\bar{A}_{\bar{\mathcal{G}}} = [\bar{a}_{ij}]_{i=0,\dots,N; j=0,\dots,N}$ with $\bar{a}_{ij} = a_{ij}$ $(i = 1, \dots, N, j = 1, \dots, N)$, $\bar{a}_{0j} = 0$ $(j = 0, 1, \dots, N)$, and $\bar{a}_{i0} = b_i$, $(i = 1, 2, \dots, N)$. A sequence of edges $(v_{i_1}, v_{i_2}), (v_{i_2}, v_{i_3}), \dots, (v_{i_{k-1}}, v_{i_k})$ is called a path from node v_{i_1} to node v_{i_k}. The extended graph $\bar{\mathcal{G}}$ is called connected if there exists a path from node v_0 (leader manipulator) to node v_i, $i = 1, 2, \dots, N$ (follower manipulator).

Lemma 3.1 *If the directed extended graph $\bar{\mathcal{G}}$ is connected, then $L_{\mathcal{G}} + B_{\mathcal{G}}$ is a positive definitive matrix.*

Proof The proof is similar to the proof of Lemma 3 in [8], so is omitted here. □

The following assumptions are made throughout this chapter.

Assumption 3.1 The topology $\bar{\mathcal{G}}$ of the multi-manipulator system is fixed: $\bar{A}_{\bar{\mathcal{G}}}$ is a constant matrix.

Assumption 3.2 The external disturbance τ_{di} in (3.1) is bounded by a given constant Δ_i: $\|\tau_{di}\| \le \Delta_i$.

Assumption 3.3 The leader manipulator's joint is time-invariant: q_0 is a constant vector.

The control objective is to develop a distributed leader-follower approach for the multi-manipulator system. The controller of each follower manipulator only requires the information of the connected neighbor manipulators. And the joints of all the follower manipulators can be regulated at the value of leader manipulator's joint. In the other word, all the manipulators in the multi-manipulator system have the same joint configuration eventually.

3.2.2 Radial Basis Function Neural Network

In control engineering, neural networks are usually employed as the function approximator to emulate the unknown ideal control signal. Due to the "linear-in-the-weight" property, the radial basis function neural network (RBFNN) is a good candidate for this purpose. In this chapter, the RBFNN shown in Fig. 3.1 is employed to approximate the continuous function, $h(Z) : \mathbb{R}^n \to \mathbb{R}^n$, as follows

$$h_{nn}(Z) = W^T S(Z), \tag{3.3}$$

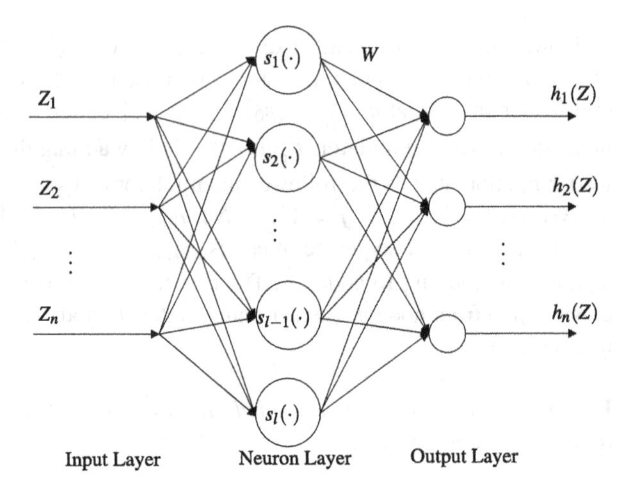

Fig. 3.1 The structure of the RBF neural network

where the input vector $Z \in \Omega \subset \mathbb{R}^n$, weight matrix $W \in \mathbb{R}^{l\times n}$, l denotes the number of neurons, and $S(Z) = [s_1(Z), \ldots, s_l(Z)]^T$ with

$$s_i(Z) = \exp\left[\frac{-(Z-\mu_i)^T(Z-\mu_i)}{\sigma_i^2}\right], \quad i = 1, 2, \ldots, l$$

where $\mu_i \in \mathbb{R}^n$ is the center of the receptive field and σ_i is the width of the Gaussian function.

It has been proven that the above RBFNN can approximate any smooth function over a compact set $\Omega_Z \subset \mathbb{R}^n$ to arbitrarily accuracy. That is, for any positive constant ε_N, given a large number of neurons l, there exists the ideal weight matrix W^* such that

$$h(Z) = {W^*}^T S(Z) + \varepsilon, \tag{3.4}$$

where ε is the bounded function approximation error satisfying $|\varepsilon| < \varepsilon_N$ in Ω_Z.

It is noted that the ideal matrix W^* is only quantity required for analytical purpose. For real applications, its estimation $\hat{W}$ is used for the practical function approximation. The estimation of $h(Z)$ can be given by

$$\hat{h}(Z) = \hat{W}^T S(Z). \tag{3.5}$$

To analyze the system performance, the following lemma is useful.

Lemma 3.2 *Let function $V(t) \geq 0$ be a continuous function defined $\forall t \geq 0$ and bounded, and $\dot{V}(t) \leq -\gamma V(t) + \kappa$, where γ and κ are positive constants, then*

$$V(t) \leq V(0)e^{-\gamma t} + \frac{\kappa}{\gamma}(1 - e^{-\gamma t}).$$

Proof See the proof of Lemma 1.1 in [25]. □

3.3 Controller Design

It is noted that the manipulator dynamics defined by (3.1) is second-order. An efficient way to deal with higher-order system is to employ the backstepping method. The backstepping scheme is featured by designing partial Lyapunov functions and auxiliary controllers for the subsystems, and integrating these individual controllers into the actual controller by "back stepping" through the system and then reassembling it from its component subsystems [26].

It is also noted that, in the real applications, the manipulator dynamics terms, $M_i(q_i)$, $V_i(q_i, \dot{q}_i)$, $G_i(q_i)$, may contain the uncertain parameters, such as the link mass and link inertia, etc. The effects of uncertain parameters should be considered in the controller design. Actually, in the proposed controller, the specific structures of these dynamics matrices are not necessarily known either.

Step 1: According to the "nearest neighbor rule", the auxiliary joint velocity $\dot{q}_{di}$ of the ith follower manipulator can be designed as follows

$$\dot{q}_{di} = -\eta\left(\sum_{j=1}^{n} l_{ij}q_j + b_i(q_i - q_0)\right), \quad i = 1, 2, \ldots, N, \tag{3.6}$$

where l_{ij} is the ith row and the jth column element of the Laplacian matrix $L_{\mathcal{G}}$; $\eta > 0$ is the constant control gain.

Equation (3.6) can be written in the following compact form that

$$\begin{aligned} \dot{q}_d &= -\eta\left((L_{\mathcal{G}} + B_{\mathcal{G}}) \otimes I_n\right) q + \eta(B_{\mathcal{G}} \otimes 1_n)(1_N \otimes q_o), \\ &= -\eta\left((L_{\mathcal{G}} + B_{\mathcal{G}}) \otimes I_n\right)(q - 1_N \otimes q_o) \end{aligned} \tag{3.7}$$

where $q_d = (q_{d1}^T, q_{d2}^T, \ldots, q_{dN}^T)^T \in \mathbb{R}^{nN}$, and $q = (q_1^T, q_2^T, \ldots, q_N^T)^T \in \mathbb{R}^{nN}$.

Define the error signal $e_i = \dot{q}_i - \dot{q}_{di}$. According to (3.1), the dynamics of e_i can be obtained as follows

$$M_i(q_i)\dot{e}_i + V_i(q_i, \dot{q}_i)e_i = \tau_i - M_i(q_i)\ddot{q}_{di} - V_i(q_i, \dot{q}_i)\dot{q}_{di} - G_i(q_i) - \tau_{di}. \tag{3.8}$$

Step 2: Design the real torque controller τ_i which makes e_i as small as possible.

Let $f_i(z_i) = M_i(q_i)\ddot{q}_{di} + V_i(q_i, \dot{q}_i)\dot{q}_{di} + G_i(q_i)$ where $z_i = (q_i^T, \dot{q}_i^T, \dot{q}_{di}^T, \ddot{q}_{di}^T)^T$. According to the description in Sect. 3.2.2, for any $\varepsilon_{Ni} > 0$, there exists the ideal weight matrix W_i^* such that $f_i(z_i) = W_i^{*T} S_i(z_i) + \varepsilon_i$ with $\|\varepsilon_i\| \leq \varepsilon_{Ni}$ in a compact region Ω_{z_i}. It will be shown in Remark 3.3 that such a region Ω_{z_i} indeed exists. Then, the proposed torque controller τ_i for the ith manipulator is given as follows

$$\tau_i = -k_i e_i + \hat{f}(z_i) - \delta_{Mi}\tanh\left(\frac{2k_u\delta_{Mi}e_i}{\varsigma_i}\right) + \frac{1}{\eta}\dot{q}_{di}, \quad i = 1, 2, \ldots, N, \tag{3.9}$$

where $k_i > 0$ is the control gain; $\hat{f}(z_i) = \hat{W}_i^T S_i(z_i)$ is the RBFNN estimation of $f_i(z_i)$; $\delta_{Mi}\tanh((2k_u\delta_{Mi}e_i)/(\varsigma_i))$ is the robustness signal with $k_u = 0.2785$; $\varsigma_i > 0$

has the effect on the control performance; The robustness gain δ_{Mi} is chosen to satisfy that $\delta_{Mi} \geq \varepsilon_{Ni} + \Delta_i$. It is easy to prove that

$$\delta_{Mi} e_i^T \tanh\left(\frac{2k_u \delta_{Mi} e_i}{\varsigma_i}\right) \geq 0, \qquad \delta_{Mi} \|e_i\| - \delta_{Mi} e_i^T \tanh\left(\frac{2k_u \delta_{Mi} e_i}{\varsigma_i}\right) \leq \varsigma_i. \tag{3.10}$$

Combining (3.8) and (3.9) yields that

$$\begin{aligned} &M_i(q_i)\dot{e}_i + V_i(q_i, \dot{q}_i)e_i \\ &\quad = -k_i e_i - \tilde{W}_i^T S_i(z_i) - \delta_{Mi} \tanh\left(\frac{2k_u \delta_{Mi} e_i}{\varsigma_i}\right) - \tau_{di} - \varepsilon_i + \frac{1}{\eta}\dot{q}_{di}, \end{aligned} \tag{3.11}$$

where $\tilde{W}_i = W_i^* - \hat{W}_i$.

By the projection algorithm, the updating law for the RBFNN weight matrix $\hat{W}_i$ $(i = 1, 2, \ldots, N)$ is derived as follows

$$\dot{\hat{W}}_i = \begin{cases} -\beta_{wi} S_i(z_i) e_i^T, \quad \text{if } \mathrm{Tr}\,(\hat{W}_i^T \hat{W}_i) < W_{\max i}, \\ \quad \text{or if } \mathrm{Tr}\,(\hat{W}_i^T \hat{W}_i) = W_{\max i} \text{ and } e_i^T \hat{W}_i^T S_i(z_i) \geq 0; \\ -\beta_{wi} S_i(z_i) e_i^T + \beta_{wi} \frac{e_i^T \hat{W}_i^T S_i(z_i)}{\mathrm{Tr}\,(\hat{W}_i^T \hat{W}_i)} \hat{W}_i, \\ \quad \text{if } \mathrm{Tr}\,(\hat{W}_i^T \hat{W}_i) = W_{\max i} \text{ and } e_i^T \hat{W}_i^T S_i(z_i) < 0; \end{cases} \tag{3.12}$$

where $\beta_{wi} > 0$ is the adaption gain; $W_{\max i}$ is a given positive constant for limiting the neural network weight matrix $\hat{W}_i$; $W_{\max i}$ is selected to satisfy $\mathrm{Tr}\,((W_i{}^*)^T W_i{}^*) \leq W_{\max i}$; and the initial neural network weight matrix $\hat{W}_i(0)$ should satisfy that

$$\mathrm{Tr}\left(\hat{W}_i^T(0)\hat{W}_i(0)\right) \leq W_{\max i}. \tag{3.13}$$

Remark 3.1 By the definition of adjacency matrices $L_{\mathcal{G}}$, $l_{ij} \neq 0$ if and only if there is information exchange between follower manipulators i and j. For the diagonal matrix $B_{\mathcal{G}}$, $b_i \neq 0$ if and only if the ith follower manipulator can receive the information of leader manipulator. Therefore, by observing the proposed controller defined by (3.9) and the parameter updating law defined by (3.12), it is easy to see that the proposed adaptive controller for the ith manipulator only uses the information of its connected neighbor manipulators. Hence, the proposed algorithm is distributed.

Remark 3.2 Compared with the previous work [17], the uncertain parameters in the manipulator's dynamics do not have to satisfy the "linearity-in-parameters" assumption. The proposed controller can be designed without knowing the specific values of matrices $M_i(q_i)$, $V_i(q_i, \dot{q}_i)$ and $G_i(q_i)$.

3.4 Performance Analysis

Lemma 3.3 *If the updating law of neural network weight matrix is defined by* (3.12), *and the initial value of neural network weight matrix satisfies* (3.13), *then*

$$\forall t \geq 0, \quad \mathrm{Tr}\left(\hat{W}_i^T(t)\hat{W}_i(t)\right) \leq W_{\max i}, \quad i = 1, 2, \ldots, N. \tag{3.14}$$

Proof First, a useful property of the trace operator is stated as follows

$$a^T b = \mathrm{Tr}\left(ab^T\right), \quad \forall a, b \in \mathbb{R}^n. \tag{3.15}$$

To prove $\mathrm{Tr}(\hat{W}_i^T(t)\hat{W}_i(t)) \leq W_{\max i}$, let $L_{wi} = \mathrm{Tr}(\hat{W}_i^T \hat{W}_i)$. By (3.12) and (3.15), the following cases are considered:

- when $L_{wi} < W_{\max i}$, the conclusion has already held;
- when $L_{wi} = W_{\max i}$ and $e_i^T \hat{W}_i^T S_i(z_i) \geq 0$,

$$\frac{dL_{wi}}{dt} = 2\,\mathrm{Tr}\left(\hat{W}_i^T \dot{\hat{W}}_i\right) = -2\beta_i e_i^T \hat{W}_i^T S_i(z_i) < 0;$$

- when $L_{wi} = W_{\max i}$ and $e_i^T \hat{W}_i^T S_i(z_i) < 0$,

$$\begin{aligned}
\frac{dL_{wi}}{dt} &= 2\,\mathrm{Tr}\left(\hat{W}_i^T \dot{\hat{W}}_i\right) \\
&= -2\,\mathrm{Tr}\left(\beta_i \hat{W}_i^T S_i(z_i) e_i^T\right) + 2\,\mathrm{Tr}\left(\hat{W}_i^T \beta_i \frac{e_i^T \hat{W}_i^T S_i(z_i)}{\mathrm{Tr}(\hat{W}_i \hat{W}_i^T)} \hat{W}_i\right) \\
&= -2\beta_i e_i^T \hat{W}_i^T S_i(z_i) + 2\beta_i e_i^T \hat{W}_i^T S_i(z_i) \\
&= 0.
\end{aligned}$$

Hence, if the initial neural network weight matrix $\hat{W}_i(0)$ satisfies condition (3.13), then $\mathrm{Tr}(\hat{W}_i^T(t)\hat{W}_i(t)) \leq W_{\max i}$, $i = 1, 2, \ldots, N$, always holds. Therefore, the approximation error $\|\tilde{W}_i\|_F = \|W_i^* - \hat{W}_i\|_F \leq \|\hat{W}_i\|_F + \|W_i^*\|_F = 2\sqrt{W_{\max i}}$ is also bounded. □

Theorem 3.1 *If the topology $\bar{\mathcal{G}}$ of multi-manipulator system is connected, by using the proposed distributed adaptive controller defined by* (3.9) *and* (3.12), *all the joints of follower manipulators could be regulated at the value of the leader manipulator's joint, and the regulation error can be reduced as small as desired by appropriately choosing the controller parameters.*

Proof Define the regulation error $r_i = q_i - q_o$. According to Assumption 3.3 and (3.7), it can be obtained that

$$\dot{r}_i = \dot{q}_i = (\dot{q}_{di}) + e_i = -\eta \left(\sum_{j=1}^{n} l_{ij} q_j + b_i (q_i - q_0) \right) + e_i \tag{3.16}$$

Or in the compact form,

$$\dot{r} = -\eta \left((L_{\mathcal{G}} + B_{\mathcal{G}}) \otimes I_n \right) r + e. \tag{3.17}$$

where $r = (r_1^T, r_2^T, \ldots, r_N^T)^T$ and $e = (e_1^T, e_2^T, \ldots, e_N^T)^T$.

Consider the following Lyapunov candidate function

$$E = E_1 + E_2 + E_3, \tag{3.18}$$

where

$$E_1 = \frac{1}{2} r^T ((L_{\mathcal{G}} + B_{\mathcal{G}}) \otimes I_n) r, \tag{3.19}$$

$$E_2 = \frac{1}{2} \sum_{i=1}^{N} \left(e_i^T M_i(q_i) e_i \right), \tag{3.20}$$

$$E_3 = \frac{1}{2} \sum_{i=1}^{N} \mathrm{Tr} \left(\frac{\tilde{W}_i^T \tilde{W}_i}{\beta_{wi}} \right). \tag{3.21}$$

By (3.17), the time derivative of E_1 is

$$\begin{aligned} \frac{dE_1}{dt} &= r^T ((L_{\mathcal{G}} + B_{\mathcal{G}}) \otimes I_n) \dot{r} \\ &= -\eta r^T ((L_{\mathcal{G}} + B_{\mathcal{G}}) \otimes I_n)^2 r + r^T ((L_{\mathcal{G}} + B_{\mathcal{G}}) \otimes I_m) e. \end{aligned} \tag{3.22}$$

By (3.7), it follows that

$$r^T ((L_{\mathcal{G}} + B_{\mathcal{G}}) \otimes I_m) e = -\frac{1}{\eta} e^T \dot{q}_d. \tag{3.23}$$

By Property 3.1, (3.10) and (3.11), the time derivative of E_2 is

$$\begin{aligned} \frac{dE_2}{dt} &= \frac{1}{2} \sum_{i=1}^{N} \left(2 e_i^T M_i(q_i) \dot{e}_i + e_i^T \dot{M}_i(q_i) e_i \right) \\ &= \frac{1}{2} \sum_{i=1}^{N} \left(-2 e_i^T V_i(q_i, \dot{q}_i) e_i + e_i^T \dot{M}_i(q_i) e_i \right) - \sum_{i=1}^{N} \left(k_i e_i^T e_i \right) \\ &\quad + \sum_{i=1}^{N} \left(-e_i^T \tilde{W}_i^T S_i(z_i) - \delta_{Mi} e_i^T \tanh\left(\frac{2 k_u \delta_{Mi} e_i}{\varsigma_i} \right) \right. \end{aligned}$$

$$
\begin{aligned}
&- e_i^T (\tau_{di} + \varepsilon_i) + \frac{1}{\eta} e_i^T \dot{q}_{di} \Big) \\
\leq & - \sum_{i=1}^{N} \left(k_i e_i^T e_i \right) - \sum_{i=1}^{N} \left(e_i^T \tilde{W}_i^T S_i(z_i) \right) + \sum_{i=1}^{N} \left(\frac{1}{\eta} e_i^T \dot{q}_{di} \right) \\
&+ \sum_{i=1}^{N} \left(\|e_i\| (\|\tau_{di}\| + \|\varepsilon_i\|) - \delta_{Mi} e_i^T \tanh \left(\frac{2k_u \delta_{Mi} e_i}{\varsigma_i} \right) \right) \\
\leq & - \sum_{i=1}^{N} \left(k_i e_i^T e_i \right) - \sum_{i=1}^{N} \mathrm{Tr} \left(\tilde{W}_i^T S_i(z_i) e_i^T \right) + \sum_{i=1}^{N} \varsigma_i + \sum_{i=1}^{N} \left(\frac{1}{\eta} e_i^T \dot{q}_{di} \right).
\end{aligned} \tag{3.24}
$$

By (3.12), it follows that

$$
\frac{dE_3}{dt} = - \sum_{i=1}^{N} \mathrm{Tr} \left(\frac{1}{\beta_{wi}} \tilde{W}_i^T \dot{\hat{W}}_i \right). \tag{3.25}
$$

By (3.12), it follows that

- when $\dot{\hat{W}}_i = -\beta_i S_i(z_i) e_i^T$, then

$$
\mathrm{Tr} \left(\tilde{W}_i^T \left(\frac{1}{\beta_i} \dot{\hat{W}}_i + S_i(z_i) e_i^T \right) \right) = 0.
$$

- when $\dot{\hat{W}}_i = -\beta_i S_i(z_i) e_i^T + \beta_i \frac{e_i^T \hat{W}_i^T S_i(z_i)}{\mathrm{Tr}(\hat{W}_i^T \hat{W}_i)} \hat{W}_i$, then $\mathrm{Tr}(\hat{W}_i^T \hat{W}_i) = W_{\max i}$ and $e_i^T \hat{W}_i^T S_i(z_i) < 0$. It can be obtained that

$$
\mathrm{Tr} \left(\tilde{W}_i^T \left(\frac{1}{\beta_i} \dot{\hat{W}}_i + S_i(z_i) e_i^T \right) \right) = \frac{e_i^T \hat{W}_i^T S_i(z_i)}{\mathrm{Tr}(\hat{W}_i^T \hat{W}_i)} \mathrm{Tr} \left(\tilde{W}_i^T \hat{W}_i \right).
$$

It is also noted that

$$
\begin{aligned}
\mathrm{Tr} \left(\tilde{W}_i^T \hat{W}_i \right) &= \mathrm{Tr} \left(\tilde{W}_i^T W_i^* \right) - \mathrm{Tr} \left(\tilde{W}_i^T \tilde{W}_i \right) \\
&= \frac{1}{2} \mathrm{Tr} \left(\tilde{W}_i^T W_i^* \right) + \frac{1}{2} \mathrm{Tr} \left(\tilde{W}_i^T W_i^* \right) - \frac{1}{2} \mathrm{Tr} \left(\tilde{W}_i^T \tilde{W}_i \right) \\
&\quad - \frac{1}{2} \mathrm{Tr} \left(\tilde{W}_i^T \tilde{W}_i \right) \\
&= \frac{1}{2} \mathrm{Tr} \left(W_i^* W_i^{*T} \right) - \frac{1}{2} \mathrm{Tr} \left(\hat{W}_i W_i^{*T} \right) - \frac{1}{2} \mathrm{Tr} \left(\tilde{W}_i \hat{W}_i^T \right) \\
&\quad - \frac{1}{2} \mathrm{Tr} \left(\tilde{W}_i \tilde{W}_i^T \right) \\
&= \frac{1}{2} \mathrm{Tr} \left(W_i^{*T} W_i^* \right) - \frac{1}{2} \mathrm{Tr} \left(\tilde{W}_i^T \tilde{W}_i \right) - \frac{1}{2} \mathrm{Tr} \left(\hat{W}_i^T \hat{W}_i \right) \\
&\leq 0,
\end{aligned}
$$

where the facts, $\mathrm{Tr}(\hat{W}_i^T\hat{W}_i) = W_{\max i} \geq \mathrm{Tr}(W_i^{*T}W_i^*)$ and $\mathrm{Tr}(\tilde{W}_i^T\tilde{W}_i) \geq 0$, have been used. Then it is easy to prove that $\mathrm{Tr}(\tilde{W}_i^T(\frac{1}{\beta_i}\dot{\hat{W}}_i + S_i(z_i)e_i^T)) \geq 0$.

From the above two cases, the following result can be obtained

$$\mathrm{Tr}\left(\tilde{W}_i^T\left(\frac{1}{\beta_i}\dot{\hat{W}}_i + S_i(z_i)e_i^T\right)\right) \geq 0. \tag{3.26}$$

By (3.22), (3.24), (3.25) and (3.26), it follows that

$$\begin{aligned}\frac{dE(t)}{dt} &= \frac{dE_1(t)}{dt} + \frac{dE_2(t)}{dt} + \frac{dE_3(t)}{dt} \\ &\leq -\eta r^T\left((L_{\mathcal{G}} + B_{\mathcal{G}}) \otimes I_n\right)^2 r - \sum_{i=1}^N \left(k_i e_i^T e_i\right) + \varsigma,\end{aligned} \tag{3.27}$$

where $\varsigma = \sum_{i=1}^N \varsigma_i$.

Since the topology $\bar{\mathcal{G}}$ of multi-manipulator system is connected, $L_{\mathcal{G}} + B_{\mathcal{G}}$ is a symmetric positive definite matrix. Then inequality (3.27) can be relaxed as follows by Property 3.1

$$\begin{aligned}\frac{dE(t)}{dt} &\leq -\eta r^T\left((L_{\mathcal{G}} + B_{\mathcal{G}}) \otimes I_n\right)^2 r - \sum_{i=1}^N \left(k_i e_i^T e_i\right) + \varsigma, \\ &\leq -\eta\frac{\lambda_{\min}((L_{\mathcal{G}} + B_{\mathcal{G}}) \otimes I_n)}{\lambda_{\max}(((L_{\mathcal{G}} + B_{\mathcal{G}}) \otimes I_n)^2)} r^T\left((L_{\mathcal{G}} + B_{\mathcal{G}}) \otimes I_n\right) r \\ &\quad - \sum_{i=1}^N\left(\frac{k_i}{m_{i2}} e_i^T M_i(q_i) e_i\right) - \sum_{i=1}^N\left(\chi\,\mathrm{Tr}\left(\frac{1}{\beta_{wi}}\tilde{W}_i^T\tilde{W}_i\right)\right) \\ &\quad + \sum_{i=1}^N\left(\frac{4\chi W_{\max i}}{\beta_i}\right) + \varsigma \\ &\leq -2\chi E(t) + \sum_{i=1}^N\left(\frac{4\chi W_{\max i}}{\beta_i}\right) + \varsigma,\end{aligned} \tag{3.28}$$

where

$$\chi = \min\left\{\eta\frac{\lambda_{\min}((L_{\mathcal{G}} + B_{\mathcal{G}}) \otimes I_n)}{\lambda_{\max}(((L_{\mathcal{G}} + B_{\mathcal{G}}) \otimes I_n)^2)}, \frac{k_1}{m_{12}}, \frac{k_2}{m_{22}}, \ldots, \frac{k_N}{m_{N2}}\right\}.$$

According to Lemma 3.2, it can be obtained that

$$E(t) \leq E(0)e^{-2\chi t} + \sum_{i=1}^N\left(\frac{2W_{\max i}}{\beta_i}\right) + \frac{\varsigma}{2\chi}. \tag{3.29}$$

For any $\kappa > 0$, choose η, β_i and k_i such that $\sum_{i=1}^{N}(\frac{2W_{\max i}}{\beta_i}) + \frac{\varsigma}{2\chi} \le \frac{\kappa}{2}$. Then,

$$\forall t \ge T = \frac{1}{2\chi} \ln\left(\frac{2E(0)}{\kappa}\right), \quad E(t) \le \kappa.$$

Since

$$E(t) \ge \frac{\lambda_{\min}((L_{\mathcal{G}} + B_{\mathcal{G}}) \otimes I_n)}{2} \|r\|^2 + \frac{1}{2}\sum_{i=1}^{N} m_{i1}\|e_i\|^2, \tag{3.30}$$

then

$$\begin{aligned} \|q_i(t) - q_0\|^2 = \|r_i(t)\|^2 \le \|r(t)\|^2 &\le \frac{2}{\lambda_{\min}((L_{\mathcal{G}} + B_{\mathcal{G}}) \otimes I_n)} E(t) \\ &\le \frac{2\kappa}{\lambda_{\min}((L_{\mathcal{G}} + B_{\mathcal{G}}) \otimes I_n)}, \quad \forall t \ge T. \end{aligned} \tag{3.31}$$

Therefore, all the joints of follower manipulators could be regulated at the value of the leader manipulator's joint, and the regulation error can be reduced as small as desired. □

Remark 3.3 According to (3.29) and (3.30), $\forall i = 1, 2, \ldots, N$,

$$\|q_i(t) - q_0\|^2 \le \frac{2}{\lambda_{\min}((L_{\mathcal{G}} + B_{\mathcal{G}}) \otimes I_n)} \left(E(0) + \sum_{i=1}^{N}\left(\frac{2W_{\max i}}{\beta_i}\right) + \frac{\varsigma}{2\chi}\right), \quad t \ge 0,$$

and

$$\|\dot{q}_i(t) - \dot{q}_{di}\|^2 = \|e_i(t)\|^2 \le \frac{2}{m_{1i}} \left(E(0) + \sum_{i=1}^{N}\left(\frac{2W_{\max i}}{\beta_i}\right) + \frac{\varsigma}{2\chi}\right), \quad t \ge 0.$$

By (3.7), it follows that $\|q_i(t)\|$ and $\|\dot{q}_i(t)\|$ $(i = 1, 2, \ldots, N)$ are bounded. Therefore, the neural network input $z_i = (q_i^T, \dot{q}_i^T, \dot{q}_{di}^T, \ddot{q}_{di}^T)^T$ can be bounded in a compact region Ω_{z_i}.

3.5 Simulation Example

In this section, a simulation example is given to demonstrate the effectiveness of the proposed method. The manipulator's dynamics defined by (3.1) is simulated by Matlab "ode15s" method.

Consider the multi-manipulator system composed of six three-link manipulators. The sketch of each three-link manipulator is shown in Fig. 3.2, and the Denavit-Hartenberg parameters are given in Table 3.1. The physical parameters of each manipulator are the same: link lengths $l_1 = 2$ m, $l_2 = 1$ m, $l_3 = 1.5$ m; link masses $m_1 = 2$ kg, $m_2 = 1$ kg, $m_3 = 1.5$ kg. All links are modeled as thin uniform rods.

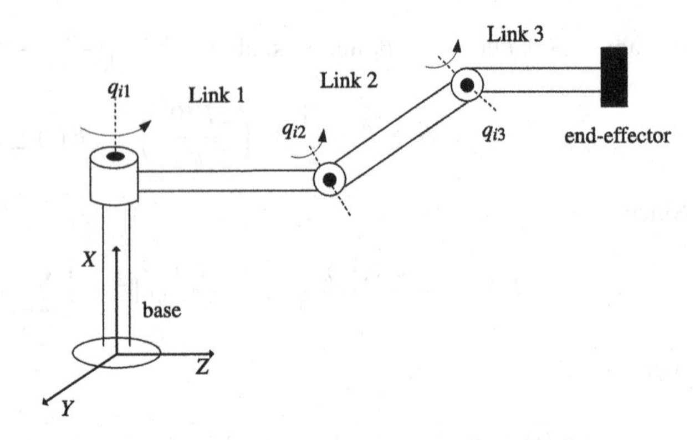

Fig. 3.2 A three-link revolute manipulator

Table 3.1 The Denavit and Hartenberg parameters of the three-link manipulator

Link i	θ_i (rad)	a_i (rad)	α_i (m)	d_i (m)
1	q_{i1}	$\pi/2$	2	0
2	q_{i2}	0	1	0
3	q_{i3}	0	1.5	0

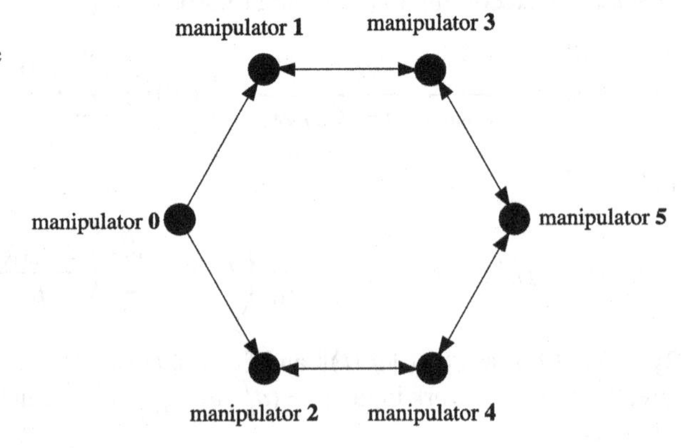

Fig. 3.3 The information exchange structure among the manipulators

The leader manipulator is indexed by 0, the five followers are indexed by 1, 2, 3, 4, 5, respectively. The information exchange among manipulators is shown in Fig. 3.3 with the following adjacency matrices

$$A_{\mathcal{G}} = \begin{bmatrix} 0.0 & 0.0 & 0.3 & 0.0 & 0.0 \\ 0.0 & 0.0 & 0.0 & 0.2 & 0.0 \\ 0.3 & 0.0 & 0.0 & 0.0 & 0.4 \\ 0.0 & 0.2 & 0.0 & 0.0 & 0.8 \\ 0.0 & 0.0 & 0.4 & 0.8 & 0.0 \end{bmatrix};$$

$$B_{\mathcal{G}} = \mathrm{diag}(0.9, 0.7, 0.0, 0.0, 0.0).$$

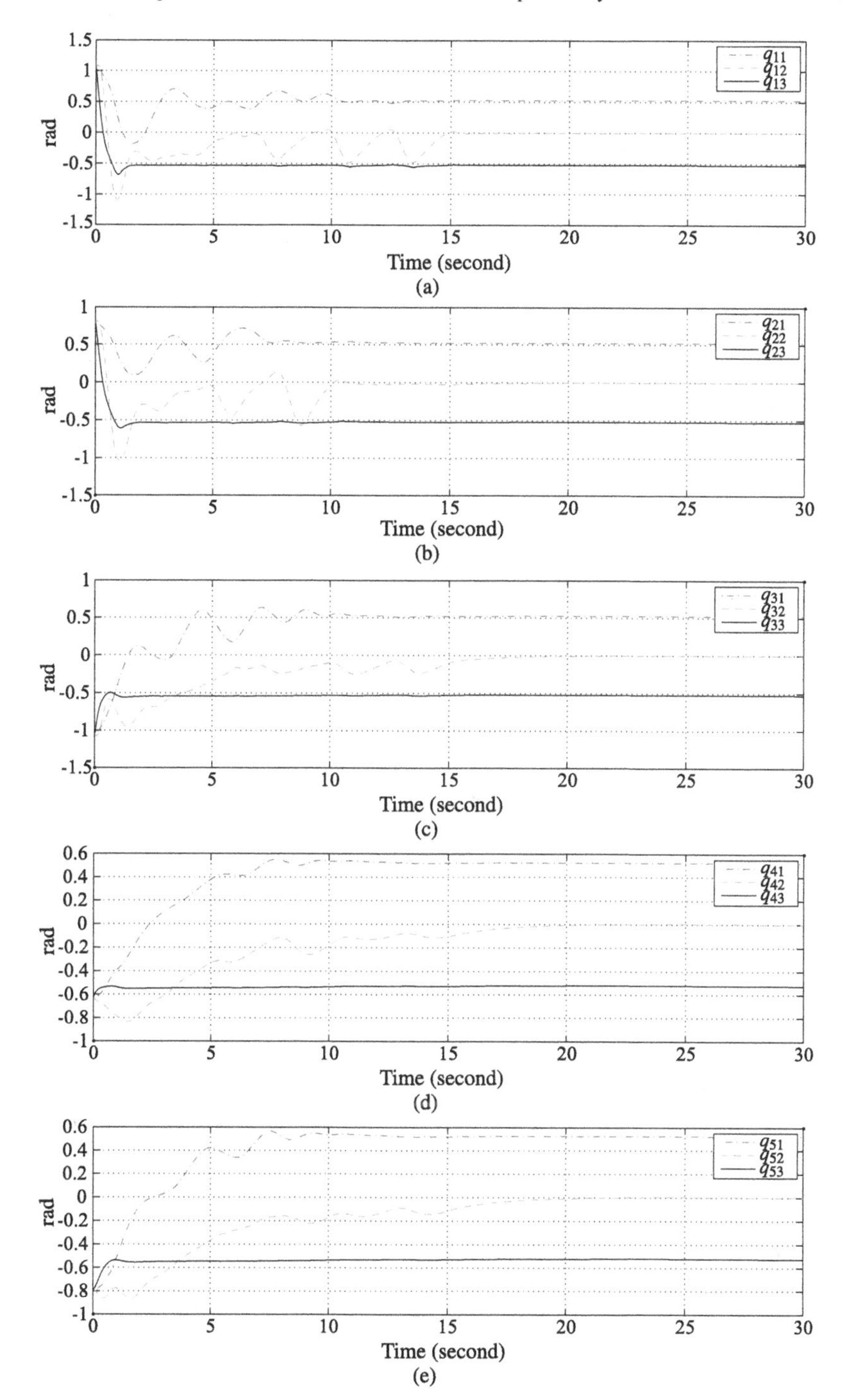

Fig. 3.4 The profiles of joint positions of follower manipulators (**a**) manipulator 1; (**b**) manipulator 2; (**c**) manipulator 3; (**d**) manipulator 4; (**e**) manipulator 5

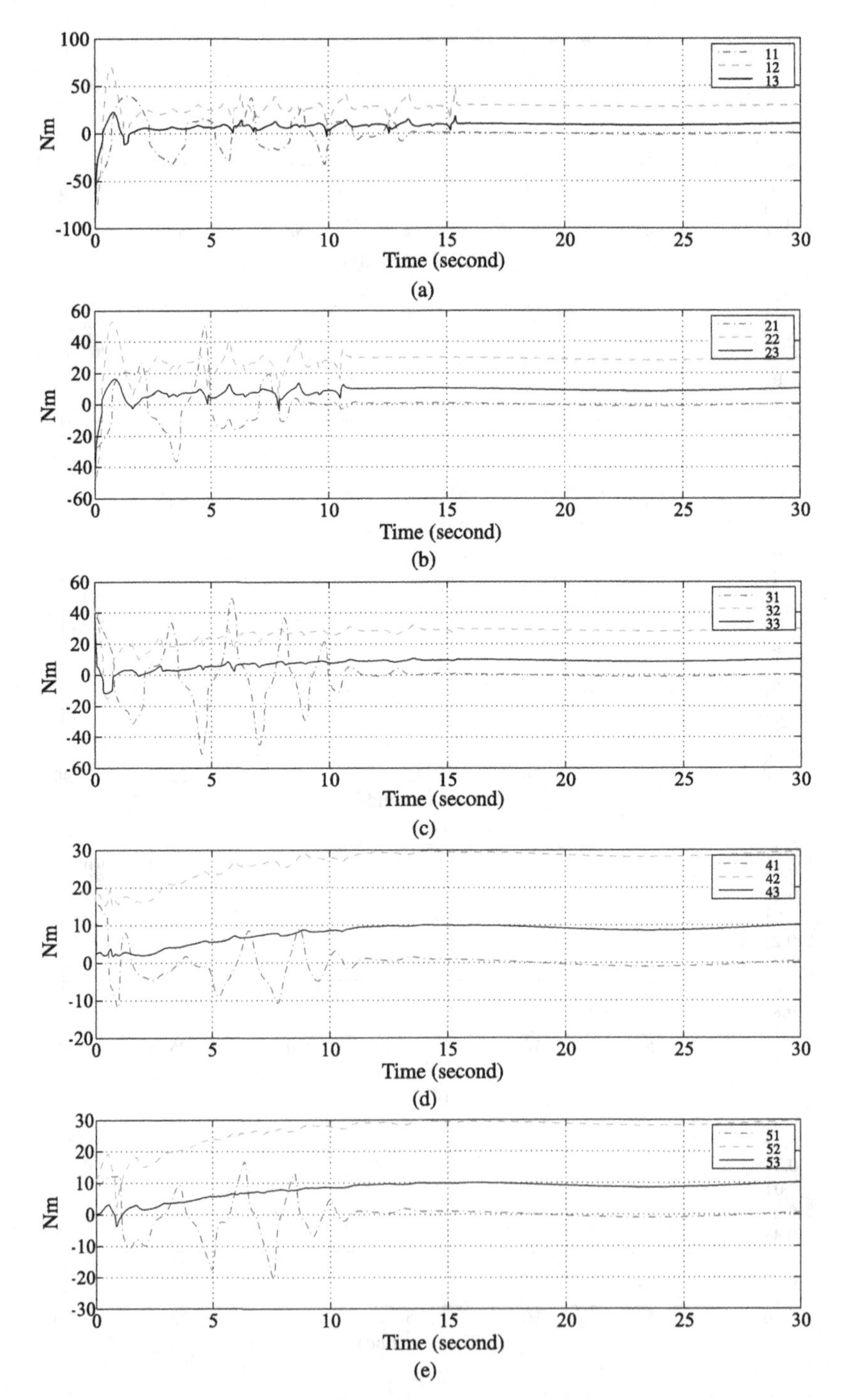

Fig. 3.5 The profiles of input torques of follower manipulators (**a**) manipulator 1; (**b**) manipulator 2; (**c**) manipulator 3; (**d**) manipulator 4; (**e**) manipulator 5

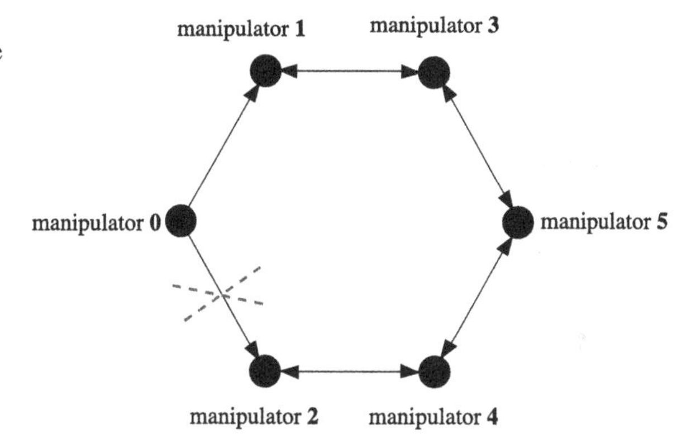

Fig. 3.6 The information exchange structure among the manipulators with link loss

The initial joint positions of follower manipulators are set as $q_1(0) = (\pi/3, \pi/3, \pi/3)^T$ rad, $q_2(0) = (\pi/4, \pi/4, \pi/4)^T$ rad, $q_3(0) = (-\pi/3, -\pi/3, -\pi/3)^T$ rad, $q_4(0) = (-\pi/5, -\pi/5, -\pi/5)^T$ rad, $q_5(0) = (-\pi/4, -\pi/4, -\pi/4)^T$ rad, respectively. And, initial joint velocity of follower manipulators are $\dot{q}_1(0) = \dot{q}_2(0) = \dot{q}_3(0) = \dot{q}_4(0) = \dot{q}_5(0) = (0, 0, 0)^T$ rad/s. The constant leader manipulator's joint position is $q_0 = (\pi/6, 0, -\pi/6)^T$ rad.

The configurations of RBFNN are the same for all the follower manipulators. The number of neurons is 16. τ_{di} is modeled by $\tau_{di} = (\sin(t/3), \sin(t/3), \sin(t/3))^T$. The controller parameters are: $\eta = 4$, $k_i = 10$, $\delta_{Mi} = 10$, $\varsigma_i = 0.2$, $\beta_{wi} = 500$, $W_{\max i} = 1500$, and the initial RBFNN weight matrix $\hat{W}_i(0)$ is chosen to be the zero matrix ($i = 1, \dots, 5$). The trajectories of joint positions of follower manipulators are provided in Fig. 3.4. The profiles of input torques of follower manipulators are given in Fig. 3.5. According to the simulation results, it can be seen that all the follower manipulators' joints can be regulated at the value of the leader manipulator's joint, which shows the satisfactory performance of the proposed neural-network-based adaptive controller.

Consider another case where the communication link from the leader manipulator to follower manipulator 2 loses after $t = 10$ s, as shown in Fig. 3.6. It can be seen that, after the link loss, the new topology is connected. Therefore, the proposed controller can still work well. The simulation is re-conducted with the same controller configuration in the previous example. The simulation results are given in Figs. 3.7 and 3.8. It can be seen that the control objective can be achieved satisfactorily too.

3.6 Conclusion

This chapter proposes a distributed neural-network-based adaptive approach for the coordinated control of multi-manipulator systems. The principle of controller design is based on the multi-agent theory. However, most work so far on the coordination of multi-agent system has assumed the agent has the exact dynamics.

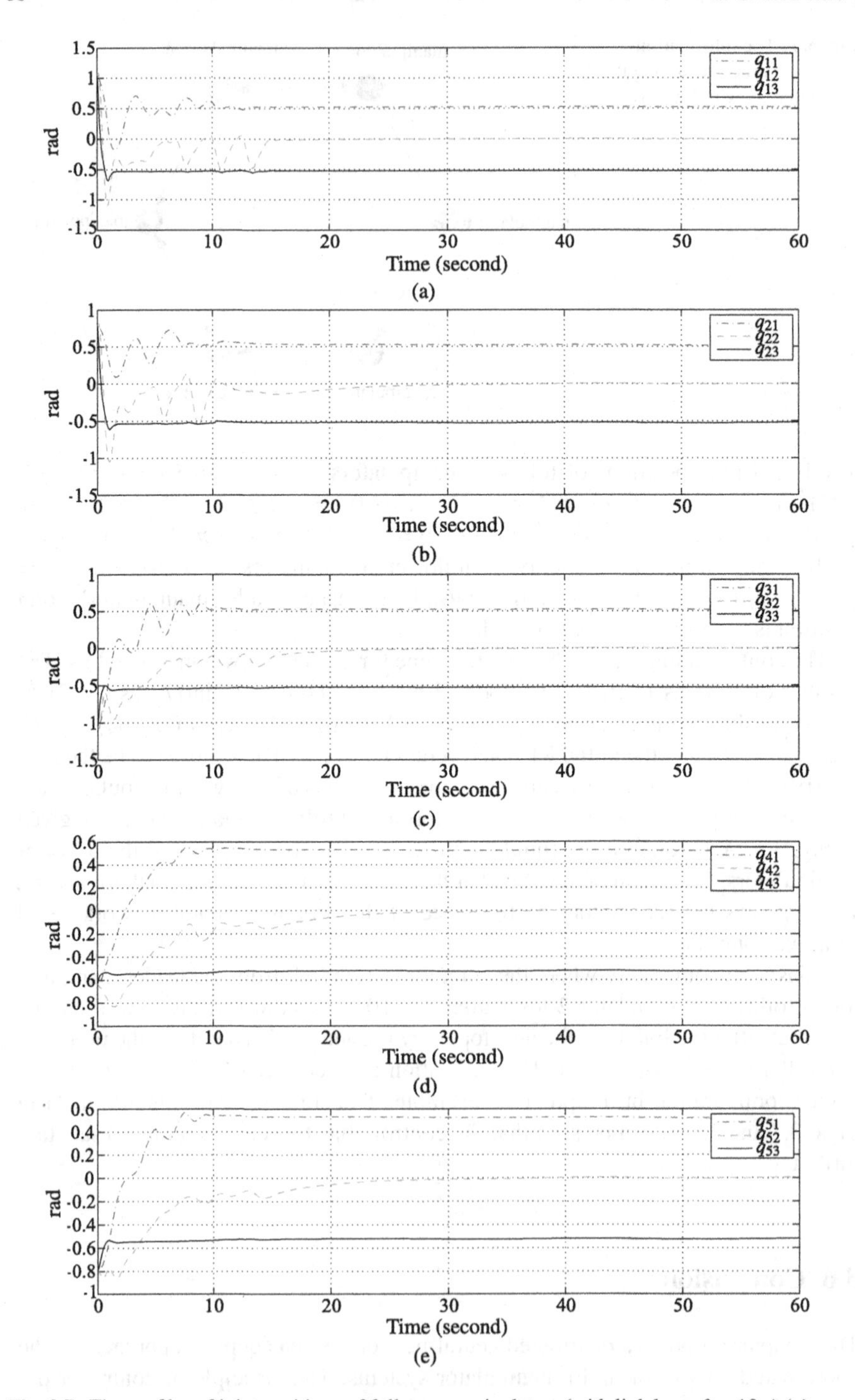

Fig. 3.7 The profiles of joint positions of follower manipulators (with link loss after 10 s) (**a**) manipulator 1; (**b**) manipulator 2; (**c**) manipulator 3; (**d**) manipulator 4; (**e**) manipulator 5

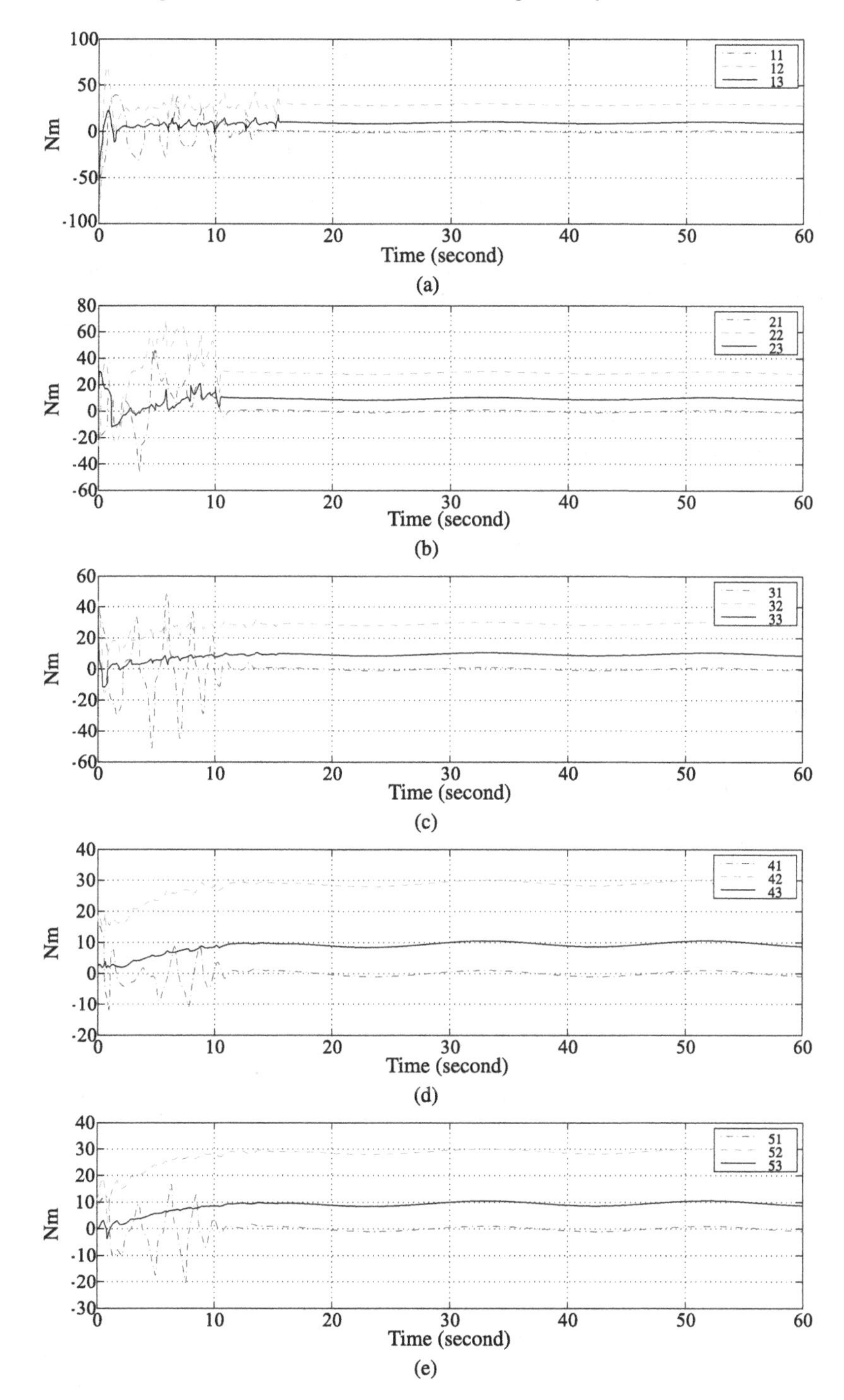

Fig. 3.8 The profiles of input torques of follower manipulators (with link loss after 10 s) (**a**) manipulator 1; (**b**) manipulator 2; (**c**) manipulator 3; (**d**) manipulator 4; (**e**) manipulator 5

Unfortunately, due to the imprecision measurement of manipulator parameters and interactions between manipulator and different environments, it is consequently difficult to obtain the exact dynamics model. The effects of uncertain manipulator dynamics are taken into account, and the RBFNN is employed to approximate the uncertain dynamics model. It is noted that the neural network has no off-line learning phase, and its weight matrix is updated on-line by the projection method. The approximation error and external disturbances in the dynamics model are counteracted by using the robustness signal. According to the theoretical analysis, it is shown that all the follower manipulators' joints can be regulated at the value of the leader's joint, and the regulation error can be reduced as small as desired by appropriately choosing controller parameters. At last, it is noted that, by the result provided in [27], the proposed controller can be extended to the following two cases, where the information exchange among manipulators is not mutual (adjacency matrix $A_{\mathcal{G}}$ is not symmetric) and where the leader manipulator's joint is time-varying.

Acknowledgements This work was supported in part by the National Natural Science Foundation of China (Grants 60725309, 60775043 and 60805038) and the National Hi-Tech R&D Program (863) of China (Grant 2009AA04Z201).

References

1. Luh, J.Y.S., Zheng, Y.F.: Constrained relations between two coordinated industrial robots for motion control. Int. J. Robot. Res. **6**, 60–70 (1987)
2. Jadbabaie, A., Lin, J., Morse, A.S.: Coordination of groups of mobile autonomous agents using nearest neighbor rules. IEEE Trans. Autom. Control **48**, 988–1001 (2003)
3. Olfati-Saber, R., Murray, R.M.: Consensus problems in networks of agents with switching topology and time-delays. IEEE Trans. Autom. Control **49**, 1520–1533 (2004)
4. Ren, W., Beard, R.W.: Consensus seeking in multiagent systems under dynamically changing interaction topologies. IEEE Trans. Autom. Control **50**, 655–661 (2005)
5. Cheng, L., Hou, Z.G., Tan, M.: Observer-based consensus protocol for linear multi-agent systems. IEEE Trans. Autom. Control (under review)
6. Moreau, L.: Stability of multiagent systems with time-dependent communication links. IEEE Trans. Autom. Control **50**, 169–182 (2005)
7. Hou, Z.G., Cheng, L., Tan, M.: Decentralized robust adaptive control for multi-agent system consensus problem using neural networks. IEEE Trans. Syst. Man Cybern., Part B, Cybern. **39**, 636–647 (2009)
8. Hong, Y., Hu, J., Gao, L.: Tracking control for multi-agent consensus with an active leader and variable topology. Automatica **42**, 1177–1182 (2006)
9. Shi, H., Wang, L., Chu, T.: Virtual leader approach to coordinated control of multiple mobile agents with asymmetric interactions. Physica D **213**, 51–65 (2006)
10. Ren, W.: Multi-vehicle consensus with a time-varying reference state. Syst. Control Lett. **56**, 474–483 (2007)
11. Hu, J., Hong, Y.: Leader-following coordination of multi-agent systems with coupling time delays. Physica A **374**, 853–863 (2007)
12. Ma, C.Q., Li, T., Zhang, J.F.: Leader-following consensus control for multi-agent systems under measurement noises. In: Proceedings of IFAC World Congress, pp. 1528–1533 (2008)
13. Ren, W., Beard, R.W., Atkins, E.M.: A survey of consensus problems in multi-agent coordination. In: Proceedings of American Control Conference, pp. 1859–1864 (2005)

14. Olfati-Saber, R., Fax, J.A., Murray, R.M.: Consensus and cooperation in networked multi-agent systems. Proc. IEEE **95**, 215–233 (2007)
15. Cheng, L., Hou, Z.G., Tan, M., Liu, D., Zou, A.M.: Multi-agent based adaptive consensus control for multiple manipulators with kinematic uncertainties. In: Proceedings of IEEE International Symposium on Intelligent Control, pp. 189–194 (2008)
16. Cheng, L., Hou, Z.G., Tan, M.: Decentralized adaptive consensus control for multi-manipulator system with uncertain dynamics. In: Proceedings of IEEE International Conference on Systems, Man, and Cybernetics, pp. 2712–2717 (2008)
17. Cheng, L., Hou, Z.G., Tan, M.: Decentralized adaptive leader-follower control of multi-manipulator system with uncertain dynamics. In: Proceedings of The 34th Annual Conference of the IEEE Industrial Electronics Society, pp. 1608–1613 (2008)
18. Narendra, K.S., Parthasarathy, K.: Identification and control of dynamical systems using neural networks. IEEE Trans. Neural Netw. **1**, 4–27 (1990)
19. Polycarpou, M.M., Ioannou, P.A.: Identification and control of nonlinear systems using neural network models: design and stability analysis. University of Southern California Technical Report 91-09-01 (1991)
20. Polycarpou, M.M.: Stable adaptive neural control scheme for nonlinear systems. IEEE Trans. Autom. Control **41**, 447–451 (1996)
21. Lewis, F.L., Jagannathan, S., Yesildirek, A.: Neural Network Control of Robot Manipulators and Nonlinear Systems. Taylor & Francis, New York (1998)
22. Ge, S.S., Lee, T.H., Harris, C.J.: Adaptive Neural Network Control of Robotic Manipulators. World Scientific, Singapore (1998)
23. Cheng, L., Hou, Z.G., Tan, M.: Adaptive neural network tracking control for manipulators with uncertain kinematics, dynamics and actuator model. Automatica **45**, 2312–2318 (2009)
24. Farrell, J.A., Polycarpou, M.M.: Adaptive Approximation Based Control: Unifying Neural, Fuzzy and Traditional Adaptive Approximation Approaches. Wiley-Interscience, Hoboken (2006)
25. Ge, S.S., Wang, C.: Adaptive neural control of uncertain MIMO nonlinear systems. IEEE Trans. Neural Netw. **15**, 674–692 (2003)
26. Kanellakopoulos, I., Kokotovic, P.V., Morse, A.S.: Systematic design of adaptive controllers for feedback linearizable systems. IEEE Trans. Autom. Control **36**, 1241–1253 (1991)
27. Cheng, L., Hou, Z.G., Tan, M.: Neural-network-based adaptive leader-following control for multi-agent systems with uncertainties. IEEE Trans. Neural Netw. (under review)

Chapter 4
A New Framework for View-Invariant Human Action Recognition

Xiaofei Ji, Honghai Liu, and Yibo Li

Abstract An exemplar-based view-invariant human action recognition framework is proposed to recognize the human actions from any arbitrary viewpoint image sequence. In this framework, human action is modelled as a sequence of body key poses (*i.e.*, exemplars) which are represented by a collection of silhouette images. The human actions are recognized by matching the observed image sequence to predefined exemplars, in which the temporal constraints are imposed in the exemplar-based Hidden Markov Model (HMM). Furthermore, a new two-level recognition framework is introduced to improve the discrimination capability for the similar human actions. The aim of the first level recognition is to decide an equivalent set in which the testing action is included instead of directly achieving the final recognition results. In the second level, the weighted contour shape feature is used to calculate the observation probability to discriminate the similar actions. The proposed framework is evaluated in a public dataset and the results show that it not only reduces computational complexity, but it is also able to accurately recognize human actions using single cameras. Besides it is verified that the weighted contour shape feature is effective to differentiate the similar arm-related actions.

4.1 Introduction

Human action recognition from video is an important and challenging research topic in computer vision with many potential applications involving human mo-

X. Ji (✉) and H. Liu
Intelligent Systems and Biomedical Robotics Group, School of Creative Technologies, The University of Portsmouth, Eldon Building, Portsmouth PO1 2DJ, UK
e-mail: xiaofei.ji@port.ac.uk; honghai.liu@port.ac.uk

X. Ji and Y. Li
School of Automation, Shenyang Institute of Aeronautical Engineering, No. 37 Daoyi South Avenue, Shenyang 110136, China
e-mail: lyb20040612@yahoo.com.cn

H. Liu et al. (eds.), *Robot Intelligence*,
Advanced Information and Knowledge Processing,
DOI 10.1007/978-1-84996-329-9_4,

Fig. 4.1 A surveillance scene in dataset for CAVIAR project

tion understanding such as visual surveillance, content based video retrieval, athletic performance analysis, human-robot interaction, *etc.* There has been significant amount of research in action recognition in recent years. Unfortunately, most of human action recognition methods are constrained with the assumptions of view dependence, *i.e.*, actors have to face a camera or to be parallel to a viewing plane [6, 7, 13, 18, 33, 47]. Such requirements on view dependence are difficult, sometimes impossible, to achieve in realistic scenarios. The actions are often observed from arbitrary camera viewpoints, for instance as shown in Fig. 4.1, so it is desired that the recognition algorithm exhibit some view invariance. That is to say that analysis methods remain unaffected by different viewpoints of camera. An action should remain recognizable while the viewpoint of the camera is changing. The viewpoint issue has been one of the bottlenecks for research development and practical implementation of human motion analysis, which has driven growing number of research groups to pay more attention to the research related to the view-invariant issue [15].

A large number of attempts and research progress on removal of the effect on human motion analysis methods had been reported in recent years [11, 17, 24, 26, 27, 34, 43, 45, 48]. Those methods can be classified into two categories: template-based methods and state-space methods.

Template-based methods focus on extracting low-level image from image sequence and converting it into a static shape pattern or a special motion feature pattern which are then compared to features that are pre-extracted from a set of action template for recognition [1]. Parameswaran and Chellappa [28, 29] handled the problem of view-invariant action recognition based on point-light displays by investigating 2D and 3D invariant theory. They obtained a convenient 2D invariant representation by decomposing and combining the patches of a 3D scene, as shown in Fig. 4.2.

Rao *et al.* [31] presented a view-invariant computational representation of human action to capture dramatic changes in the speed and direction of a motion trajectory, which was presented by spatio-temporal curvature of a 2D trajectory. Furthermore a representative spatio-temporal action volumes (STV) was proposed by Yilmaz

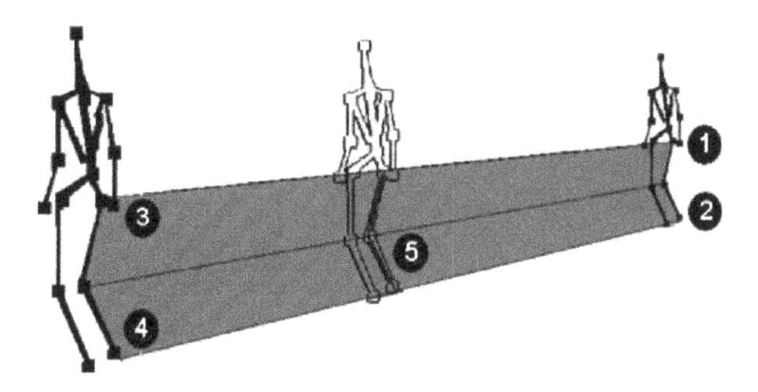

Fig. 4.2 Geometrical invariants can be computed from five points that lie in a plant [28]

Fig. 4.3 Space-time shapes of "jumping-jack", "walk", "run" actions [5]

and Shah [46] to achieve view-invariant action recognition. On the basis of this work, a novel 4D action feature (4D-AFM) was presented for recognizing actions from arbitrary camera views [44]. The above methods are all based on the assumption that point correspondences are available in parts of images. Interest points are likely to be unreliable in cases of smooth surfaces, motion singularities and low-quality videos, so their applications are limited to some special occasions. Another approach was proposed by Blank and Gorelick [5] that represented human actions as three-dimensional shapes induced by the silhouettes in the space-time volume, as shown in Fig. 4.3. This method extracts space-time features that do not require computing point correspondence. This method is not fully view-invariant, however it is robust to large changes in viewpoint (up to 54 degrees).

A novel framework is proposed that fuses multiple features including a quantized vocabulary of local spatio-temporal volumes and a quantized vocabulary of spin-images to improve the action recognition. The results demonstrate that fusion of multiple features helps in achieving improved performance [20]. The template based approach is a two-stage method. It first directly investigates view-invariant action representations, then considers the action recognition as a classification problem. The key to template-matching approaches is to find the vital and robust feature sets. The advantages of template-based methods are the low computational cost and the simple implementation, however they are usually more sensitive to noise and variance of movement duration [15].

Meantime, the methods based on state-space models, *e.g.*, Hidden Markov Models (HMMs), have been widely applied to express the temporal relationships inherent in human actions [2, 12, 23, 30, 35]. These methods usually define each static human posture as a state of the model. These states are connected by certain probabilities, and any motion sequence is considered as a tour going through

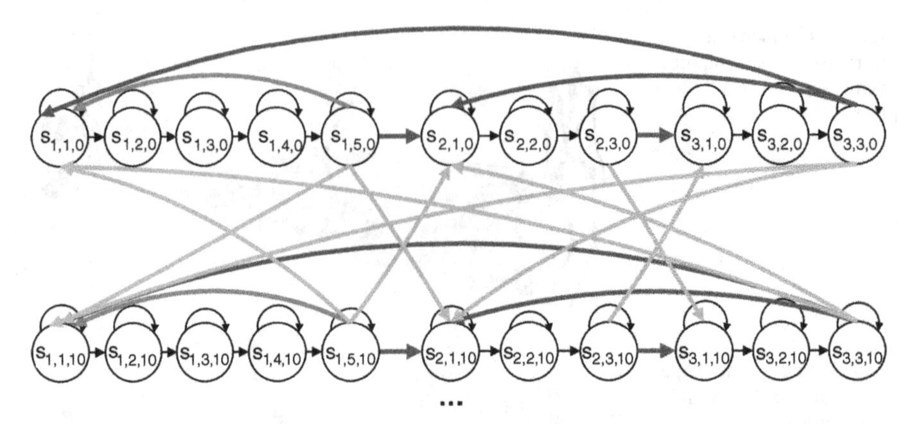

Fig. 4.4 Action graph models [22]

various states of these static poses. Probabilities are computed through these tours, and maximum values are selected as the criteria for human action classification and recognition [1]. State-space methods usually utilize the results of 3D pose estimation as input to achieve view-invariant action recognition. Lv *et al.* [21] decomposed a high dimensional 3D joint space into a set of feature spaces, in which each feature corresponds to the motion of a single joint or combination of related multiple joints. In the learning process, for each feature, the dynamics of each action class are learned with one HMM. In the recognition process, given an image sequence, the observation probability is computed in every HMM to recognize each action class, where an AdaBoost scheme is formed to detect and recognize the feature. The proposed algorithm is effective in that the results are convincing in recognizing 22 actions on a large number of motion capture sequences as well as several annotated and automatically tracked sequences. Lv and Nevatia [22] presented an example-based view-invariant action recognition system that explored the use of contextual constraints. Those constraints were inherently modelled by a novel action graph model representation called *Action Net*, as shown in Fig. 4.4. Each link in the action net specified the possible transition of the key poses within an action class or across different action classes. This approach was demonstrated on challenging video sets consisting of 15 complex action classes. Owing to the complexity of the action net, modelling transitional probability for each link is not applicable in practice. So, this action net representation neglects the transitional probability.

A similar work on exemplar-based HMMs was proposed for view-invariant human motion analysis [42]. This model can account for dependencies between three dimensional exemplars, *i.e.*, representative pose instances and image cues. Inference is then used to identify the action sequence that best explains the image observations. This work uses a probabilistic formulation instead of the deterministic linked action graph introduced in [22], so it can handle uncertainties inherent to actions performed by different people and different styles. However the learning process is relatively complex. HMM and its variants have been widely used to recognition

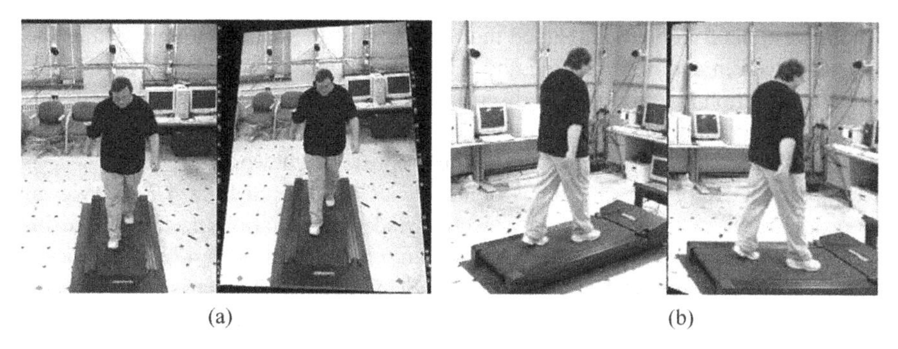

(a) (b)

Fig. 4.5 Original image and transformed image for frontal: (**a**) "rear-diagonal", (**b**) "diagonal" [34]

problems such as modelling human motions. However assumption of independence is usually required in such generative models, which makes the methods unsuitable for accommodating multiple overlapping features or long-range dependences among observations.

Apart from the above-mentioned approaches, there is another kind of methods based on image normalization. Those methods pre-process the image sequence before abstract the human motion feature. In order to remove viewpoint effect, those methods directly transform all observations into a canonical coordinate frame, then template matching method and state space method can be used to recognize the human actions. Normally the actual motion direction must be detected in advance by using the detected body parts or walking direction. Then matching takes place after the observations have been normalised. For example, Kale *et al.* [16] proposed a method for view invariant gait recognition, in which a person is walking along a straight line (*i.e.*, make a constant angle with the images plane), then a side-view is synthesized by homography. The viewpoint invariance is also achieved by projecting all the images onto the ground plane [32]. A method was presented by Rogez *et al.* [34] that selects a 2D viewpoint-insensitive model (made of shape and stick figure), then uses the 3D principal directions of man-made environments and the direction of motion to transform both 2D Model and input images to a common frontal view, as shown in Fig. 4.5. Though these approaches can remove the viewpoint effect directly, a problem with them is that all results completely depend on the robustness of the body orientation estimation. Furthermore the computational cost is significantly high.

As the above discussion, the tradeoffs have to be handled between computational cost and recognition performance in the view-invariant human action recognition. On the basis of the works [22, 42], we propose a simplified view-invariant human action recognition framework using exemplar-based HMMs. In our framework, each human action is modelled by a sequence of static key poses, which are represented by a set of 2D silhouette images captured from multiple camera viewpoints. Action recognition is achieved by using Viterbi search on the exemplar-based HMMs. The silhouette distance signal is abstracted as shape feature of silhouette image, which is efficiently obtained. Hence the reduction of the computational complexity is achieved in action modelling and recognition. Furthermore, the second level action recognition is introduced to discriminate the similar human actions by using the weighted contour shape features in the observation probability. It is helpful to improve the discrimination capability of the proposed framework.

The remainder of this chapter is organized as follows. Section 4.2 overviews the proposed framework. Exemplar selection and representation are introduced in Sect. 4.3. Action modelling and recognition are proposed in Sect. 4.4. The experimental results are presented and discussed in Sect. 4.5. The chapter is concluded in Sect. 4.6 with analysis on future research challenges and directions.

4.2 Overview of the Proposed Approach

Human actions involve both spacial (the body pose in each time step) and temporal (the transition of the body poses over time) characters in their representations. The actual appearance of the spacial-temporal representation varies significantly with camera viewpoint changes. A simplified action representation was utilized in our method only using remarkable key poses instead of including body poses in all frames [22, 26]. For example, a cross arm action can be represented by using three key poses, *e.g.*, stand, raise two arms, cross two arms, as shown in Fig. 4.6. This representation cannot only reduce the computational complex, but also deal with the variance in execution styles of the actions.

To accommodate variants of human action in appearance due to camera viewpoint, the action model should include appearances at different camera viewpoints. A exemplar-based view-invariant action recognition method is proposed. The framework of the proposed method is shown in Fig. 4.7, the details are provided in the following:

1. Key pose extraction and representation: the key poses of each action class are extracted from a given small set of action sequences from Inria Xmas Motion Acquisition Sequences (IXMAS) dataset [43] by clustering the 3D visual hull representations. This dataset provides the 3D visual hull representations which have been computed using a system of 5 calibrated cameras. Then the 3D key poses are projected into multiple-view 2D silhouette images using the camera projection principle under the assumption that only the orientation of an actor

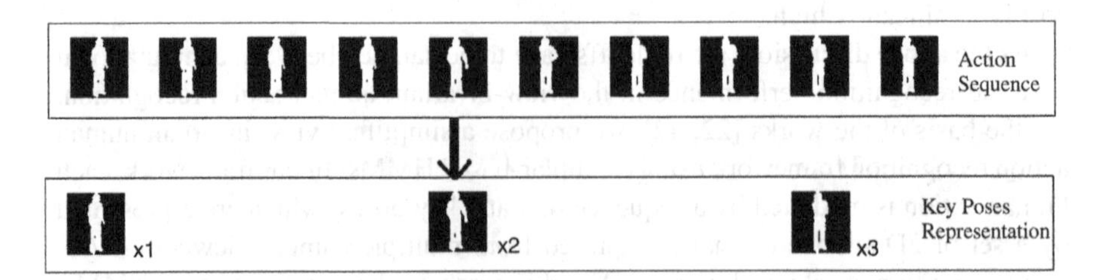

Fig. 4.6 The key pose representation of human action

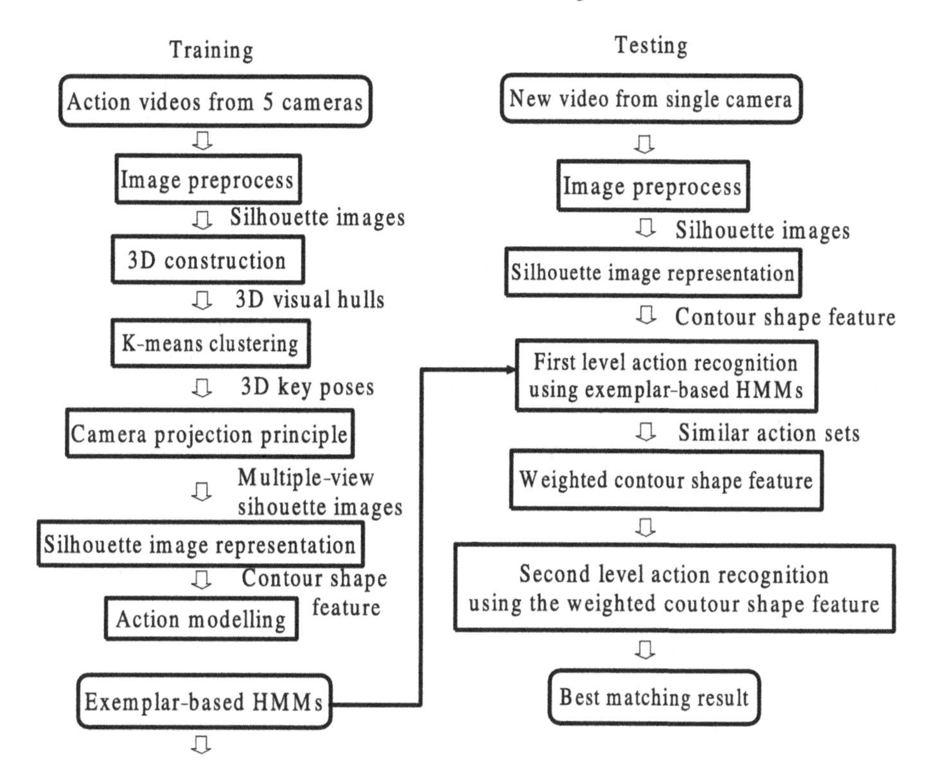

Fig. 4.7 The framework of proposed approach

around the vertical axis is variable. Contour shape feature is utilized to represent those silhouette images.

2. Action representation and model learning: each human action class is modelled by an exemplar-based HMM. Each state in the HMMs accords with a key pose exemplar which is represented by a collection of silhouette images observed from 36 different viewpoints. The dynamics, (*i.e.*, transfer matrices) are learned by utilizing the expectation maximization (EM) algorithm from the given action sequences. At the same time, the contextual constraints are imposed in the transfer matrices to reduce the searching scope.
3. The first level action recognition: the exemplar-based HMMs are used to find the most likely sequence of actions seen from a single viewpoint video which is obtained by applying the Viterbi algorithm in recognition phrase. At each frame the observation probability is computed based on shape similarity which is achieved by calculating the Euclidean distance between the observation and the exemplar. In this process, we force the camera viewpoint to remain constant or change smoothly from one key pose to the next consecutive pose.
4. The second level action recognition: according to some similar actions, especially arm-related actions, it is possible to get the wrong recognition results in the first level recognition. In order to efficiently discriminate those similar ac-

tions, the categories of those similar arm-related actions are revalidated by the second level action recognition. In this process, the weighted contour shape features is introduced in calculating the observation probability to obtain the best matching result.

4.3 Exemplar Selection and Representation

In the proposed framework, each class of human action is modelled as a sequence over a set of 3D key poses, the exemplars, which are described by multiple view silhouette images. It is evident that collecting multiple view human pose images from real experimental conditions is a difficult task. So we directly obtain multiple view pose dataset by projecting 3D key poses into multiple view 2D silhouette images by using the camera projection principle where the camera parameters are known [42]. 3D key poses are extracted from a small set of action sequences by k-means clustering.

4.3.1 Key Pose Extraction

There are some popularly used methods of extracting human key poses to describe human action, including motion capture data based motion energy minima and maxima [22], image based optical flow magnitude of foreground pixels extremum [26], 3D pose representation based k-means clustering and wrapper [42]. As we know, the key point in the process of key pose extraction is to obtain as much key poses as possible for a special action class and to keep as much distance as possible between key poses of different actions. In the framework, firstly, we manually select a small set of human action sequences for each action class from IXMAX dataset which is a multiple-actor and multiple-action dataset. In this dataset human poses in every frame are represented in 3D visual hulls that have been computed using a system of 5 calibrated cameras as shown in Fig. 4.8.

Fig. 4.8 3D visual hull presentation

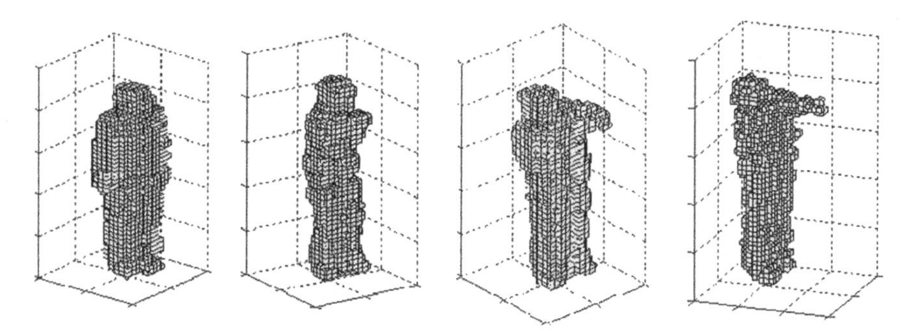

Fig. 4.9 The exemplars for an action punch

Key poses are extracted for each action by clustering those given 3D visual hull representations. We extracted 4 exemplars for each action by using k-means method, which is sufficient to accurately recognize those actions. The exemplars of the action punch are shown in Fig. 4.9.

4.3.2 2D Silhouette Image Generation

A camera is a mapping between a 3D world (object space) and a 2D image. Let (X_w, Y_w, Z_w) be the 3D coordinates of a point in the world coordinate system. Its projection on the image plane (u, v) is given by:

$$\begin{bmatrix} u \\ v \\ 1 \end{bmatrix} = \begin{bmatrix} \alpha_x & 0 & u_0 & 0 \\ 0 & \alpha_y & v_0 & 0 \\ 0 & 0 & 1 & 0 \end{bmatrix} \begin{bmatrix} R & t \\ 0^T & 1 \end{bmatrix} \begin{bmatrix} X_w \\ Y_w \\ Z_w \\ 1 \end{bmatrix} = M_1 M_2 \begin{bmatrix} X_w \\ Y_w \\ Z_w \\ 1 \end{bmatrix}. \tag{4.1}$$

M_1 is the internal parameter of the camera. M_2 is the external parameter of the camera that is decided by the camera position in world coordinate. R is the rotation matrix, t is the transition vector.

Due to the fact that a camera is a mapping device between a 3D world (object space) and a 2D image, 3D visual hull presentations are projected into multiple-view 2D silhouette images by using the camera projection principle. It is only considered that the orientation of a person around the vertical axis is variable. The orientation angle is discretized into 36 equally spaced angles within $[0, 2\pi]$, the multiple views for a given 3D key pose are shown in Fig. 4.10; multiple-view silhouette images of a given key pose are provided in Fig. 4.11.

The multiple-view silhouette images are centred and normalized in order to contain as much foreground as possible, it leads to the fact that the motion shape is not distorted and all input frames are equal dimensions.

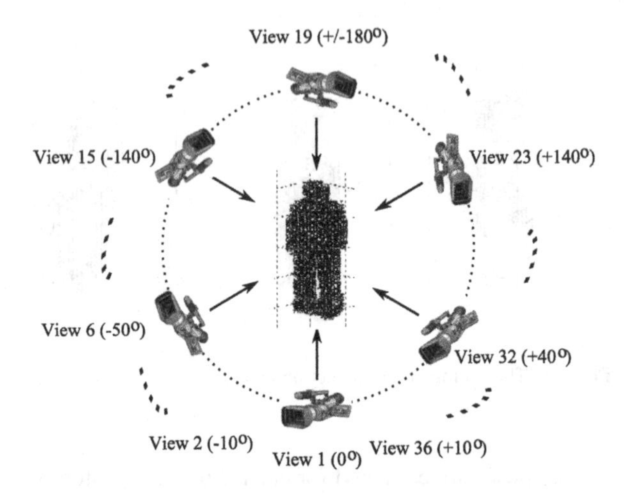

Fig. 4.10 Multiple views for a given key pose

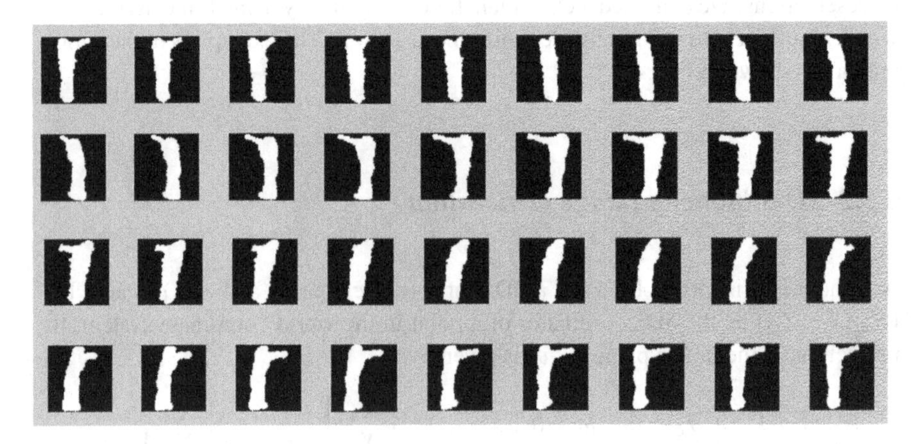

Fig. 4.11 Multi-view silhouette images of a given key pose

4.3.3 Contour Shape Feature

There are some representative shape features of the silhouette image in the previous chapters, such as shape context descriptor [4], width feature [8], block-based feature [40] and orientation code [39]. We describe the silhouette images using the silhouette distance matrix, in which it not only can capture both structural and dynamic information for an action, but also it can be efficiently obtained [10]. An example of the distance signal is shown in Fig. 4.12, which is generated by calculating the Euclidean distance between the centroid of the mass points and each edge point of the silhouette images in clockwise direction.

In order to obtain image rotation invariance, firstly the principle axis of the silhouette is computed before computing shape feature, the calculation process is described as follows:

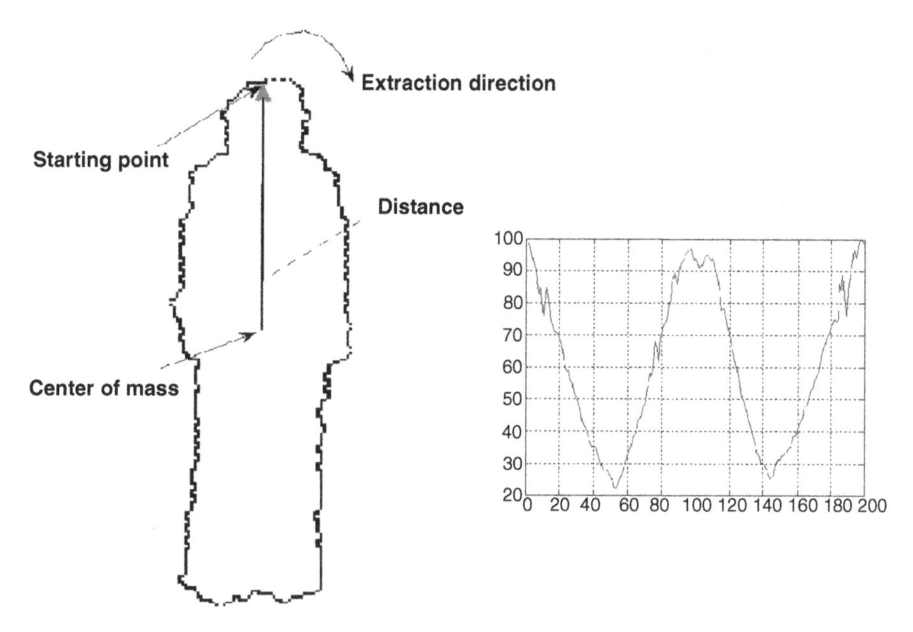

Fig. 4.12 Contour shaper feature of the silhouette image

1. The centroid of silhouette image at time t, is

$$\mu_t = \left(\frac{1}{P} \sum_{j=1}^{P} V_{t,j} \right) \tag{4.2}$$

   V_t is the coordinate representation of the silhouette image at time t, P is according to the total number of the silhouette image points.
2. The covariance matrix, is

$$\left(\frac{1}{p-1} \right) \sum_{j=1}^{P} \left(V_{t,j} - \mu_t \right) * \left(V_{t,j} - \mu_t \right)^T . \tag{4.3}$$

3. Calculate the eigenvalues and eigenvectors of above covariance matrix. The direction according to the maximal eigenvalue is the principle axis of the silhouette image [3].

Before computing shape feature, the rotation angle of the principle axis is compensated so that the principal axis is vertical. That means the silhouette image is vertical. Furthermore, in order to achieve the scale invariant feature, the contour of the silhouette image is uniformly re-sampled to 200 edge points, and the distance is normalized into [0, 100], as shown in Fig. 4.12.

By utilizing above method, we selected four exemplars for every action, and built up a multiple-view dataset for those exemplars. Meantime the representation of those multiple-view silhouettes is scale and rotation invariance that is a very useful feature for action modelling and recognition.

4.4 Action Modelling and Recognition

Human actions evolve dynamically over time, so temporal models such as HMMs and their variants have been widely applied to model human actions [2, 12, 23, 30, 35]. In our method, each action class is modelled from multiple-person and multiple-view datasets by learning an exemplar-based HMM. In order to achieve view-invariant human action recognition, each state in this graph model corresponds to one human static key pose, which is represented as a collection of contour shape features of multiple-view 2D silhouette images. Action recognition is achieved by using the standard HMM algorithm, *i.e.*, the maximum a posteriori estimate, to find the most likely sequence of actions seen in a single viewpoint video. Observed from the arbitrary viewpoint, some silhouette images in different human actions look very similar. That will lead to wrong recognition results. So in our framework, after the first level action recognition, some similar actions will be revalidated by using the weighted contour shape feature in the observation probability.

4.4.1 Exemplar-based Hidden Markov Model

The exemplar-based HMM has been used in action recognition to solve the problem that the space of observations is not Euclidean [12, 42]. In this case, the mean and variance cannot be defined. The novelty of the exemplar-based HMM is that mixture density functions are not entered on arbitrary means values, but centred on prototypical data instances, the exemplars.

A representative graphical model is shown in Fig. 4.13. An action class is modelled as a hidden state sequence Q, *e.g.*, a motion sequence in a pose space which follows a first order Markov chains over time. An exemplar space is specified by a set of exemplars, $X = \{x_t, t = 1, 2, \ldots, M\}$. The exemplars are intermediate observations that are being emitted by the underlying process Q. The final observation y_t at time t derives from a geometric transformation of exemplars, $y_t \approx T_{\alpha_t} x_t$. T_{α_t} is a geometric transformation with parameter α_t.

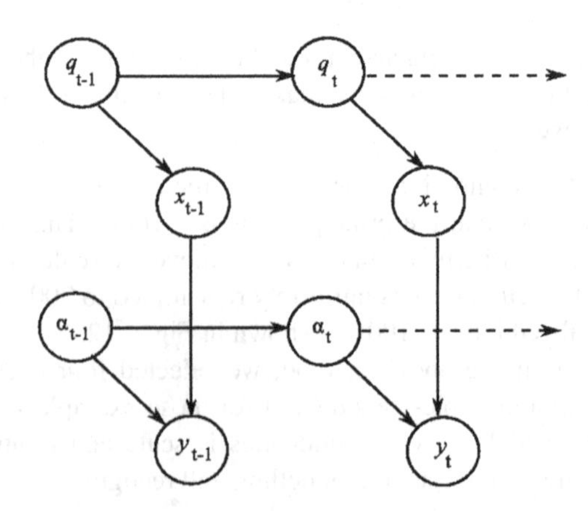

Fig. 4.13 Graphical model for exemplar-based tracking [12, 38]

Learning this probability model involves learning the exemplars from the training set, learning the dynamics, *i.e.*, transfer matrix in the form of $P(x_t|x_{t-1})$, and learning the exemplar probability given the state $P(x_t|q_t)$ [12, 38].

4.4.2 Action Modelling

In our method, each action class is modelled by an exemplar-based HMM. Exemplars estimation is no longer coupled with the HMMs parameter estimation. Exemplars previously are selected by using the method that has been introduced in Sect. 4.3. Then the exemplars are represented by a set of contour shape feature of multiple view 2D silhouette images, which don't deterministically link to motion states Q. So there is no coupling between the states and the exemplars. Under this condition, exemplar probability given the state $P(x_t|q_t)$ need not be estimated. Only the dynamic *i.e.*, transfer matrix $P(x_t|x_{t-1})$, need to be learned by using the traditional EM approach, in which each exemplar can be treated as a discrete symbol.

The EM procedure iterates is as follows:

1. E-step Using the current parameter values and training data, the expected sufficient statistics(ESS) are estimate.
2. M-step The parameters are re-estimate using the ESS, which for HMMs is achieved by normalizing the ESS into conditional probabilities.

Note that efficient implementations of the EM iteration usually use the forward-backward algorithm to estimate the ESS.

Actually learning the dynamics in the form of $P(x_t|x_{t-1})$ needs a large number of training data. In order to deal with this problem, special action temporal constraints are imposed in the dynamics *i.e.*, $P(x_t|x_{t-1})$. Such constraints represent two aspects: first, the key poses in an action should occur according to some specific order, for example, during the process of sitting down, pose "went down" must happen between the pose "standing" and the pose "sitting down". Some transitions from one key pose to another one is impossible or no meaning in the real human actions. Second, change in actor's orientation should be smooth. That is to say, the camera viewpoint should remain constant or change smoothly from one key pose to the next consecutive pose [22]. In our method, we used left-right HMMs to impose some constraints on the dynamics which lead to better generalization since there are less transition parameters to consider.

4.4.3 Action Recognition

Separate action models H_c are learned for each action class $c \in 1, \dots, C$. Given a set of observations $Y = \{y_1, y_2, \dots, y_T\}$, the task is to calculate the probability of that observation sequence given each of the action models. Then the sequence of observation Y can be recognized with maximum of the probability. The algorithm detail is as follows:

The actual observation y is the detected edge features at each new input frame. The observation probability is a probabilistic function of the current state of the action model *i.e.*, $p(y|x)$ is defined as

$$p(y|x=i)=\frac{1}{Z_i}\exp(-d(y,x_i)^2/\sigma_i^2) \tag{4.4}$$

where d is a distance function between the observation and the exemplar, *e.g.* Euclidean distance or more specialized distance such as the chamfer distance. In our method the Euclidean distance is employed. The variance σ_i and the normalization constant Z_i are selected as the method proposed in [12], *i.e.*, $\sigma_i=\sigma, i=1,\dots,M$ (M is the number of exemplars).

Given a sequence of observation Y, using Bayes theorem the posterior of a class $c\in 1,\dots,C$ is

$$p(H_c|Y)=\frac{p(Y|H_c)p(H_c)}{p(Y)}. \tag{4.5}$$

Because $p(Y)$ is independent of the class, it follows that

$$p(H_c|Y)\approx p(Y|H_c)p(H_c). \tag{4.6}$$

Then the sequence of observation Y can be recognized with respect to the maximum a posteriori estimate:

$$p(Y)=\arg\max_c p(Y|H_c)p(H_c). \tag{4.7}$$

The joint probability of observation sequence $p(Y|H_c)$ can be obtained by the Viterbi algorithm, which is efficiently based on dynamic programming. The prior probabilities $p(H_c)$ is used as an uniformly distributed prior in the framework.

4.4.4 Action Category Revalidation

After the previous action recognition, some similar actions (*e.g.* check watch and cross arm, wave hand and scratch head) may get wrong recognition results. In order to improve the recognition rate, the categories of those similar arm-related actions are revalidated by the second level action recognition that introduces weighted contour shape features in the observation probability. That is, in our framework, the action recognition is a two level process as shown in Fig. 4.14. The aim of the first level process of the recognition is to decide an action set(including multiple similar actions) in which the testing action is included instead of directly achieving the final action recognition results. In the second level, weighted contour shape feature is used to discriminate the similar actions that are included in the same action set to obtain the best matching result [12].

After the first level action recognition, we know the action set in which the similar actions are included. In the same action set, the most of parts of the action silhouettes

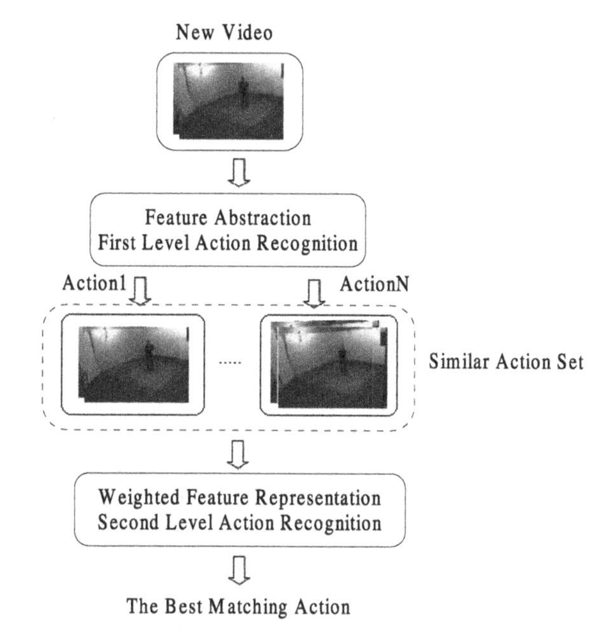

Fig. 4.14 The flowchart of the two levels action recognition

look very similar from a given camera viewpoint, some difference exists in some relative small parts. For example, the head, torso, crura parts of the body are likely to be similar in different silhouette images, while the arm parts of the body are very different at different silhouette images, especially in the arm-related action. Since the arm part is represented by only a small number of coutour shape features with respect to the whole silhouette image, it is possible that the matching between the observation and exemplar is decided by the major body parts. However, it is desired that the feature of arm parts is crucial to calculate the matching probability since these parts will be more discriminative between different silhouette images of similar actions.

To achieve this goal, different weights are assigned to different feature points in each silhouette image, that has been used in [12]. Therefore each exemplars x_t is represented as a set of feature locations as well as a set of weights ω_t. According to each feature of exemplar, the relatively big weights are assigned to the feature of arm-related parts and the smaller weights are assigned to the other parts. It is very easy to achieve in our framework because our contour shape feature is scaled and rotate invariant and that has been normalised to the equal dimensions.

In order to efficiently discriminate those similar actions, weighted contour shape feature is introduced in the observation probability, as shown in (4.8):

$$p(y|x=i) = \frac{1}{Z_i} \exp(-\omega_i * d(y, x_i)^2 / \sigma_i^2). \tag{4.8}$$

According to the similar actions in the same action set, we discriminate them by advising the observation probability of above HMMs to obtain the best matching results. In our framework, the second level action recognition is only used to recognize the arm-related similar actions. It is ineffective to discriminate the actions that are involved by whole human body.

4.5 Experiments

We demonstrate the proposed framework on a public dataset, the multiple view IXMAS dataset. It contains 12 actions (*i.e.*, check watch, cross arms, scratch head, sit down, get up, turn around, walk in a circle, wave a hand, punch, kick, point, pick up), and each of them was performed 3 times by 12 actors (5 females and 7 males). Each action is recorded by 5 cameras with the frame rate of 23 fps. In this dataset, actor orientations are arbitrary since no specific instruction is given during the acquisition, as shown in Fig. 4.15. So this dataset is very suitable for testing the effectiveness of the view-invariant action recognition algorithm.

Human silhouette extraction from image sequence is relatively easy for current vision techniques. How to achieve the silhouette images is not considered in our framework. We directly use the human silhouette images of the observation sequences that are provided with the dataset. The quality of the silhouette image is general good but many leaks and intrusions are also presented due to imperfect background subtraction. So morphological close and open operations are applied to the silhouette image in order to deal with noise, as shown in Fig. 4.16. However, not all the defects can be repaired, as shown in Fig. 4.17.

It is a challenging task to recognize the actions from this dataset. Human actions are not absolutely consistent when they perform a given action. The same action looks quite different when observed from different camera viewpoints, in addition

Fig. 4.15 The IXMAS database [43]

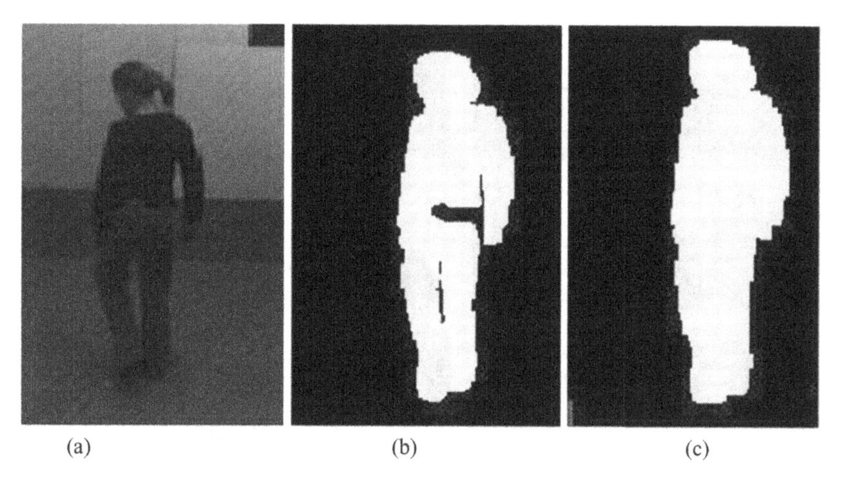

Fig. 4.16 An example of defect in the provided silhouette image, (**a**) observe image, (**b**) the silhouette image, (**c**) the image after repairing

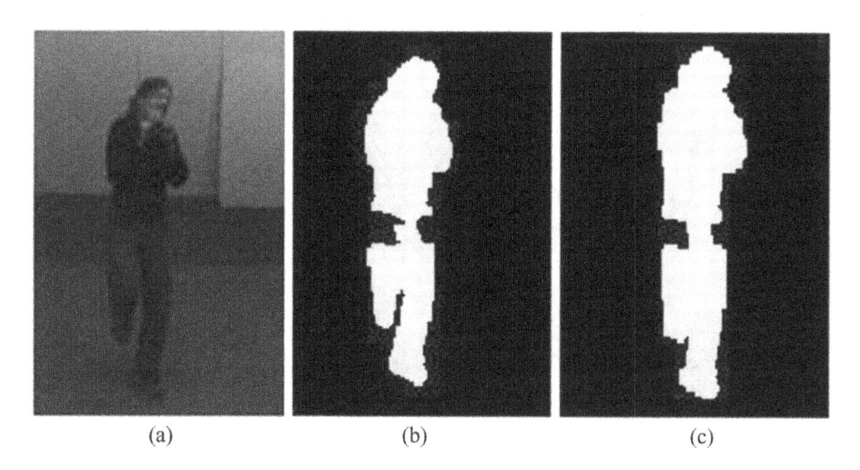

Fig. 4.17 Other example of defect in the provided silhouette image, (**a**) observe image, (**b**) the silhouette image, (**c**) the image after repairing

that the same action executed multiple times by the same person, or by different persons will exhibit variation as shown in Fig. 4.18.

Since male and female actors' execution styles are significantly different, we chose 10 actions (*i.e.*, check watch, cross arms, scratch head, sit down, get up, turn around, walk in a circle, wave a hand, punch, kick), performed by 5 female actors, each 3 times, and viewed by 4 cameras (except top camera) as training and testing objects in our experiment. The action sequences were all manually segmented in advance, so no action segmentation was considered. 4 actors were used for exemplar extraction and model learning each time, and another one was used to test the models. Finally the average recognition rate was calculated.

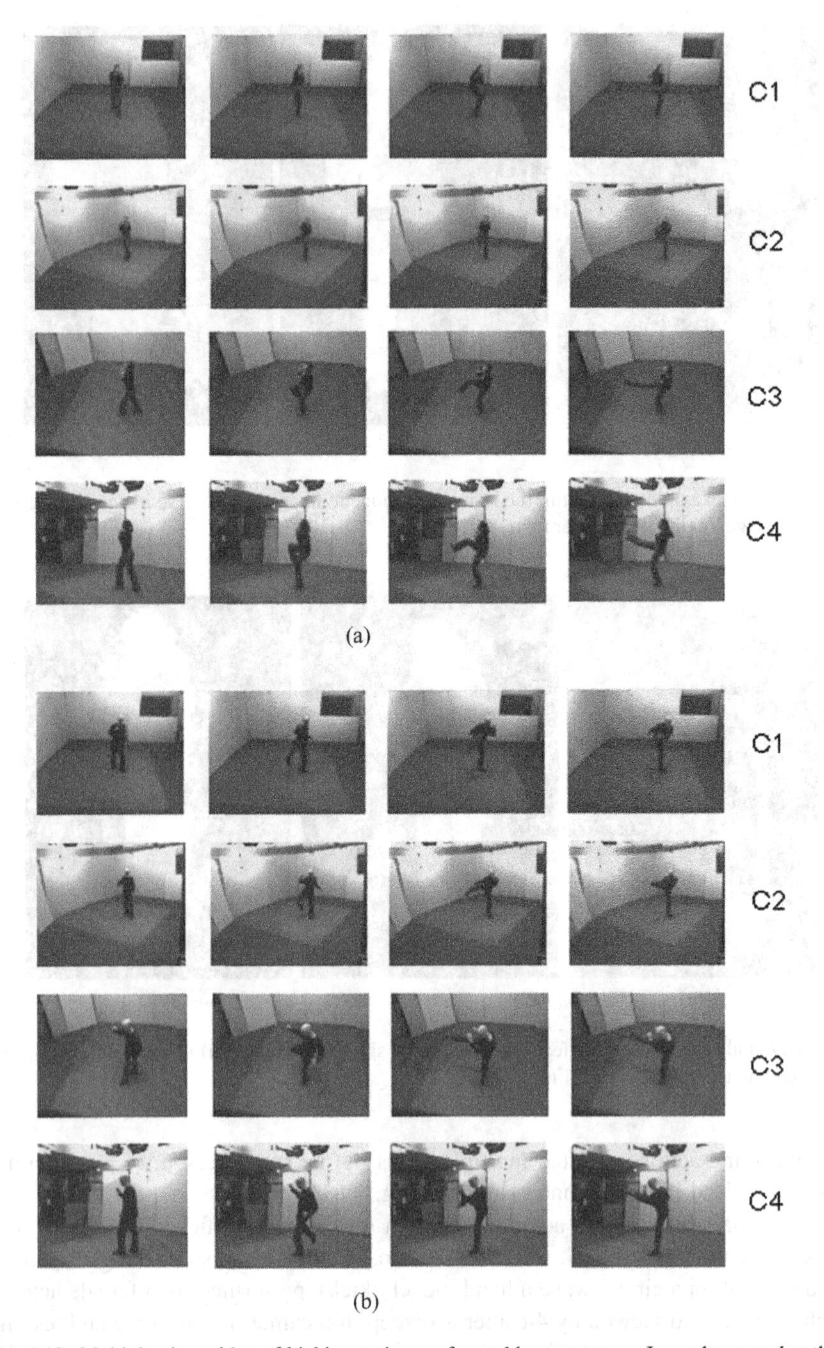

Fig. 4.18 Multiple view video of kicking action performed by two actors. It can be seen that the same action may be look quite different when being observed from different camera viewpoint. Variation also exists when the same action is performed by different actors. (**a**) Kicking action of Alba. (**b**) Kicking action of Andreas

Table 4.1 The recognition rate

Action	Recognition rate (%)
check watch	78.0
cross arms	80.0
scratch head	75.0
sit down	86.7
get up	85.0
turn around	71.7
walk in a circle	70.0
wave a hand	76.7
punch	83.3
kick	81.7
overall	78.8

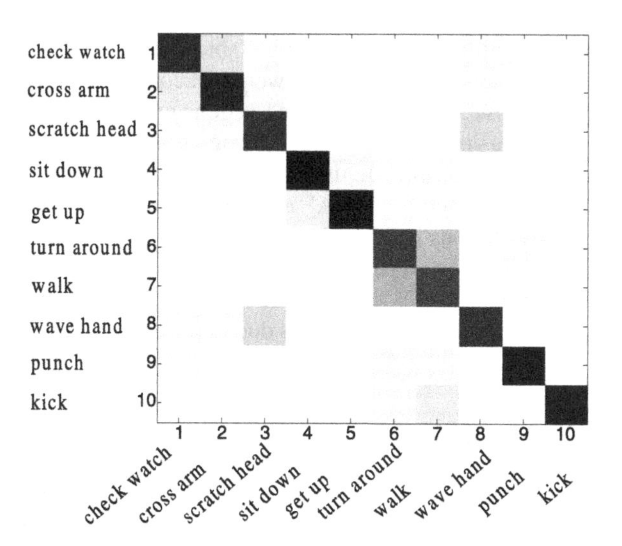

Fig. 4.19 Confusion matrix for recognition

We extracted 4 exemplars for each action, which is sufficient to accurately recognize those actions. One exemplar-based HMM is used to model each action class, then a sequence of observation is recognized by using the maximum a posteriori estimate. The recognition rates for each action class are listed in Table 4.1 after the first level action recognition. The confusion matrix is provided in Fig. 4.19 to show the effectiveness of the method.

The results showed that, our system achieved a satisfying average recognition rate of 78.8% by using a single camera. Our system runs at 15 frames/second, so it has potential to implement the proposed work into the real time human motion recognition. Furthermore, among those 10 actions, "sit down" and "get up" were the easiest actions to recognize because they were more remarkable than the other actions. Some arm related actions such as "check watch", "cross arm", "scratch

Table 4.2 The recognition rate

Action	Recognition rate (%)
check watch	81.7
cross arms	85.0
scratch head	81.7
wave a hand	83.8
overall	83.1

head" and "wave hand" got relatively low recognition rate in that some silhouette images of the key poses were very similar from a single camera.

The confusion matrix showed that action "check watch" and "cross arm", "scratch head" and "wave hand" were very easy to be confused. So after the first level recognition, actions "check watch" and "cross arm", "scratch head" and "wave hand", were separately defined as the similar action set. We revalidated those similar actions by using the weighted contour shape features in the observation probability. The recognition rates are listed in Table 4.2.

The results showed that the weighted contour shape feature was effective to recognize those similar arm-related actions even under the condition of the arbitrary camera viewpoint. The average recognition rate rises to 83.1%. Weighted feature is useful to discriminate the similar actions that just involve the small parts action. Although the actions "walk" and "turn around" are with high confusion possibility, those actions involve almost the whole body action. It is no sense to revalidate them by using the weighted contour shape features. There were some view-invariant works proposed and tested on the IXMAS dataset, however it is difficult to directly compare our results with others due to different experimental settings.

4.6 Conclusion

An exemplar-based view-invariant action recognition framework has been proposed and tested on a challenge dataset (IXMAS dataset) in this chapter. The novelty of the method is two-fold: (a) a simplified two-level exemplar-based action recognition framework was proposed for view-invariant action recognition, (b) weighted contoure shape feature was utilized to recognize the similar human actions. Experimental results have demonstrated that the proposed framework can effectively recognize human actions performed by different people and different actions types in nearly real time without the camera viewpoint constraints. Furthermore, it was verified that the weighted contour shape feature was effective to discriminate the similar arm-related actions. Our future works are targeted as follows:

1. Single feature is weak to recognize the complex human actions. Different human action features have various discriminative abilities, such as silhouettes, shapes, appearances, optical flow *etc*. It is necessary to study the characterise of those cues and fuse multiple features of action to improve the algorithm's effectiveness and robustness [8, 14, 20, 40].

2. Assumption of independence is usually required in HMMs, which makes the method unsuitable for accommodating multiple overlapping features or long-range dependences among observations. Researchers have been attempting to introduce conditional random fields (CRFs) to overcome the independence assumption between observations in human motion analysis [25, 36, 37, 41, 49]. Thereby it is interesting to investigate how to effectively combine CRFs and HMMs to deal with the viewpoint issue in human motion analysis [24, 40].
3. At present, human behaviour understanding is still restricted to simple and special action patterns and special scenes. Therefore, research on semantic description of human behaviours in complex unconstrained scenes still remains an open issue. Research on behaviour patterns constructed by fusing fuzzy qualitative description for unknown scenes is another future research direction [9, 19].

Acknowledgements The authors would like to thank Dr. Weinland *et al.* for kindly providing the INRIA IXMAS dataset.

References

1. Aggarwal, J., Cai, Q.: Human motion analysis: a review. Comput. Vis. Image Underst. **73**(3), 428–440 (1999)
2. Ahmad, M., Lee, S.: Hmm-based human action recognition using multiview image sequences. Proc. Int. Conf. Pattern Recognit. **1**, 263–266 (2006)
3. Anderson, D., Luke, R.H., Keller, J.M., Skubic, M., Rantz, M.J., Aud, M.A.: Modeling human activity from voxel person using fuzzy logic. IEEE Trans. Fuzzy Syst. **17**, 39–49 (2009)
4. Belongie, S., Malik, J., Puzicha, J.: Shape matching and object recognition using shape contexts. IEEE Trans. Pattern Anal. Mach. Intell. **24**(4), 509–522 (2002)
5. Blank, M., Gorelick, L., Shechtman, E., Irani, M., Basri, R.: Actions as space-time shapes. Proc. IEEE Conf. Comput. Vis. **2**, 1395–1402 (2005)
6. Bobick, A., Davis, J.: The recognition of human movement using temporal templates. IEEE Trans. Pattern Anal. Mach. Intell. **23**(3), 257–267 (2001)
7. Chen, H., Chen, H., Chen, Y., Lee, S.: Human action recognition using star skeleton. In: Proc. the 4th ACM International Workshop on Video Surveillance and Sensor Networks, pp. 171–178 (2006)
8. Cherla, S., Kulkarni, K., Kale, A., Ramasubramanian, V.: Towards fast, view-invariant human action recognition. In: Proc. IEEE Conf. Computer Vision and Pattern Recognition, pp. 1–8 (2008)
9. Chan C.S., Liu H.: Fuzzy qualitative human analysis. IEEE Trans. Fuzzy Syst. **17**(4), 851–862 (2009)
10. Dedeoğlu, Y., Töreyin, B., Güdükbay, U., Çetin, A.: Silhouette-based method for object classification and human action recognition in video. In: Proc. European Conf. Computer Vision, pp. 62–77 (2006)
11. Elgammal, A., Lee, C.: Inferring 3D body pose from silhouettes using activity manifold learning. In: Proc. IEEE Conf. Computer Vision and Pattern Recognition, pp. 681–688 (2004)
12. Elgammal, A., Shet, V., Yacoob, Y., Davis, L.: Learning dynamics for exemplar-based gesture recognition. In: Proc. IEEE Conf. Computer Vision and Pattern Recognition, vol. 1, pp. 571–578 (2003)
13. Fathi, A., Mori, G.: Action recognition by learning mid-level motion features. In: Proc. IEEE Conf. Computer Vision and Pattern Recognition, pp. 1–8 (2008)
14. Hu, Y., Cao, L., Lv, F., Yan, S., Gong, Y., Huang, T.S.: Action detection in complex scenes with spatial and temporal ambiguities. In: Proc. IEEE Conf. Computer Vision, pp. 1–8 (2009)

15. Ji, X., Liu, H.: Advances in view-invariant human motion: a review. IEEE Trans. Syst. Man Cybern., Part C, Appl. Rev. **40**, 13–24 (2010)
16. Kale, A., Chowdhury, A., Chellappa, R.: Towards a view invariant gait recognition algorithm. In: Proc. IEEE Conf. Advanced Video and Signal Based Surveillance, pp. 143–150 (2003)
17. Lee, C., Elgammal, A.: Simultaneous inference of view and body pose using torus manifolds. Proc. Int. Conf. Pattern Recognit. **3**, 489–494 (2006)
18. Li, H., Lin, S., Zhang, Y., Tao, K.: Automatic video-based analysis of athlete action. In: Proc. IEEE Conf. Image Analysis and Processing, pp. 205–210 (2007)
19. Liu, H.: A fuzzy qualitative framework for connecting robot qualitative and quantitative representations. IEEE Trans. Fuzzy Syst. **16**(6), 1522–1530 (2008)
20. Liu, J., Ali, S., Shah, M.: Recognizing human actions using multiple features. In: Proc. IEEE Conf. Computer Vision and Pattern Recognition, pp. 1–8 (2008)
21. Lv, F., Nevatia, R.: Recognition and segmentation of 3-D human action using HMM and multi-class AdaBoost. In: Proc. European Conf. Computer Vision, vol. 4, pp. 359–372 (2006)
22. Lv, F., Nevatia, R.: Single view human action recognition using key pose matching and Viterbi path searching. In: Proc. IEEE Conf. Computer Vision and Pattern Recognition, pp. 1–8 (2007)
23. Mori, T., Segawa, Y., Shimosaka, M., Sato, T.: Hierarchical recognition of daily human actions based on Continuous Hidden Markov Models. In: Proc. IEEE Conf. Automatic Face and Gesture Recognition, pp. 779–784 (2004)
24. Natarajan, P., Nevatia, R.: View and scale invariant action recognition using multiview shape-flow models. In: Proc. IEEE Conf. Computer Vision and Pattern Recognition, pp. 1–8 (2008)
25. Ning, H., Xu, W., Gong, Y., Huang, T.: Latent pose estimator for continuous action recognition. In: Proc. European Conf. Computer Vision, pp. 1–7 (2008)
26. Ogale, A., Karapurkar, A., Aloimonos, Y.: View-invariant modeling and recognition of human actions using grammars. Proc. IEEE Conf. Comput. Vis **5**, 115–126 (2005)
27. Ong, E., Micilotta, A., Bowden, R., Hilton, A.: Viewpoint invariant exemplar-based 3D human tracking. Comput. Vis. Image Underst. **104**(2–3), 178–189 (2006)
28. Parameswaran, V., Chellappa, R.: View invariants for human action recognition. In: Proc. IEEE Conf. Computer Vision and Pattern Recognition, vol. 2, pp. 83–101 (2003)
29. Parameswaran, V., Chellappa, R.: View independent human body pose estimation from a single perspective image. In: Proc. IEEE Conf. Computer Vision and Pattern Recognition, vol. 2, pp. 16–22 (2004)
30. Patrick, P., Vand Geoff Svetha, W.: Tracking as recognition for articulated full body human motion analysis. In: Proc. IEEE Conf. Computer Vision and Pattern Recognition, pp. 1–8 (2007)
31. Rao, C., Yilmaz, A., Shah, M.: View-Invariant representation and recognition of actions. Int. J. Comput. Vis. **50**(2), 203–226 (2002)
32. Ribeiro, P., Santos-Victor, J., Lisboa, P.: Human activity recognition from video: modeling, feature selection and classification architecture. In: Proc. Int Workshop. Human Activity Recognition and Modelling, pp. 1–10 (2005)
33. Rittscher, J., Blake, A., Roberts, S.: Towards the automatic analysis of complex human body motions. Image Vis. Comput. **20**(12), 905–916 (2002)
34. Rogez, G., Guerrero, J., Martınez, J., Orrite, C.: Viewpoint independent human motion analysis in man-made environments. In: Proc. British Machine Vision Conference, pp. 659–668 (2006)
35. Shi, Y., Bobick, A., Essa, I.: Learning temporal sequence model from partially labeled data. In: Computer Vision and Pattern Recognition, pp. 1631–1638 (2006)
36. Sminchisescu, C., Kanaujia, A., Metaxas, D.: Conditional models for contextual human motion recognition. Comput. Vis. Image Underst. **104**(2–3), 210–220 (2006)
37. Sutton, C., McCallum, A., Rohanimanesh, K.: Dynamic conditional random fields: factorized probabilistic models for labeling and segmenting sequence data. J. Mach. Learn. Res. **8**, 693–723 (2007)
38. Toyama, K., Blake, A.: Probabilistic tracking with exemplars in a metric space. Int. J. Comput. Vis. **48**, 9–19 (2002)

39. Ullah, F., Kaneko, S.: Using orientation codes for rotation-invariant template matching. Pattern Recognit. **37**(2), 201–209 (2004)
40. Wang, L., Suter, D.: Recognizing human activities from silhouettes: motion subspace and factorial discriminative graphical model. In: Proc. IEEE Conf. Computer Vision and Pattern Recognition, pp. 1–8 (2007)
41. Wang, S., Quattoni, A., Morency, L., Demirdjian, D., Darrell, T.: Hidden conditional random fields for gesture recognition. In: Proc. IEEE Conf. Computer Vision and Pattern Recognition, vol. 2, pp. 1521–1527 (2006)
42. Weinland, D., Grenoble, F., Boyer, E., Ronfard, R., Inc, A.: Action recognition from arbitrary views using 3D exemplars. In: Proc. IEEE Conf. Computer Vision, pp. 1–7 (2007)
43. Weinland, D., Ronfard, R., Boyer, E.: Free viewpoint action recognition using motion history volumes. Comput. Vis. Image Underst. **104**, 249–257 (2006)
44. Yan, P., Khan, S., Shah, M.: Learning 4D action feature models for arbitrary view action recognition. In: Proc. IEEE Conf. Computer Vision and Pattern Recognition, vol. 12, pp. 1–8 (2008)
45. Yang, Y., Hao, A., Zhao, Q.: View-invariant action recognition using interest points. In: Proc. Int. Conf. Multimedia Information Retrieval, pp. 305–312 (2008)
46. Yilmaz, A., Shah, M.: Actions as objects: a novel action representation. In: Proc. IEEE Conf. Computer Vision and Pattern Recognition, pp. 984–989 (2005)
47. Yu, H., Sun, G., Song, W., Li, X.: Human motion recognition based on neural network. In: Proc. IEEE Conf. Communications, Circuits and Systems, vol. 2, pp. 977–982 (2005)
48. Yu, S., Tan, D., Tan, T.: Modelling the effect of view angle variation on appearance-based gait recognition. In: Proc. Asian Conf. Computer Vision, vol. 1, pp. 807–816 (2006)
49. Zhang, J., Gong, S.: Action categorization with modified hidden conditional random field. Pattern Recognit. **43**, 197–203 (2010)

Chapter 5
Using Fuzzy Gaussian Inference and Genetic Programming to Classify 3D Human Motions

Mehdi Khoury and Honghai Liu

Abstract This research introduces and builds on the concept of Fuzzy Gaussian Inference (FGI) (Khoury and Liu in Proceedings of UKCI, 2008 and IEEE Workshop on Robotic Intelligence in Informationally Structured Space (RiiSS 2009), 2009) as a novel way to build Fuzzy Membership Functions that map to hidden Probability Distributions underlying human motions. This method is now combined with a Genetic Programming Fuzzy rule-based system in order to classify boxing moves from natural human Motion Capture data. In this experiment, FGI alone is able to recognise seven different boxing stances simultaneously with an accuracy superior to a GMM-based classifier. Results seem to indicate that adding an evolutionary Fuzzy Inference Engine on top of FGI improves the accuracy of the classifier in a consistent way.

5.1 Introduction

This study presents a novel machine learning technique tested in the application domain of behaviour understanding, that is to say the recognition and description of actions and activities from the observation of human motions. The process of behaviour understanding is usually performed by comparing observations to models inferred from examples using different learning algorithms. Such techniques presented in [3] and [4] can be used either in the context of template matching [5], state-spaces approaches [6], or semantic description [7]. Our application domain is focused on sport, and more precisely, boxing. We have discarded template matching as it is generally more susceptible to noise, variations of the time intervals of the movements, and is viewpoint dependent [4]. We are not interested in a pure

M. Khoury (✉) and H. Liu
Intelligent Systems and Biomedical Robotics Group, School of Creative Technologies, The University of Portsmouth, Eldon Building, Portsmouth PO1 2DJ, UK
e-mail: mehdi.khoury@port.ac.uk; honghai.liu@port.ac.uk

H. Liu et al. (eds.), *Robot Intelligence*,
Advanced Information and Knowledge Processing,
DOI 10.1007/978-1-84996-329-9_5,

semantic description as we need to analyse and evaluate a boxing motion in a relatively detailed way. We therefore focus on identifying static states during a motion (i.e., state-spaces approach). Conventionally, machine learning techniques in use for solving such problems vary from dynamic Time Warping [8], to Hidden Markov Models [9], Neural Networks [10], Principal Component Analysis [11], or variations of HMM or NN such as Coupled Hidden Markov Models [9], Variable-Length Markov Models [12], Fuzzy HMM [13], or Time-Delay Neural Networks [14]. This chapter introduces a different method that allows us to build from learning samples fuzzy qualitative models corresponding to different states. An automated way to generate fuzzy membership function is proposed [1]. It is applicable to biologically "imprecise" human motion, by mapping an estimation of centroid and range from a cumulative normal distribution to a membership function. Refined qualitative statement are then extracted from membership scores using Genetic Programming instead of using a standard Mangmani-typed rule-based system like in [15]. First the human skeletal representation in use will be described, then the process by which stances are recognized (Guard, Jab, Cross, Lower Cross, Lower Jab, Right Hook, Left Hook, Lower Left Hook, and Right Uppercut) with fuzzy membership functions, then some mathematical properties of this technique, some informations on the Genetic Programming based rule extraction process, and finally, experimental results will be presented and discussed.

5.2 Human Skeletal Representation

There exists a wide range of ways to represent the human body in the study of human motion. In kinesiology or biomechanics, models of the human body are based on precise 3-D joint data, kinematic analysis, and analytical dynamics (forces and torques for a movement are also of interest [16]). In this experiment, such detailed information might not be needed, especially considering that models like the International Society of Biomechanics Joint-Coordinate system [17, 18] or the Tilt-Twist representation system [19] are still incomplete and computationally expensive. The next step of our work will be focused not only on motion recognition but also performance analysis. There might be a need to assess the correctness of a given motion. This means that focusing on watching the displacements of the end-effectors of a chain of links (for example watching the trajectory of the end of a foot and hands instead of the whole body) to understand a motion [5, 20] is not sufficient. It is important to be able to decompose the motion into subcomponents such as the rotations of individual joints. Therefore, the representation system must be able to keep track of multiple joints rotations. For this research, representation system in use is the widely spread .BVH motion capture format [21] in which a human skeleton is formed of skeletal limbs linked by rotational joints (see Fig. 5.1). It uses Euler angles to quantify rotations of joints having three Degrees of Freedom. This system is not perfect (Gimbal Lock is a possible issue), but allows to gather data easily when using motion capture while keeping track of subcomponents such

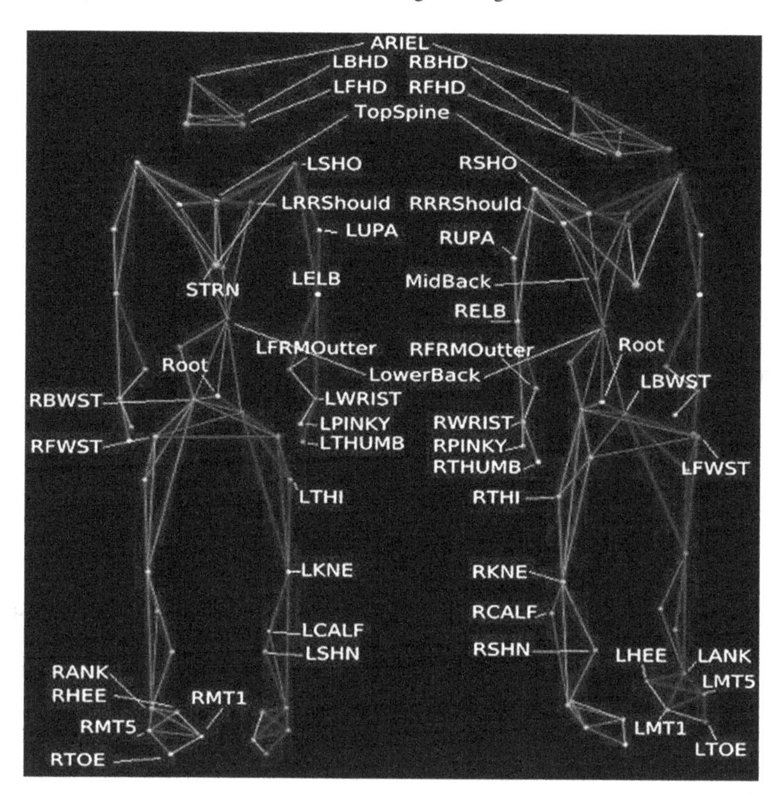

Fig. 5.1 Optical markers—resulting BVH representation

as the rotations of individual joints. The following choices and assumptions are made:

- Knowing that motion capture data cannot give absolutely exact skeletal displacements of the joints [22] due to soft tissues movements, this work simply seeks to use it to obtain an approximation which would be good enough to characterize the motion.
- The body is simplified to nineteen main joints and it is assumed that this number is sufficient to characterize and understand the general motions of a human skeleton performing boxing combinations. (Human being can after all identify a motion from the observation of a limited number of points moving in space.) Reference to the psychology study...
- Each joint is seen as having three degrees of freedom. The rotations of such joints are represented by Euler ZXY angles. A joint rotation is therefore characterized by three rotation angles Z, X and Y given in degrees by the .BVH motion capture format sampled at the speed of 120 frames per second.

In practice, for every frame, our observed data takes the shape of a nineteen-by-three matrix describing ZXY Euler Angles for all nineteen joints in a simplified human skeletal representation. In other words, 57 continuous variables (each between 0 and 360) characterize a stance at any time.

5.3 The Learning Method

To learn to recognize a stance, a model needs to be extracted (i.e., a fuzzy membership function) for this stance from learning data. This stance is later identified during a motion by evaluating the membership score of the observed data with respect to the learned model. This chapter will first describe the model itself (the notion of fuzzy membership function). It will then describe the novel process by which this model is generated: Fuzzy Gaussian Inference. Finally, it will show how the degree of membership of observed data to a given template is computed.

5.3.1 Model Description: the Fuzzy Membership Function

The fuzzy linguistic approach introduced by Zadeh [23] allows us to associate a linguistic variable such as a "guard" stance with linguistic terms expressed by a fuzzy membership function. Using a trapezoid fuzzy-four-tupple (a, b, α, β) which defines a function that returns a degree of membership in [0, 1] (see Fig. 5.2 and (5.1)) seems to be more interesting as there is a good compromise between precision and computational efficiency (compared with, for example, the triangular membership function):

$$\mu(x) = \begin{cases} 0, & x < a - \alpha, \\ \alpha^{-1}(x - a + \alpha), & x \in [a - \alpha\ a], \\ 1, & x \in [a\ b], \\ \beta^{-1}(b + \beta - x), & x \in [b, b + \beta], \\ 0, & x > b + \beta. \end{cases} \tag{5.1}$$

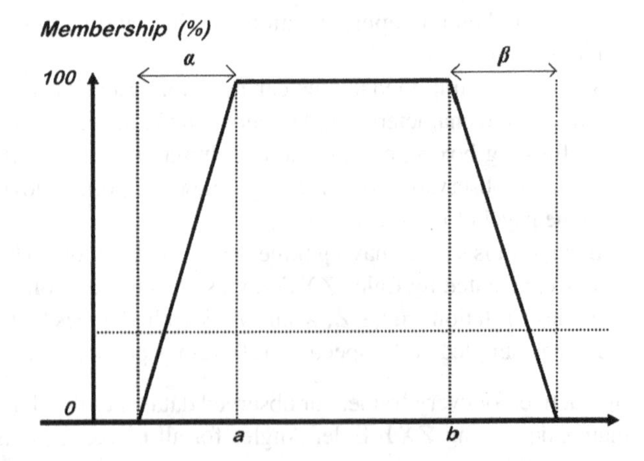

Fig. 5.2 Fuzzy-4-tuple membership function

5.3.2 Model Generation: Fuzzy Gaussian Inference

Frames identified as "Guard" of membership equal to one are used as learning samples. The identification of these example data is made by a system similar to Reverse Rating [24], which is to say that, in our case, an expert (a human observer) is asked to do the following: identify a group of frames whose motion indicates a stance that possesses the degree 1.0 of membership in the fuzzy set "Guard". Once these learning data are obtained, a fuzzy membership function can be generated. Many kinds of procedures for the automated generation of membership functions can be found in the literature. Fuzzy Clustering [25], Inductive Reasoning [26], Neural Networks [27] and Evolutionary Algorithms [28] have been used, among others, to build such functions. Estimation of S-Function from Histograms [29] has also been done, some of it based on the optimization of fuzzy entropy [30]. So far, one downside of such techniques has been the difficulty to link the notion of fuzzy membership to the notion of probability distribution. One noticeable attempt to link both concepts in the generation of membership functions has been done by Frantti [31] in the context of mobile network engineering. Unfortunately, this approach is relatively limited as the minimum and maximum of the observed data are the absolute limits of the membership function. As a consequence, such a system ignores motions which are over the extremum of the learning range of the examples. This work presents a method that overcomes this problem by introducing a function that maps the probability that values fall within a given cumulative normal distribution to a degree of membership. This relies on the assumption that, for a population of samples representing a given motion, the Z, X and Y Euler angles characterizing the motion tend to be normally distributed. Assuming that the space of known boxing motions is informationally structured by these hidden Gaussian Distributions, there is a need to build fuzzy membership functions that map to these underlying structures. The mapping from probability distribution to membership score is done by examining the range and center of density of the learning data for one specific motion. In our experiment, there is a limited number of motion capture learning samples of a given stance (let us say a defensive posture called "Guard"). Looking at each Euler angle Z, X, Y for every joint j for this type of motion, it can be observed that, in our training sample, each Euler Angle e in each joint has a global minimum and maximum. This range is defined between minimum and maximum of the learning sample as the range $\delta(e, j)$ of degree of membership one in the fuzzy set "Guard". Knowing the size of our training sample, it is possible to estimate how much *the range* of our learning sample represents compared to *the range* of all possible guards. For example, if the range of our sample represents around 68.2% of the maximum range of all possible guards ($\gamma = 0.682$), then there is a degree of membership 1 for two standard deviations (one on each side) on the population maximum range. This means that the rest of the distribution that will have membership inferior to one will take three remaining standard deviations on each side. To summarize the salient points of our method, considering the range and center of density of the learning sample, the shape of a fuzzy membership function will be defined by the following four factors:

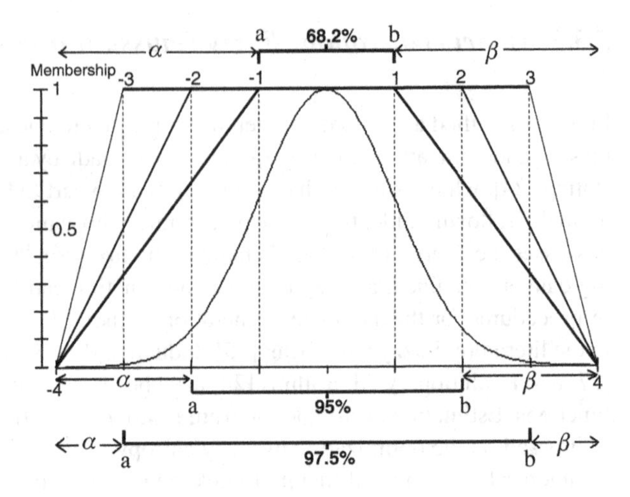

Fig. 5.3 Influence of the cumulative normal distribution parameter on the shape of the fuzzy membership function

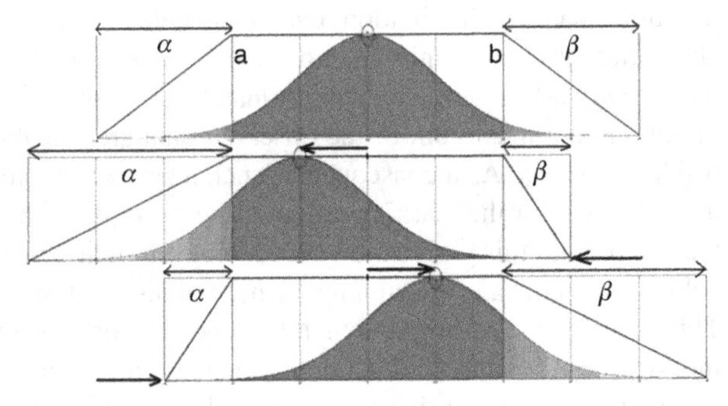

Fig. 5.4 Moving the centroid shifts the distribution and deforms the fuzzy membership function

- The maximum number of standard deviations covered by the fuzzy membership function. In this example, the maximum range is approximated by assuming that it is four standard deviations away in both directions from the mid-point of the range of membership one. This will define the length of the base of the trapezoid shape.
- Depending on the cumulative normal distribution evaluation defining the parameter γ, a portion of the four standard deviations representing the total range will be allocated to the membership-one-range and the remaining part will be allocated to the lower membership degrees. This will define the length of the top part of the trapezoid shape (see Fig. 5.3).
- The average of the means is extracted out of each learning sample. This will correspond to the centroid of the data samples of membership one.
- While the distance $|(b + \beta) - (a - \alpha)|$ will be constant, $a - \alpha$ and $b + \beta$ will be shifted to the side proportionally to the way the centroid is shifted from the midpoint (see Fig. 5.4 and (5.3)). This will shift the base of the trapezoid shape to either side.

For example, if the centroid is at the same position with the middle of the membership-one-range $\delta(e, j)$, and this range is evaluated as representing 95% of the maximum theoretical range, then our fuzzy membership function will be symmetric ($\alpha = \beta = 2$ standard deviations on each side of the membership-one-range). The centroid c and the constant μ are such that:

$$|c - a| = \mu \times |b - a|. \tag{5.2}$$

Let this range be evaluated as representing 95% of the global theoretical range, then the fuzzy membership function would be shifted to the left such that:

$$\begin{cases} \alpha = (1 - \mu) \times (\alpha + \beta), \\ \beta = \mu \times (\alpha + \beta). \end{cases} \tag{5.3}$$

5.3.3 Membership Evaluation

Our observed data take the shape of a nineteen-by-three matrix describing ZXY Euler Angles for all nineteen joints. One evaluates how close this matrix is from a "Guard" stance by calculating the degree of membership of every Euler Angle in every joint (we have previously built a fuzzy-4-tuple corresponding to the "Guard" stance for every one of these Euler angles), and then, an average membership score is computed. This approach could probably be improved in the near future by introducing weighted average for certain joints (for example, the position of the elbow might be more important than the position of the knee when in guard). If a frame has a high membership score for several fuzzy sets, an order of preference of these sets can be established by comparing the Euclidean distance of the observed data to the centroid of each fuzzy set.

5.4 Mathematical Properties

Fuzzy Gaussian Inference (FGI) does not have the problem linked to dimensionality reduction of methods such as PCA as we keep the initial number of dimensions when building the model. The method decompose what would normally be a Gaussian Mixture of a number x of m-dimensional Normal distributions into $x \times m$ Fuzzy Membership Functions (see Fig. 5.5). In this study nineteen 3-dimensional rotation continuous data are used to produce $19 \times 3 = 57$ fuzzy membership functions.

The flexibility of a machine learning method is generally determined by how successfully it can be applied to different application domains. Empirically speaking, making use of supervised machine learning techniques generally involves testing a data sample with different parameter values in order to reach an optimal combination leading to a maximized performance of the given system. Two of the contributing factors to the degree of usability for such methods are the number of parameters in use and the sensitivity the system exhibits to slight variations in parameters values. In other words, if our classifier is parameter dependant like most machine learning techniques, we want to know what is the relationship between the parameters, and how do variations in these parameters influence the overall system performance.

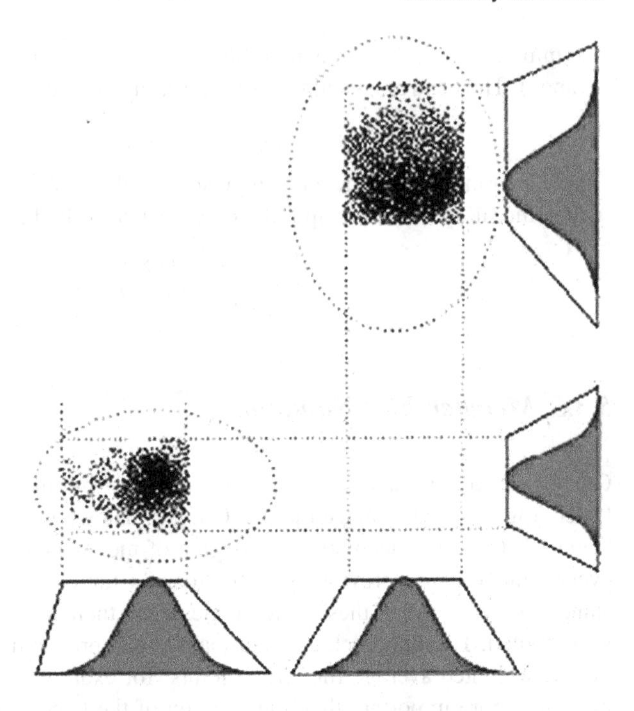

Fig. 5.5 How FGI decomposes Gaussian mixture models

Fuzzy Gaussian Inference is based on two parameters which, combined with input data, produce a classification with a certain degree of accuracy. The first parameter is the evaluation of the "relative size" of our sample. Intuitively it could be defined as the ratio of the correct "guard" movements that the learning sample represents over the range of all possible correct "guard" movements. This number would be a percentage lying in the interval $]0, 0.999936657516[$ where the maximum range considered is 4 standard deviations in both directions. This ratio is transformed into a z-score n. To be more precise, this ratio represents the average over $x \times m$ dimensions of the area under the bell curve between $\mu - n\sigma$ and $\mu + n\sigma$ in terms of the cumulative normal distribution function ϕ given by:

$$\phi(n) - \phi(-n) = 2\phi(n) - 1 = \mathrm{erf}(n/\sqrt{2}) \tag{5.4}$$

where erf() is defined as the error function such that:

$$\mathrm{erf}(x) = \frac{2}{\sqrt{\pi}} \int_0^x e^{-t^2} dt. \tag{5.5}$$

The z-score n can therefore be deduced from the parameter by using the inverse error function. The second parameter is a ratio representing the membership threshold in use with the classifier. A membership threshold of 0.95 means for example that we are interested in identifying all frames which have a membership score $\geq 95\%$ of the fuzzy membership function "Guard". When classifying different types of movements, for a given specific accuracy, there seems to be a mathematical relationship between the parameter ϕ and the membership threshold t. Let g be the function that maps the parameter ϕ (an estimation of the relative-size of the learning sample) to

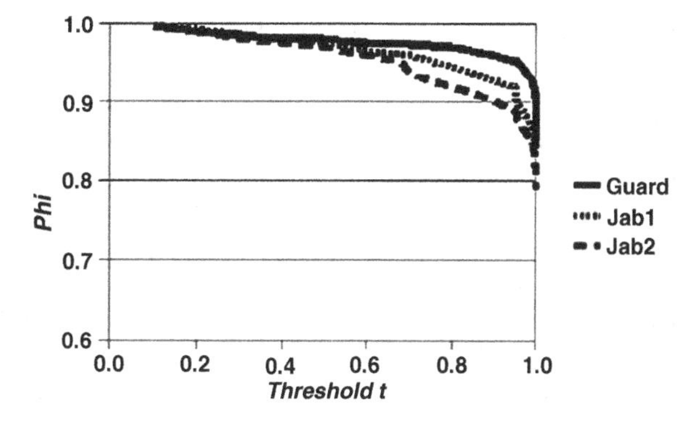

Fig. 5.6 Function mapping the relative-size ϕ to the threshold t

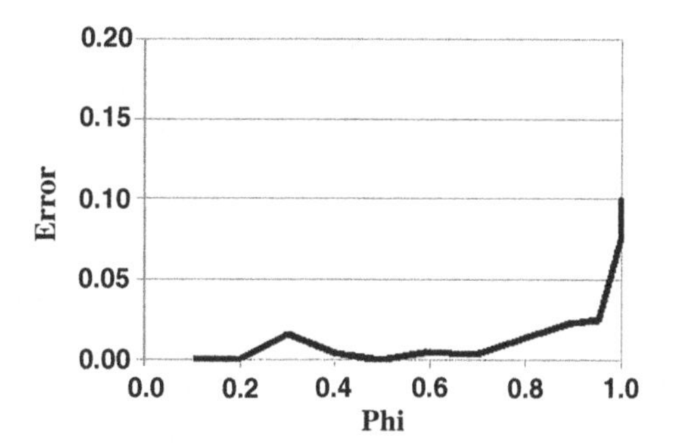

Fig. 5.7 Function mapping the error to over-estimation of ϕ in the "Guard" learning sample

the membership threshold t for a given accuracy such that: $g(\phi) = t$. One can observe that for any parameter ϕ, it seems that: $\dot{g}(\phi) < 0$, that is to say that the slope of the function g is always negative. For a given accuracy, the threshold t seems to vary as a function of ϕ following a general curve with an equation of the form:

$$t = \delta + 1/(\gamma \times \log \phi) \tag{5.6}$$

where δ and γ are constants linked to the dataset considered. Figure 5.6 shows different plots of the function g mapping the relative-size parameter ϕ on the x-axis to the threshold t on the y-axis using three different data sets. Using the concept of elasticity to evaluate if the threshold t is ϕ-elastic, it becomes noticeable that the elasticity is poor when using a very high ϕ value (superior to 0.95). The maximum elasticity is obtained when ϕ is between 40 and 95%. This means that in our data set, the variations of the ϕ parameter are more likely to influence the threshold t if ϕ is kept between 0.4 and 0.95. Regarding the relationship between accuracy and parameters, the accuracy seems to falter with higher values of ϕ. This makes sense because, our sample being of limited size, over-estimating its relative-size will damage the accuracy of the classifier. The loss in accuracy is determined as a function of the relative-size parameter ϕ. When classifying a guard, the error is rising with over-estimation of ϕ up to a maximum of 10% which is relatively reasonable (see Fig. 5.7).

5.5 Extracting Fuzzy Rules Using Genetic Programming

FGI classifies frames into identified motions using membership scores. However, this classification is done frame by frame and is not time-based. Neither does it take into account previous or following motions. In this context, refining the qualitative output using some kind of context-aware fuzzy-rules becomes the next logical step. Rules are inferred using evolution, and more specifically Genetic Programming. For the purpose of this research, a Strongly-Typed Genetic Programming open-source distribution was built [32]. The genetic programming (GP) system evolves rules of the type *If Then Replace X by Y* that are applied to qualitative output of each frame. The GP terminal and function sets contain the following elements:

- Operators that return the frames that belong to motions of a given duration. e.g. *is_short* expresses a duration of less than 5 frames, *is_medium* expresses a duration between 5 and 19 frames, *is_long* expresses a duration of more than 19 frames. These duration figures seemed to be empirically the most suited to the observed data set.
- Operators that return the groups of frames that belong to the first, second and third best membership scores of a motion e.g. *membership_1(jab)* return groups of frames with the best membership score for a motion as a jab, *membership_2(left_hook)* return groups of frames with the second best membership score for a motion as a left hook, *membership_3(lower_left_hook)* return groups of frames with the third best membership score for a motion as a lower left hook.
- Operators that return the groups of frames that have specific previous motions e.g. *left_2(guard, jab)* returns groups of frames preceded in order by a guard and then a jab motion, and *right_3(hook, guard, jab)* returns groups of frames succeeded by a hook, a guard and then a jab motion.
- Logical operators e.g. and, or, not
- An *If Then Replace X by Y* statement that use the previous operators to identify groups of frames and replace their best motion membership score by a different one, e.g. *If Then Replace X by Y(membership_2(guard), jab, cross)* replace the "jab" first membership score with "cross" in groups of frames defined by a "guard" second best membership score.

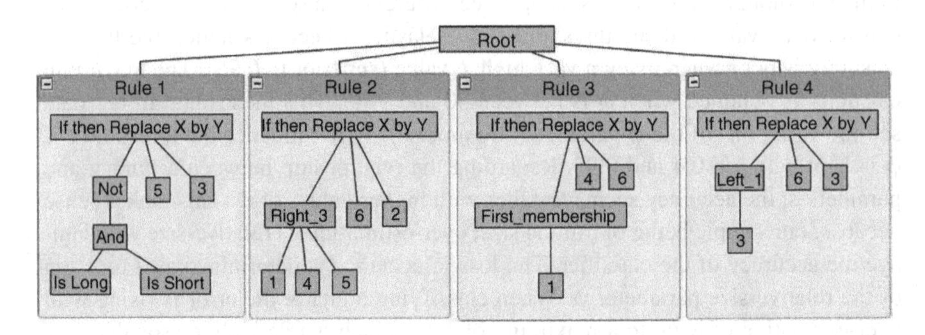

Fig. 5.8 A typical set of rules generated by GP

One individual consists of four interconnected *If Then Replace X by Y* rules. The GP parameters are:

- A population size of 1000 individuals, with a maximum number of 400 generations per evaluation.
- Tournament selection of size seven and selection probability 0.8.
- The probabilities of crossover, mutation, and reproduction are respectively 0.5, 0.49 and 0.01.

The fitness function simply sums the number of frames that have a different qualitative output from the frames present in a "perfect" sequence as defined by a human observer. So if y is the total number of frames observed, Δ is one unit that expresses a difference of classification on one frame between FGI and the human observer, then the fitness F is such that:

$$F = \sum_{r=1}^{y} \Delta_r. \tag{5.7}$$

The Koza operators such as the tree building "ramped half and half" algorithm, and operators such as crossover, mutation are all modified with a Strongly-Typed flavour. In practice, this means that the structure of all the individuals generated will be defined by a set of rules. These rules associate for each parent node an ordered set of children nodes. To be more precise, each parent node maps to a list of possible children nodes which can be either function nodes or terminal nodes as shown in Table 5.1.

5.6 Experiment and Results

5.6.1 Apparatus

The motion capture data are obtained from a Vicon Motion Capture Studio with eight infra-red cameras. The motion recognition is implemented in MATLAB 2007 on a single machine: a PC with an Intel core duo 2 GHz with 2 Gigs of RAM. An additional MATLAB toolbox [33] is also used for extracting Euler Angles from .BVH files.

5.6.2 Participants

Three male subjects, aged between 18 and 21, of light to medium-average size (167 cm to 178 cm) and weight (59 to 79 kg), all practising boxing in competition at the national level. None of them presented any abnormal gait. Optical Markers were placed in a similar way on each subject to ensure a consistent motion capture (see Fig. 5.9).

Table 5.1 Grammar rules defining the individuals produced with strongly typed GP

Parent nodes	Associated children nodes	
	Function nodes	Terminal nodes
Root	If Then Replace X by Y	Empty
If Then Replace X by Y	Membership_1	Is_Short
	Membership_2	Is_Medium
	Membership_3	Is_long
	Left_1	Mvt Type 1
	Left_2	Mvt Type 2
	Left_3	Mvt Type 3
	Right_1	Mvt Type 4
	Right_2	Mvt Type 5
	Right_3	Mvt Type 6
	And	Mvt Type 7
	Or	
	Not	
Membership_1	Empty	Mvt Type 1
Membership_2		Mvt Type 2
Membership_3		Mvt Type 3
Left_1		Mvt Type 4
Left_2		Mvt Type 5
Left_3		Mvt Type 6
Right_1		Mvt Type 7
Right_2		
Right_3		
And	Membership_1	Is_Short
Or	Membership_2	Is_Medium
	Membership_3	Is_long
	Left_1	
	Left_2	
	Left_3	
	Right_1	
	Right_2	
	Right_3	
	And	
	Or	
	Not	

Table 5.1 (Continued)

Parent nodes	Associated children nodes	
	Function nodes	Terminal nodes
Not	Membership_1	Is_Short
	Membership_2	Is_Medium
	Membership_3	Is_long
	Left_1	
	Left_2	
	Left_3	
	Right_1	
	Right_2	
	Right_3	
	And	
	Or	

5.6.3 Procedure

The motion capture data is obtained from several subjects performing each boxing combination four times. There are twenty-one different boxing combinations, each separated by a guard stance. These are performed at two different speeds (medium-slow and medium fast). The boxing combinations are using in different order basic boxing stances. There are in total nine quite precisely defined basic stances (the level of precision needed to identify such motions is non-negligible). These are described in [34] as follow:

- Guard: a defensive position where the boxer stands with the legs shoulder-width apart and the rear foot a half-step behind the lead foot. The lead (left) fist is held vertically about six inches in front of the face at eye level. The rear (right) fist is held beside the chin and the elbow tucked against the ribcage to protect the body.
- Jab: a quick, straight punch thrown with the lead hand from the guard position. The jab is accompanied by a small, clockwise rotation of the torso and hips, while the fist rotates 90 degrees, becoming horizontal upon impact.
- Lower Jab: similar to a jab in a crouching stance.
- Cross: a powerful, straight punch thrown with the rear hand. From the guard position, the rear hand is thrown from the chin, crossing the body and travelling towards the target in a straight line. The rear shoulder is thrust forward and finishes just touching the outside of the chin. At the same time, the lead hand is retracted and tucked against the face to protect the inside of the chin. For additional power, the torso and hips are rotated counter-clockwise as the cross is thrown.
- Lower Cross: similar to a cross in a crouching stance.
- Right Hook: a semi-circular punch thrown with the lead hand to the side of the opponent's head. From the guard position, the elbow is drawn back with a horizontal fist (knuckles pointing forward) and the elbow bent. The rear hand is tucked

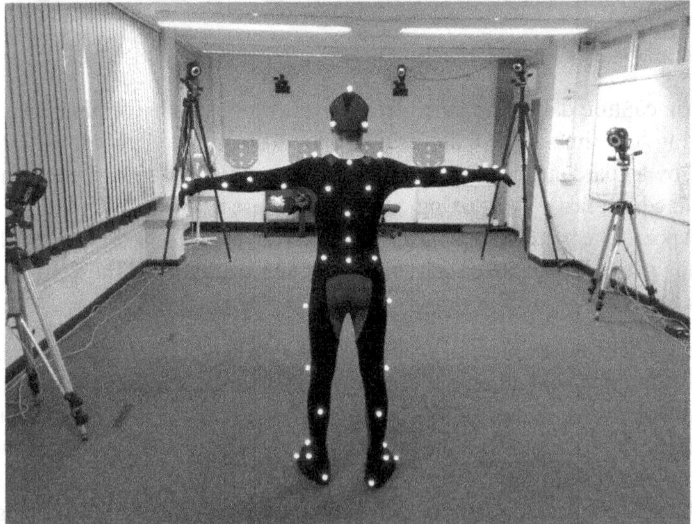

Fig. 5.9 Optical markers placement—front and back view

firmly against the jaw to protect the chin. The torso and hips are rotated clockwise, propelling the fist through a tight, clockwise arc across the front of the body and connecting with the target. At the same time, the lead foot pivots clockwise, turning the left heel outwards.

- Left Hook: similar to a Right Hook, but done with the rear hand.
- Lower Left Hook: similar to a Left Hook in a crouching stance.
- Right Uppercut: a vertical, rising punch thrown with the rear hand. From the guard position, the torso shifts slightly to the right, the rear hand drops below the

level of the opponent's chest and the knees are bent slightly. From this position, the rear hand is thrust upwards in a rising arc towards the opponent's chin or torso. At the same time, the knees push upwards quickly and the torso and hips rotate anti-clockwise and the rear heel turns outward, mimicking the body movement of the cross.

Due to time constraints, we use a subset of seven stances in the experiments involving the GP qualitative output filter: Guard, Jab, Cross, Right Hook, Left Hook, and Lower Left Hook. The fuzzy membership function template corresponding to a "Guard" stance is extracted from various samples. First all three participants are used to learn and to test how well the system recognizes some of their Guard stances. Then, an evaluation is done to see how the system cope to learn from only two participants, and test how well it recognize stances from a third different participant. The accuracy of the system is examined when learning to recognize seven different boxing stances simultaneously. At first there is an evaluation on how accurately each frame is classified individually (see Fig. 5.10). Then, a Genetic Programming fuzzy rule-based system is used to classify frames by looking at groups of frames and their relative positions (see Fig. 5.11). The inputs for each given time frame are the seven membership scores of each known move. These membership scores s_i are re-scaled (see Fig. 5.10) by fine-tuning the thresholds t_i linked to each input i the following way:

$$s_i = (s_i - t_i) \div / (1 - t_i)\,. \tag{5.8}$$

Fuzzy rules are generated using Strongly-Typed Genetic Programming (a specific Python based open source package has been built for this purpose [32]). They have an If-Then type of structure, and take as input for each group of frames the first, second and third highest membership scores. They produce as output groups of frames with modified first membership scores (see Fig. 5.11). There can be seven different types of moves, therefore seven possible qualitative outputs for a group of frames. The system generates four rules. Let $j = 1, 2, \ldots, 4$ be the identifier of a given rule. Each rule f_j can be seen as a function of the form that is applied to each group of frames k:

$$\big\{f_j : X_k \to X_k | X_k = 1, 2, \ldots, N\big\}\, N = 7, \quad \text{and} \quad j = 1, 2, \ldots, 4. \tag{5.9}$$

Each rule produces an output which is used in turn as an input for the next rule. This means that one set of 4 rules is in fact a function composition of the type:

$$\{f_1 \circ f_2 \circ f_3 \circ f_4(X_k) : X_k \to X_k | X_k = 1, 2, \ldots, N\}\, N = 7. \tag{5.10}$$

The fitness of set of rules is evaluated by looking at the accuracy the classification of groups of frames depending on their relative positions. This "context-aware" accuracy (as opposed to the "short-sighted" accuracy of an individual frame) is evaluated by summing the overall number of frames that differ from the classification made by a human observer.

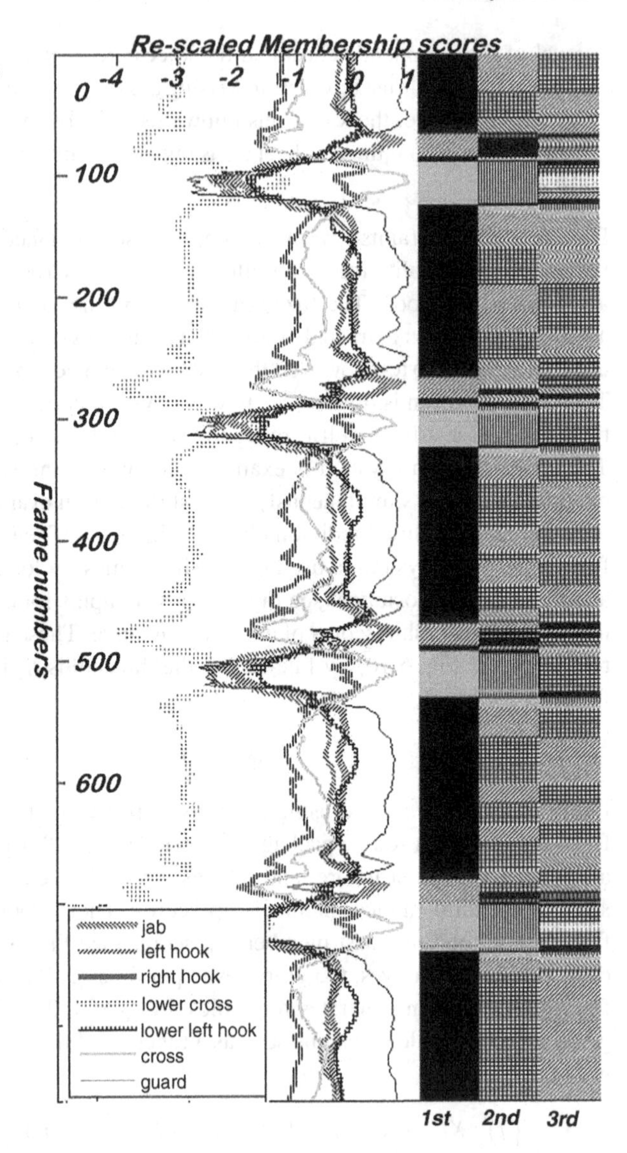

Fig. 5.10 Extraction of the three best membership scores for all frames

5.6.4 Results

An evaluation of the classifier is done by comparing its performance to a human observer. One expert identifies "Guard" frames of membership 1 and of membership 0 (non-guard frames). The number of false positives (frames identified by the expert as non-guards, but identified by the classifiers as guards) and false negatives (frames identified by an expert as guards, but classified as non-guards by the system) are taken into account. ROC analysis is used to plot the true positive rates versus the false positive rates as a function of different membership thresholds. The data are

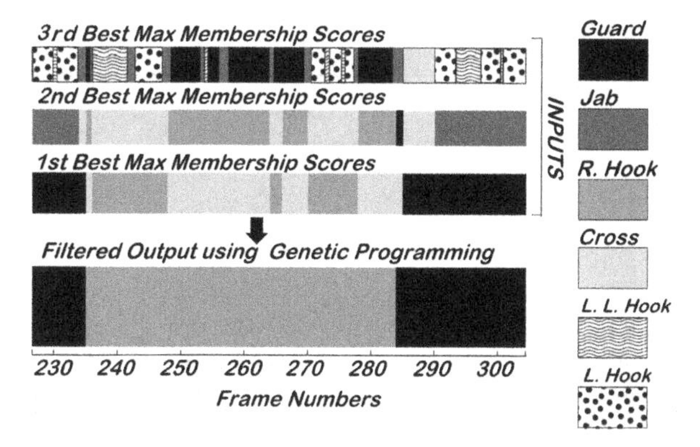

Fig. 5.11 Fuzzy rule-based improved classification of a right hook movement

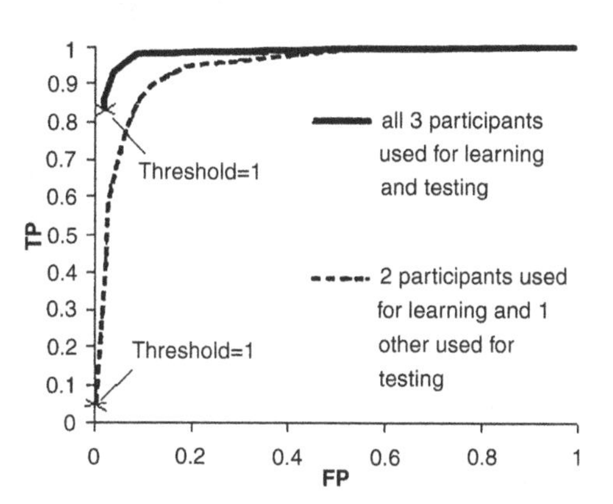

Fig. 5.12 3-fold cross-validation ROC analysis of the guard classifier

partitioned into sub-samples and tests are run using K-fold cross-validation. We present results for a 3-fold cross-validation (as we have three different participants) where one third of the data is used for learning while the rest is used for testing as shown in Fig. 5.8. The testing samples represent in total about 107000 unidentified frames. In this example, we analyse two situations:

- first situation: all participants are used for learning and testing. This means there is a greater similarity between the learning and the test samples, as the gait differences are reduced. This situation generally leads to better results in the classification task, and is the approach in use most of the time in the research area.
- second situation: two participants are used for learning and a third one with a different gait for testing (all combinations are averaged to produce a single estimation). In this case, there are greater gait differences between the learning sample and the test sample as we do not use the same subjects for learning and for testing. This approach generally leads to weaker results but also validates the generalisation power of the classifier.

As seen in both the ROC Fig. 5.12 and Table 5.2, the optimum accuracy of the classifier is 0.95 if the same participants are used for learning and testing, or 0.88 when different participants are used for learning and testing. Crisp evaluation(the

Table 5.2 Optimal accuracy of the classifier and the corresponding threshold values

3-fold validation	Accuracy	Threshold
3/3 boxers	0.95	0.997
3/3 boxers	0.906	1
2/3 boxers	0.885	0.978
2/3 boxers	0.506	1

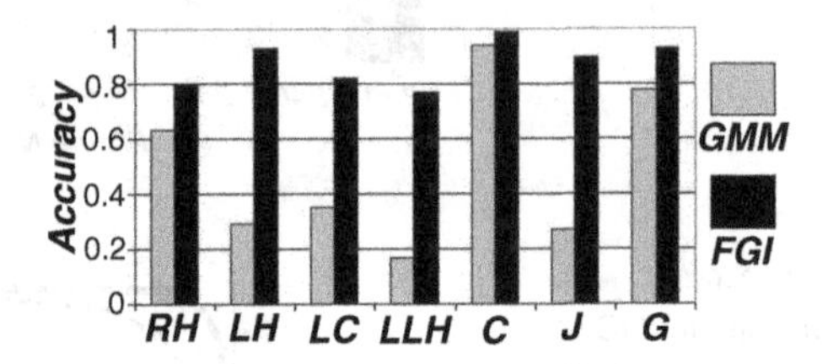

Fig. 5.13 Comparing accuracy on seven stances: GMM versus FGI

accuracy obtained for detecting frames of "Guard" membership only equal to 1.0) gives inferior results: 0.906 in the first case and 0.506 in the second case. Accuracy is defined as:

$$accuracy = \frac{tp + tn}{tp + fp + fn + tn} \tag{5.11}$$

where tp represents true positive rate, tn true negative rate, fp false positive rate, and fn false negative rate. Figure 5.13 shows a comparison between the accuracy of Fuzzy Gaussian Inference(FGI) and a standard Gaussian Mixture Models (GMM) algorithm when classifying seven different stances (Guard, Jab, Cross, Lower Cross, Right Hook, Left Hook, and Lower Left Hook).

The system can recognize nine different stances with an average accuracy of 88.68% with half/half of the data used for learning and testing, on all three participants. When these movements have very few learning examples available, the threshold is fine-tuned by decreasing it to compensate for the data sparsity. Time performance of FGI alone is measured on a laptop, using non-optimized Matlab code (average times over a thousand runs obtained using Matlab Profiler and Tic Toc tools):

- Around 0.2 milliseconds to learn one fuzzy membership function for one joint.
- Around 13.2 milliseconds to learn a full model of 57 joints.
- Around 0.05 milliseconds to compute the membership score of one joint (2.85 ms for 57 joints).

Beside looking at the individual accuracy of a given frame, we also look at the accuracy of frames depending on the context. On its own, a stance could be classified as a Right Hook movement. But if is surrounded by Cross stances, this short Cross stance corresponds in fact to the middle of a Cross movement. The relative positions of groups of frames contain crucial informations taking into account the time dimension. We compare this "context-aware" accuracy (as opposed to the "short-sighted"

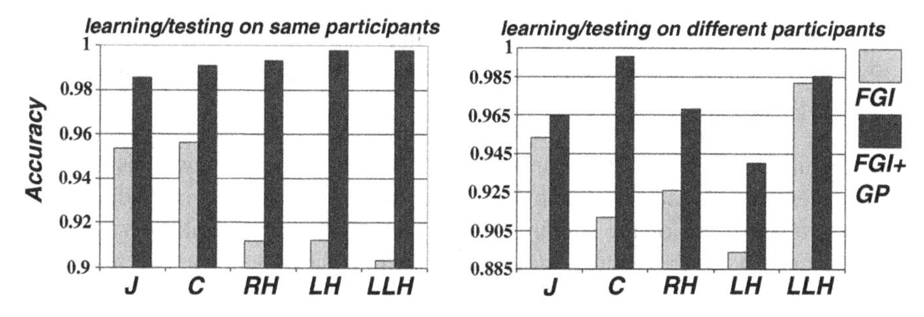

Fig. 5.14 Comparing "context-aware" accuracy: FGI versus FGI+GP

accuracy of an individual frame) of FGI to a mixture of FGI and Genetic Programming (see Fig. 5.14) over six different moves (Guard, Jab, Cross, Right Hook, Left Hook, and Lower Left Hook) performed by each one of all three individuals (data are 3-fold validated). Results show the accuracy when the same individuals are used for learning and testing and when the individuals used for learning are different from the individuals used for testing.

5.7 Discussion

The following observations can be made:

- This system is not a binary but a fuzzy classifier. The threshold value will therefore stay between 0 and 1, which might give the illusion of an "unfinished" ROC curve if the learning and test samples are similar enough. That is the case in Fig. 5.12 where the membership-one point which marks the beginning of the upper curve starts with a higher True Positive rate. This is because the learning and testing are done on the same boxers (even if the samples in use are different, these use participants with the same gait).
- The ROC curves show that, the fuzzy classifier performs better than its crisp counterpart (the one that only identifies Guards of membership one). This gain is especially noticeable when the learning and the testing data present less similarity. It has been observed that a high threshold value is needed to obtain good results. If the threshold is inferior to a membership degree of 0.8, we obtain a maximum True Positive Rate (most of known guards are correctly identified) and a minimum false Negative rate (nearly all known non-guard are identified as guards).
- A t-test shows with 95% confidence that FGI seems to perform significantly better than the GMM-based one (besides, it is worth noting that it has a general average individual frame accuracy of 87.71% while the GMM algorithm is 49% accurate).
- Hidden Markov Models have been tested but perform poorly with our small data set which seems to be insufficient in size to allow significant results. One likely explanation is that, in our natural motion data sample, when the observer determines manually the clusters defining a movement, these clusters are imposed by the observer and they do not necessarily follow the distribution that would

have resulted from standard clustering techniques such as k-means. This seems to make HMM obtain consistently infinite negative log-likelihoods due to the near-singularity of some matrices.

- Another t-test confirms with 95% confidence that the mixture of FGI and Genetic Programming performs significantly better than FGI alone, even with as little as four rules in total. Although the models performed well when the individual concerned formed part of the training group, the classifier performance worsened significantly when they were removed. Despite this phenomenon in line with previous findings [35, 36], it is worth noticing that the association of FGI and GP still shows consistently better results than FGI on its own. It takes in average less than 17 milliseconds to create a template and evaluate a membership score for one stance using FGI.
- The FGI time complexity for recognizing n stances is of the order $O(n)$. As the recognition of a stance takes less than 3 milliseconds on non-optimized Matlab code, we can expect applicability of the order of real time. This time evaluation, however, is valid on FGI alone, and does not include the generation of fuzzy logic rules to filter the qualitative output.

5.8 Conclusion

Fuzzy Gaussian Inference can be used to learn and classify boxing stances with several distinct advantages. First, there is no need for pre-processing the data (which can be a problem when using techniques such as GMM). Another good point is that static models can be obtained from very few examples. Our method also allows one to use different parameters to tailor the precision of every model to the quantity of available data. It is worth noticing that FGI alone appears to be fast enough to allow real-time recognition. The addition of Fuzzy Logic Rules created through Genetic Programming to filter the qualitative output of the system improves the "context-aware" accuracy of the classifier and makes it possible to refine the classification by taking into account the time dimension and identify combinations of moves. Having validated our method on a real-life data set, the next step is to prioritise the recognition of new moves from partial information. Future work might include a robot kinematics [37–39] representation system to deal with the occluded data.

Acknowledgements The project is funded by EPSRC Industrial CASE studentship and MM2G Ltd under grant No. 07002034. Many thanks to Portsmouth University Boxing Club and to the Motion capture Team: Alex Counsell, Geoffrey Samuel, Ollie Seymour, Ian Sedgebeer, David McNab, David Shipway and Maxim Mitrofanov.

References

1. Khoury, M., Liu, H.: Mapping fuzzy membership functions to normal distributions to understand boxing motions. In: Proceedings of UKCI 2008, De Montfort University, Leicester, UK (2008)

2. Khoury, M., Liu, H.: Fuzzy qualitative Gaussian inference: finding hidden probability distributions using fuzzy membership functions. In: IEEE Workshop on Robotic Intelligence in Informationally Structured Space (RiiSS 2009) (2009)
3. Aggarwal, J.K., Cai, Q., Liao, W., Sabata, B.: Articulated and elastic non-rigid motion: a review. In: Proceedings of the IEEE Workshop on Motion of Non-Rigid and Articulated Objects, pp. 2–14 (1994)
4. Wang, L., Hu, W., Tan, T.: Recent developments in human motion analysis. Pattern Recognit. **36**(3), 585–601 (2003)
5. Chan, C.S., Liu, H., Brown, D.J.: Recognition of human motion from qualitative normalised templates. J. Intell. Robot. Syst. **48**(1), 79–95 (2007)
6. Yamato, J., Ohya, J., Ishii, K.: Recognizing human action in time-sequential images using hidden Markov model. In: IEEE Computer Vision and Pattern Recognition, pp. 379–385 (1992)
7. Remagnino, P., Tan, T.N., Baker, K.D.: Agent orientated annotation in model based visual surveillance. In: International Conference on Computer Vision, pp. 857–862 (1998)
8. Bobick, A.F., Wilson, A.D.: A state based technique for the summarization and recognition of gesture. In: International Conference on Computer Vision, pp. 382–388 (1995)
9. Pentland, A.P., Oliver, N., Brand, M.: Coupled hidden Markov models for complex action recognition. In: Massachusetts Institute of Technology, Media Lab (1996)
10. Guo, Y., Xu, G., Tsuji, S.: Understanding human motion patterns. In: International Conference on Pattern Recognition, pp. 325–329 (1994)
11. Yacoob, Y., Black, M.: Parameterized modeling and recognition of activities. In: Sixth International Conference on Computer Vision, pp. 120–127, 1998 (1998)
12. Galata, A., Johnson, N., Hogg, D.C.: Learning variable-length Markov models of behavior. Comput. Vis. Image Underst. **81**(3), 398–413 (2001)
13. Zhang, X., Naghdy, F.: Human motion recognition through fuzzy hidden Markov model. Proceedings of the International Conference on Computational Intelligence for Modelling **02**, 450–456 (2005)
14. Lin, C.-T., Nein, H.-W., Lin, W.-C.: A space-time delay neural network for motion recognition and its application to lipreading. Int. J. Neural Syst. **9**(4), 311–334 (1999)
15. Anderson, D., Luke, R., Keller, J., Skubic, M.: Modeling human activity from voxel person using fuzzy logic. IEEE Trans. Fuzzy Syst. **17**(1), 39–49 (2009)
16. Calvert, T.W., Chapman, A.E.: Analysis and Synthesis of Human Movement, pp. 431–474 (1994)
17. Wu, G., et al.: ISB recommendation on definitions of joint coordinate system of various joints for the reporting of human joint motion, part I: ankle, hip, and spine. J. Biomech. **35**(4), 543–548 (2002)
18. Wu, G., et al.: ISB recommendation on definitions of joint coordinate systems of various joints for the reporting of human joint motion, part II: shoulder, elbow, wrist and hand. J. Biomech. **38**(5), 981–992 (2005)
19. Crawford, N.R., Yamagushi, G.T., Dickman, C.A.: A new technique for determining 3-d joint angles: the tilt/twist method. Clin. Biomech. **17**(2), 166 (1998)
20. Dragulescu, D., Tascau, M., Stanciu, D.: Kinematic and dynamic modeling of human lower limb. In: IASTED International Conference Robotics and Applications, pp. 19–22 (2001)
21. Thingvold, J.: Biovision BVH format (1999). http://www.cs.wisc.edu/graphics/Courses/cs-838-1999/Jeff
22. Favre, J., Aissaoui, R., Jolles, B.M., Siegrist, O., de Guise, J.A., Aminian, K.: 3d joint rotation measurement using mems inertial sensors: application to the knee joint. In: ISB-3D: 3-D Analysis of Human Movement, Valenciennes, France, 28–30 June 2006
23. ZADEH, L.: Fuzzy sets. Inf. Control **8**, 338–353 (1986)
24. Sanghi, S.: Determining membership function values to optimize retrieval in a fuzzy relational database. In: Proceedings of the 2006 ACM SE Conference, vol. 1, pp. 537–542 (2006)
25. Iokibe, T.: A method for automatic rule and membership function generation by discretionary fuzzy performance function and its application to a practical system. In: Proceedings of the First International Joint Conference of the North American Fuzzy Information Processing Society Biannual Conference, pp. 363–364 (1994)

26. Kim, C., Russell, B.: Automatic generation of membership function and fuzzy rule using inductive reasoning. Third International Conference on Industrial Fuzzy Control and Intelligent Systems, IFIS'93, pp. 93–96 (1993)
27. Kim, J., Seo, J., Kim, G.: Estimating membership functions in a fuzzy network model for part-of-speech tagging. J. Intell. Fuzzy Syst. **4**, 309–320 (1996)
28. Simon, D.: H infinity estimation for fuzzy membership function optimization. Int. J. Approx. Reason. **40**(3), 224–242 (2005)
29. Devi, B.B., Sarma, V.V.S.: Estimation of fuzzy memberships from histograms. Inf. Sci. **35**(1), 43–59 (1985)
30. Nieradka, G., Butkiewicz, B.S.: A method for automatic membership function estimation based on fuzzy measures. In: IFSA. Lecture Notes in Computer Science, vol. 4529, pp. 451–460. Springer, Berlin (2007)
31. Frantti, T.: Timing of fuzzy membership functions from data. Academic Dissertation, University of Oulu, Finland, July 2001
32. Khoury, M.: Pystep or python strongly typed genetic programming. Available online at: http://pystep.sourceforge.net/
33. Lawrence, N.D.: Mocap toolbox for Matlab. Available on-line at http://www.cs.man.ac.uk/~neill/mocap/
34. http://en.wikipedia.org/wiki/Boxing
35. Darby, J., Li, B., Costen, N.: Activity classification for interactive game interfaces. Int. J. Comput. Games Technol. **2008**(3), 1–7 (2008). http://dx.doi.org/10.1155/2008/751268
36. Darby, J., Li, B., Costen, N.: Human activity recognition: Enhancement for gesture based game interfaces. In: 3rd International Conference on Games Research and Development, CyberGames (2007)
37. Liu, H.: A fuzzy qualitative framework for connecting robot qualitative and quantitative representations. IEEE Trans. Fuzzy Syst. **16**(8), 1522–1530 (2008)
38. Liu, H., Brown, D.J., Coghill, G.M.: Fuzzy qualitative robot kinematics. IEEE Trans. Fuzzy Syst. **16**(3), 808–822 (2008)
39. Chang, C.S., Liu, H.: Fuzzy qualitative human motion analysis. IEEE Trans. Fuzzy Syst. **17**(4), 851–862 (2009)

Chapter 6
Obstacle Detection Using Cross-Ratio and Disparity Velocity

Huiyu Zhou, Andrew M. Wallace, and Patrick R. Green

Abstract In this chapter we consider the detection of hazards within the ground plane immediately in front to a moving pedestrian. Consecutive views of the scene are acquired by a standard video camera. Using epipolar constraints between the two views, detected features are matched to compute the camera motion and reconstruct the 3-D geometry. Assuming the ground is planar, projective invariance of the cross-ratio and the presence or absence of significant peaks in a Lomb-Scargle periodogram are used in a region-growing technique to label a triangulated mesh as obstructed or unobstructed ground plane. On the other hand, for a less feature based scene a new disparity velocity based obstacle detection scheme is presented. This scheme can be used to find image points of large disparity estimates and hence single out suspicious obstructed ground points. The experimental work shows the performance of these two algorithms in real image sequences.

6.1 Introduction

6.1.1 Background

Systems for vehicle navigation have included road following, tactical-level planning, and the avoidance of large obstacles [1–4]. For pedestrians, previous work has been directed to locate and avoid large obstacles (e.g., waste bins, lamp-posts) [5, 6], but there is no reliable system to detect small obstacles such as kerbs, small stones or uneven pavements. Developing such an aid is valuable, especially for older

H. Zhou (✉)
Queen's University Belfast, Belfast, UK
e-mail: H.Zhou@ecit.qub.ac.uk

A.M. Wallace and P.R. Green
Heriot-Watt University, Edinburgh, UK
e-mail: A.M.Wallace@hw.ac.uk; P.R.Green@hw.ac.uk

H. Liu et al. (eds.), *Robot Intelligence*,
Advanced Information and Knowledge Processing,
DOI 10.1007/978-1-84996-329-9_6,

people who have less efficiency in detecting and anticipating tripping hazards [7], typically obstacles of size ~ 3 cm at a range of ~ 3 m. Walking obstacle detection is a challenging problem as the environment is dynamic and unknown to any proposed system. In addition, the system is under real-time constraints since the 3 m's distance only leads a normal male pedestrian to walking 2–3 seconds.

Conventional systems such as the Automated Highway System [2] and the guiding robot [8] cannot be feasibly implanted into the proposed system due to the two manifolds: firstly, the former-like systems aim to solve the problem of finding small obstacles at long distances, indicating instant reactions of the systems are not demanded. Secondly, the latter-like ones put much cognitive load on the pedestrians so as to be reliably working.

Before the proposed system is configured, a realistic situation must be taken into account, which is that an ordinary CCD camera without wide angle lens (500-by-500 pixels^2) is only able to view an area with the width of 5–8 m. Suppose a potential obstacle having a size of 0.05×0.05 mm^2 is lying 3 m ahead. This implies that the obstacle in the field of view approximately occupies 5-by-5 pixels^2. As such, the detailed information on the surface of the obstacle will be too difficult to retrieve. Any attempt to detect the obstacle by analysing its shape or characteristic intensity will hence be unsuccessful. To distinguish an obstacle from the background, the height of any point on the obstacle should be identified, followed by clarifying this point's relationship to the neighbouring areas in terms of height and intensity. Based on this cue, in the first algorithm we suggest to use the perspective invariance of the cross-ratio, coupled with periodic frequency analysis, to identify safe regions in the ground plane. In a scene of less image features, we consider a map of disparity velocity that can be used to distinguish the potential obstacles from the computed 3-D points.

6.1.2 Algorithm Overview

The first proposed algorithm, illustrated in Fig. 6.1, is based on automatic feature tracking across successive frames from the calibrated single camera, followed by estimation of the ego-motion and the tracked feature positions. In stage 1, we obtain a robust estimate of the ground plane and sets of included and excluded features from the 3-D coordinates of tracked corner features, represented as a triangular mesh. It must be pointed out that the feature points are tracked through frames in combination with the dynamically extracted gait model, and may arise from reflectance and textural variations on smooth surfaces, or from small or large physical obstacles in the field of view. In stage 2 the mesh is extended by adding untracked feature points that may have been missed in stage 1 (and may be hazardous). We verify the hypotheses by additional co-planarity checks (using the simple cross-ratio invariant) and periodic analysis of the triangular regions to find additional evidence for the probable safety of those areas. Eventually a safe route on the ground plane is determined, and unsafe regions containing potential obstacles are marked.

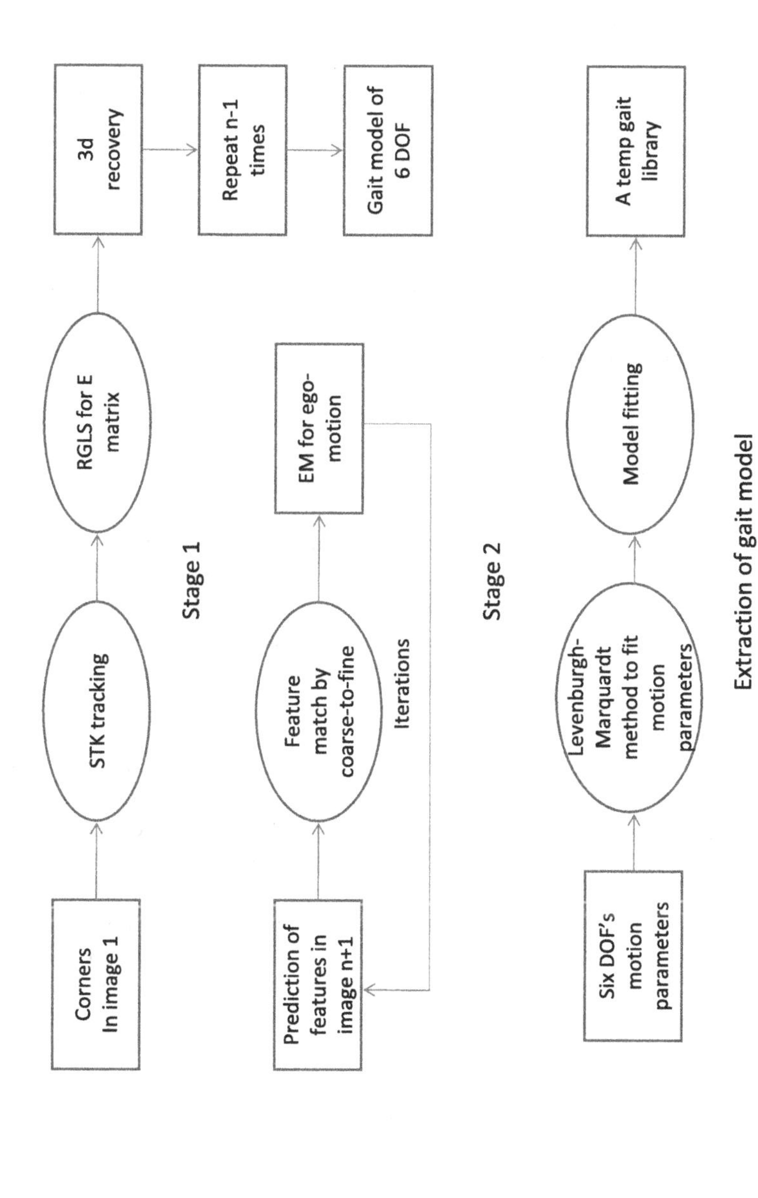

Fig. 6.1 The flow-chart of the proposed filtering algorithm

In the second proposed algorithm, the computation used to reconstruct 3-D data is the same as that of the first algorithm. After the 3-D structure is obtained, being different form the first algorithm, the second algorithm applies the disparity velocity computed from the neighboring frames to differentiate the obstacle points from the computed 3-D points. This algorithm has the advantage of effectively working in a less feature based scene.

6.2 Generation of Mesh Maps

The combination of mesh and perspective invariance modalities was pointed out as a solution to the identity of potential obstacles. The former is adopted as it provides clearly structured details within each triangle [9], whilst the latter has been well-developed to evaluate surfaces with coplanar properties [10]. The fundamental of mesh generation is therefore briefly described in the following section, and that of perspective invariance will be explored in the later application.

6.2.1 Mesh Generation

Meshes, represented by triangles or tetrahedra, are used to model the geometries of physical quantities, i.e., terrains or cups. The produce of meshes is based on the intuition that an object under investigation can be partitioned into small pieces of simple shape. This is compelled by the fact where numerical methods, such as the *finite element method* (FEM), are not able to simulate an observed object in a closed-form due to arbitrary complexities.

The Delaunay triangulation by Delaunay [11] in 1934 is a geometric structure that accompanies mesh generation after the latter was born. This technique seems unique to tackle the meshing problem due to its effectiveness. In two dimensions, the Delaunay triangulation works in such a way: given a set M of vertices (or nodes), let m_1 and m_2 are two of them. Define an *empty* circle as the one that does not enclose any vertex of M, and a *circumcircle* of the edge $m_1 m_2$ as any circle of passing through m_1 and m_2. The edge $m_1 m_2$ is thus in the Delaunay triangulation if and only if there exists an empty circumcircle of $m_1 m_2$ [12]. The edge $m_1 m_2$ is hence *Delaunay*. Figure 6.2 shows a Delaunay triangulation. The Delaunay triangulation of the set M of vertices is obviously unique. Also, it has the properties represented as follows.

Lemma 6.1 *Let N be a triangulation. If all the triangles of N are Delaunay, then all the edges of N are Delaunay, and vice versa.*

Before the second lemma is prompted, let us introduce the *flip algorithm*, which leads all the edges of the triangulation created by itself to being *Delaunay*. The flip algorithm starts with an arbitrary triangulation of M, and seeks an edge that is

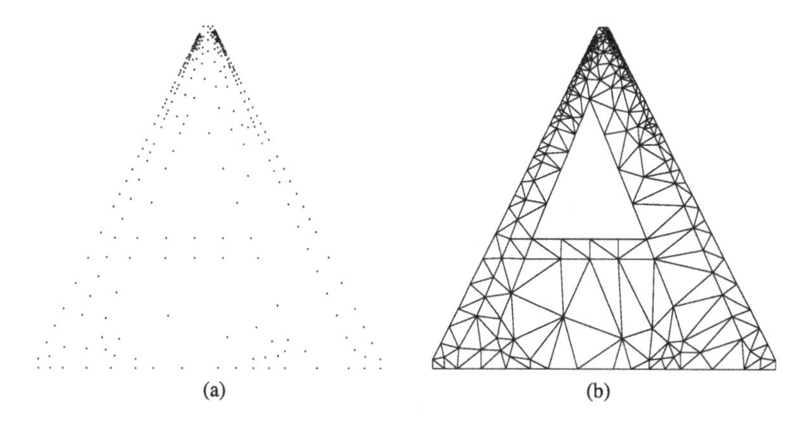

Fig. 6.2 Illustration of a Delaunay triangulation. (**a**) Vertices. (**b**) Delaunay triangulation. Vertex data courtesy of Shewchuk, J.R., University of California at Berkeley

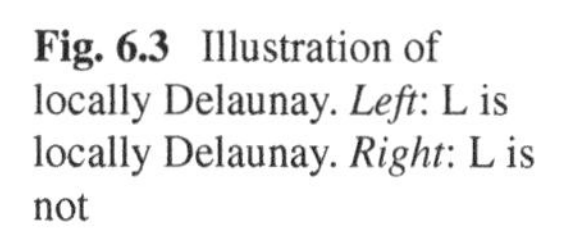

Fig. 6.3 Illustration of locally Delaunay. *Left*: L is locally Delaunay. *Right*: L is not

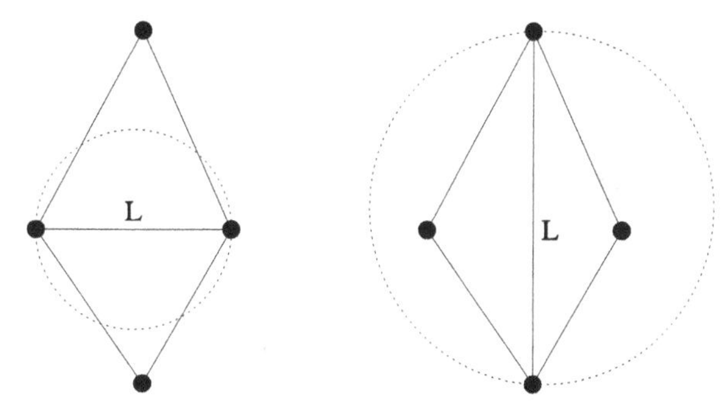

not locally Delaunay (a local Delaunay refers to the edge on the boundary of the triangulation). This is illustrated in Fig. 6.3.

If an edge has been judged by the flip algorithm to be not locally Delaunay, the edge can be flipped that means its absence from the triangulation. The triangulation at right in Fig. 6.3 is thereby converted into the one at left.

Lemma 6.2 *Let L be an edge of a triangulation of M. L is either locally Delaunay or flippable, and the edge generated by making L flipped is locally Delaunay.*

Lemma 6.3 *Let N be a triangulation, which only has edges of locally Delaunay. Thus, each edge of N is globally Delaunay.*

Lemma 6.4 *Suppose there is a triangulation out of M vertices. The flip algorithm only can stop unless a triangulation whose edges are all Delaunay after M^2 edge flips.*

Lemma 6.5 *Let M be a set of vertices that are not all collinear. Without four co-circular vertices, the Delaunay triangulation of M is the one produced by the flip algorithm.*

Lemma 6.6 *Of all triangulations of a vertex set, the Delaunay triangulation is the one that maximizes the minimum angle in the triangulation, minimizes the largest circumcircle, and minimizes the largest min-containment circle, which is the smallest circle of enclosing itself.*

These six lemmas help define the Delaunay triangulation of a vertex set, and the reader can directly conduct the proofs or alternatively refers to the literature [12]. In the proposed project, the mesh generation in two dimensions is preferred to the other dimensional for the height and the distance (to the camera) of a potential obstacle can be identified and bound within the two-dimensional meshes. For instance, a point within one of the triangles is marked at a specific height by 3-D rendering, and the distance from the point with the height to the camera position is estimated as well.

6.3 Estimation of the Ground Floor

Estimation of the ground floor is so crucial that any potential obstacle can be deduced by computing the relative height of any point in the scene in terms of that of the ground floor. This procedure is illustrated in Figs. 6.4 and 6.5, where a group of 3-D points have been retrieved by using the proposed camera system. Meanwhile, it should be noticed that the discrimination between the floor and the potential obstacle points is dependent on the pre-defined threshold.

Before the ground floor is determined, an assumption is undertaken as follows:

Assumption 6.1 The ground floor is a flat surface.

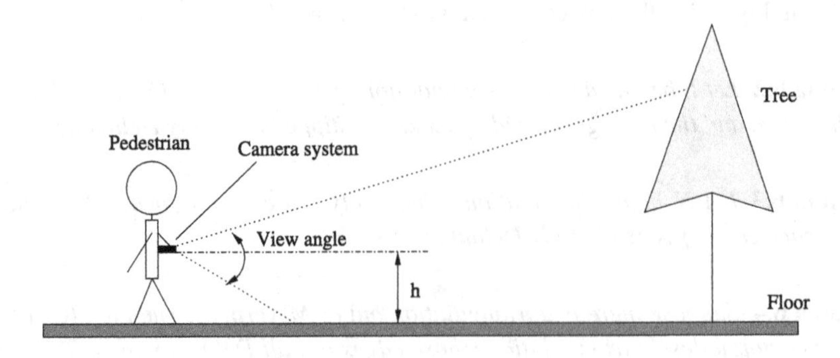

Fig. 6.4 Illustration of a pedestrian carrying the proposed camera system

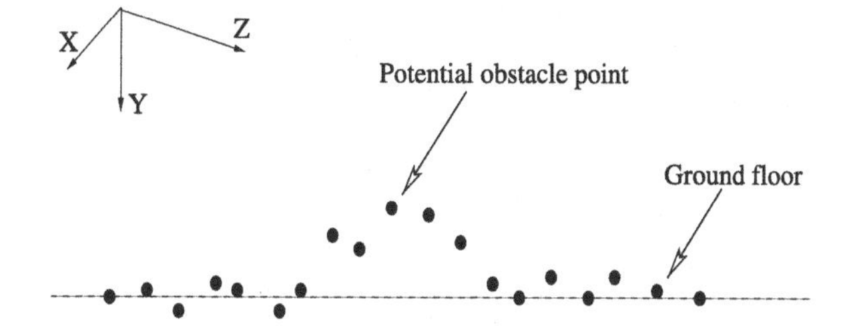

Fig. 6.5 Illustration of the floor and potential obstacle points

Li [13] and Molton [14] propose a *dynamic ground plane recalibration* (DGPR) algorithm to determine the ground floor, which normally works based on the disparity map of a cyclopean image. Let the ground disparity, D_{dis}, change linearly with cyclopean image plane position as

$$D_{dis} = au_c + bv_c + c, \tag{6.1}$$

where (u_c, v_c) is the cyclopean image coordinate, which is the mid-point of the left and the right image coordinates. A least square fit is then employed to estimate the parameter vector $[a, b, c]$. In the real application of obstacle detection, any potential obstacle will be coincident with a larger disparity than the pre-defined threshold, D_{dis}. This approach undoubtedly works in a variety of circumstances. However, due to the inaccurate correspondences and image distortions the ground plane disparity fitting cannot be well-achieved, resulting in unstable judgement on ground or obstacle points [14].

Zhou [15] raises a concept of dominant plane that is regarded as the potential ground floor. This strategy works in the case of which deserves most feature points (minimally 50%) on the flat ground, but fails when most points reside on other planes, e.g. surfaces of buildings, since these planes might be incorrectly identified as the flat ground.

Here, our aim is to develop a novel scheme, which works as a compensation of Zhou's approach [15] in the circumstance where a few feature points can be tracked. It is evident that, with a few feature points being tracked, the ground floor cannot be meshed densely, resulting in missed localization of potential obstacles due to the shortage of remarkable features on surfaces. The introduction to the framework starts from several propositions, which are described as follows.

Proposition 6.1 *On the flat ground, the points closer to the camera have higher disparities.*

Proof Referred to Fig. 6.5, the disparity, D_{dis}, is defined as $D_{dis} = fd_{int}/Z_{dis}$, where f is the focal length of the camera, d_{int} is the interocular distance, and Z_{dis} is the distance between the camera and the ground points [16]. Decreasing Z_{dis} leads to the increase of D_{dis}, indicating that the points closer to the camera have higher disparities than the others, and vice versa. The proof is complete. □

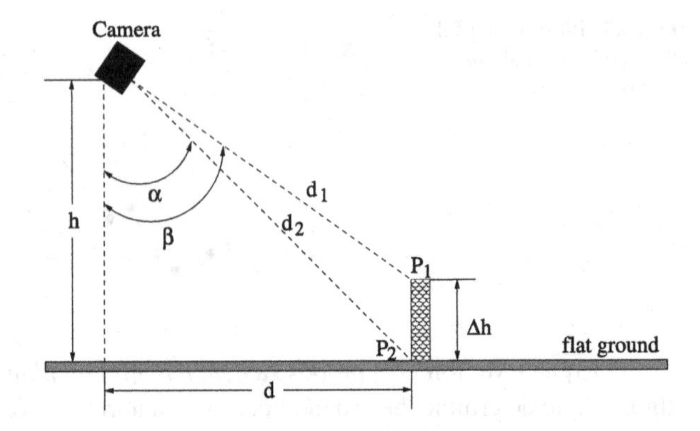

Fig. 6.6 Illustration of two points on or over the flat ground (P_1—obstacle and P_2—ground)

Proposition 6.2 *The points on obstacles, whose heights are assumed to be less than that of the camera, deserve larger disparities than those of the corresponding points on the flat ground.*

Proof Referred to Fig. 6.6, there are two points on or over the flat ground (P_1—obstacle and P_2—ground). Their distances to the camera are denoted as d_1 and d_2, respectively. Due to the height of the obstacle point, Δh, $\beta > \alpha$ that leads to $d_1 < d_2$ where $d_1 = d/\sin(\beta)$, and $d_2 = d/\sin(\alpha)$. On the revisitation of $D_{dis} = f d_{int}/Z_{dis}$, one can replace Z_{dis} by d_1 or d_2. The obstacle point, P_1, has larger disparity than that of the ground point, P_2.

□

Proposition 6.3 *The points on the flat ground but with mismatches across frames are of irregular disparities.*

Proof Replacing Z_{dis} of $D_{dis} = f d_{int}/Z_{dis}$ by $(Z_{dis} + \Delta Z_{dis})$, one gets a new disparity, D_{dis}', depending on the sign of ΔZ_{dis}. If $\Delta Z_{dis} \le 0$, then the disparity increases; otherwise it decreases. □

Proposition 6.4 *The disparity velocity of a point on or over the flat ground is limited by the upper and lower boundaries, which are defined by the moving distance of the camera.*

Proof Differentiating the both sides of $D_{dis} = f d_{int}/Z_{dis}$ in terms of time t, one has

$$\frac{\partial D_{dis}}{\partial t} \equiv -\frac{\partial Z_{dis}}{\partial t} \cdot \frac{f d_{int}}{Z_{dis}^2} + \frac{f}{Z_{dis}} \cdot \frac{\partial d_{int}}{\partial t}. \tag{6.2}$$

Due to $Z_{dis} = d/\sin(\alpha)$, referred to Fig. 6.6, the above equation is equivalent to the form as

$$\frac{\partial D_{dis}}{\partial t} \equiv -\frac{\partial (d/\sin(\alpha))}{\partial t} \cdot \frac{f d_{int}}{Z_{dis}^2} + \frac{f}{Z_{dis}} \cdot \frac{\partial d_{int}}{\partial t}. \tag{6.3}$$

Since α changes slightly (≤ 0.1 rads per frame) during motion, one obtains

$$\frac{\partial D_{dis}}{\partial t} \approx -\frac{\partial d}{\partial t} \cdot \frac{f d_{int}}{{Z_{dis}}^2 \sin(\alpha)} + \frac{f}{Z_{dis}} \cdot \frac{\partial d_{int}}{\partial t} = \left(\frac{f}{Z_{dis}} - \frac{f d_{int}}{{Z_{dis}}^2 \sin(\alpha)}\right) v_z, \quad (6.4)$$

where v_z is the gait velocity along the walking direction (here Z-axis), falling in the range $[0.0, 2.0]$ (m/s) for normal walking [17], and $\forall \alpha \in (0, \pi)$ (the viewing angle of the mobile camera). Hence, the disparity velocity has the lower and upper boundaries, which ends the proof. □

A real scene contains nearer and further feature points referred to the camera position. Imagine that a pedestrian stands on a pathway where there is no obstacle nearby before he/she starts walking. Propositions 6.1–6.3 demonstrate the *complexity* and *uncertainty* of feature points on the floor so the assumption of *no obstacle nearby* is demanding in this case! Making this assumption allows to predetermine the flat floor and its characteristics that will be used to discriminate the points on or over the floor. Consequently, such a property is enclosed as follows:

Property 6.1 *The flat ground is constructed by the feature points that are close to the camera position, and have disparity velocities bound by* (6.4).

Proof The proof can be divided into two parts, where the first one results from the assumption of no obstacle nearby leading to the residence of only ground points around the camera, and the second one has been proved in Proposition 6.4. □

To feasibly implement the above definition, two parameters must be prior defined: (1) What is the concerned area for determining the ground plane? (2) What is the lower or upper boundary for (6.4). These two parameters are *interactive* and *coherent*. To effectively define them, (6.4) is reviewed here. One expects to detect potential obstacles of 3–5 m ahead, so $Z_{dis} > 3$ m but $d_{int} < 2$ m per second. For a camera system with regular sampling rate, e.g., 25 frames per second in our case, $d_{int} < 0.08$ m during 0.04 s. Let us look at such a typical example, where the overall parameters are chosen by taking into account the configuration and motion of the real camera during 0.04 s: (1) if $f = 0.00567$ m, $Z_{dis} = 5$ m, $\alpha = 0.43\pi$, and $v_z = 1.0$ m/s, then the minimum disparity velocity will be 0.00113 m/s; (2) if $f = 0.00567$ m, $Z_{dis} = 1$ m, $\alpha = 0.25\pi$, and $v_z = 1.0$ m/s, then the maximum disparity velocity will be 0.00567 m/s. That is to say, only those points with the disparity velocities of $[0.00113, 0.00567]$ m/s are considered as the candidates for determining the ground plane.

Given a short interval such as 0.04 s, v_z is treated *constant* during two consecutive periods. Thus, (6.4) is only determined by Z_{dis}. In tests, two neighboring images are required in tracking in order to compute disparities. We then study those points whose disparity velocities are bound by $\frac{v_{dis}}{5.0} \sim \frac{v_{dis}}{2.0}$ (v_{dis} is the disparity velocity of the closest feature point to the camera). It is possible that some feature points from further areas are incorrectly contained in the candidates. A 3 or 5 m trade-off (along the walking direction z-axis) is defined to artificially "remove" the further feature points from the candidate list.

Once the candidates of the ground floor have been specified, SVD technique is used in the sense of least square in order to seek a unique solution to the Cartesian equation of the ground plane:

$$c_1X + c_2Y + c_3Z + c_4 = 0. \tag{6.5}$$

Due to the availability of more than four points in the scene the system equation is over-determined.

6.4 Identification of Safe Regions within the Ground Plane

Without loss of generality, a couple of images are chosen to show how the proposed algorithm works in achieving obstacle detection. Assume these images have been corresponded using a tracking strategy (e.g., the STK [18] or the gait-based tracker [19]). The well-established epipolor constraints are then used to further refine the corresponding of features and recover the camera and scene geometry [20]. Figure 6.7(a) shows 150 feature points extracted in the first image using the SUSAN corner detector, and (b) refers to a reduced set of features after tracking (31 in this example). Due to the prior definition of the ground floor, the tracked feature points can be considered as belonging or not to the plane on the basis of their Euclidean distances to the plane [15]. Figure 6.7(c) illustrates the 20 tracked features that are effectively included in the ground plane.

To identify safe regions in the scene, we use proximity, co-planarity and a measure of intensity variations in the triangles formed between feature points. A 2-D Delaunay triangulation of the feature points in the latest image is used to define the basic structure: the vertices of the triangles are the points in the image that have known (x, y, z) coordinates, the interiors of the triangles are regions that are on or off the ground plane.

6.4.1 Incremental Addition of Feature Points

However, we also wish to obtain a denser tessellation of the ground plane in front of the pedestrian, as this is the area of principal interest. To do this, each triangle in the tessellation is examined.

The vertices define 3 coplanar points. Using any of the three adjacent triangles which have two vertices in common, chosen at random, the third vertex of that adjacent triangle defines a fourth coplanar point. Then, using the original output from the SUSAN corner detector, several additional feature points are selected from within the original triangle, provided the magnitude of the response is above a fixed threshold. Each of these additional interior points is added to the already defined four coplanar points to define a cross-ratio, a projective invariant that is formed from five coplanar points [21].

Fig. 6.7 Feature detection, tracking and recovery of the epipolar geometry. (**a**) Corner features; (**b**) Tracked features; (**c**) Features in the ground plane. Two epipolar lines are shown in images (**a**) and (**b**)

With reference to Fig. 6.8, D_i, $i = 1, 2, 3, 4$, is a set of lines formed from a set of five points with $Q_0(x, y)$ as the common intersection. $(x_i, 0)$ and $(0, y_i)$ are the Cartesian coordinates whose origin is $(0, 0)$ with the axes X_0 and Y_0, where $Q_1(0, y_1)$, $Q_3(0, y_3)$, $Q_2(x_2, 0)$, and $Q_4(x_4, 0)$ are the other four points. From projective geometry, we can write for each line D_i under a translated Cartesian coordinate system,

$$\frac{x}{x_i} = \frac{y}{y_i}. \tag{6.6}$$

Hence:

$$x\left(\frac{1}{x_i} - \frac{1}{x_j}\right) = -y\left(\frac{1}{y_i} - \frac{1}{y_j}\right) \tag{6.7}$$

and:

$$CR = \frac{(y_3 - y_1)(y_2 - y_4)}{(y_2 - y_1)(y_3 - y_4)} = \frac{(x_3 - x_1)(x_2 - x_4)}{(x_2 - x_1)(x_3 - x_4)} \tag{6.8}$$

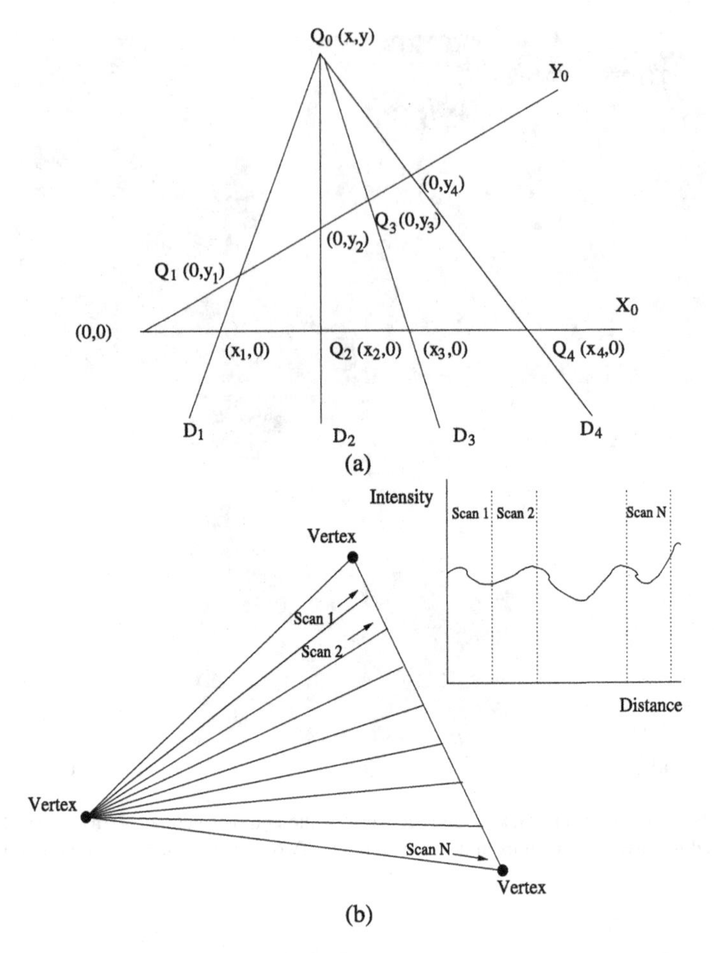

Fig. 6.8 (**a**) Cross-ratio and projective geometry in a translated Cartesian coordinate system; (**b**) Sampling within a triangular region

where *CR* is one of the cross-ratios computed. Therefore, we compute one of the six cross-ratios of the five points in the first view. Denoting (u_i, v_i) the image coordinates of Q_i, $i = 0, 1, 2, 3, 4$, in the first image coordinate system, let (u'_i, v'_i) be the coordinates of an expected point in the second image that corresponds to $Q_i(u_i, v_i)$ in the first image. Suppose we are looking for the correspondence of (u_3, v_3) and (u'_3, v'_3). Assume, in the second image, (u''_1, v''_1) is the intersection between lines Q_0Q_1 and Q_2Q_4, and (u''_3, v''_3) between lines Q_0Q_3 and Q_2Q_4. With the known epipolar constraint,

$$\mathbf{p}_2^T F \mathbf{p}_1 = 0, \tag{6.9}$$

we have

$$\alpha u'_3 - v'_3 + \beta = 0 \tag{6.10}$$

where

$$\alpha = -\frac{f_{11}u_3 + f_{12}v_3 + f_{13}}{f_{21}u_3 + f_{22}v_3 + f_{23}}, \tag{6.11}$$

$$\beta = -\frac{f_{31}u_3 + f_{32}v_3 + f_{33}}{f_{21}u_3 + f_{22}v_3 + f_{23}}, \tag{6.12}$$

and f_{jk} $(j, k = 1, 2, 3)$ refer to the components of the fundamental matrix. Hence we can express the square of the known *CR* in terms of the image coordinates in the second image:

$$CR^2 = \frac{((u_1'' - u_3'')^2 + (v_1'' - v_3'')^2)((u_2' - u_4')^2 + (v_2' - v_4')^2)}{((u_1'' - u_2')^2 + (v_1'' - v_2')^2)((u_3'' - u_4')^2 + (v_3'' - v_4')^2)}. \tag{6.13}$$

(u_3', v_3') can be deduced from (6.10), (6.13) and from the line equations Q_0Q_3 and Q_2Q_4. This gives the projected position of the additional point in the second image, *assuming it is on the ground plane and is visible.* Visibility is checked by a correlation of the intensity values around the hypothetical points matched in the two views. Coplanarity is verified by checking the independent cross-ratios, using a threshold that has the form $\frac{|crossratio_1 - crossratio_2|}{crossratio_1 + crossratio_2}$, where $crossratio_1$ and $crossratio_2$ refer to the cross-ratios in the two frames respectively.

6.4.2 Safe Path Detection

If we can confirm the addition of a further co-planar feature point, this one is added to the set of points of the mesh included in the ground plane; otherwise, it is added to the set of excluded points. A safe path through the mesh might be one that traverses the mesh from vertex to vertex within the ground plane. However, we also examine the variation of intensity in each triangle that has one of its three sides defined by the vertex to vertex boundary under consideration.

Usually a natural small obstacle (kerb, uneven paving stone, pothole, etc.) will result in a spatial distribution of intensity values that can be detected. We have experimented with the use of the *Lomb-Scargle normalized periodogram* [22] to detect structures in the images by combining a number of scans in a triangular region into a single, non-uniformly sampled scanline, as illustrated in Fig. 6.8(b). The non-uniformity arises because a fixed number of points are sampled on each of the scans which are of unequal length, and because there is irregularity when successive scans are concatenated.

Assume that there are N pixels in the combined scan, $p_i \equiv p(s_i)$, $i = 1, 2, \ldots, N$, where s_i is the distance of the point from the beginning of the combined scanline. Then the Lomb-Scargle normalized periodogram, which expresses the spectral power of the data as a function of the angular frequency ω, is defined as

$$P_N(\omega) \equiv \frac{1}{2\sigma^2} \times \left\{ \frac{[\sum_i (p_i - \overline{p}) \cos \omega (s_i - s_0)]^2}{\sum_i \cos^2 \omega (s_i - s_0)} + \frac{[\sum_i (p_i - \overline{p}) \sin \omega (s_i - s_0)]^2}{\sum_i \sin^2 \omega (s_i - s_0)} \right\}$$

where $\overline{p}$ and σ^2 are the mean and variance of the pixel intensities. s_0 is an offset that lets $P_N(\omega)$ be independent of shifting the s_i's by any constant, defined by the equation:

$$\tan(2\omega s_0) = \frac{\sum_i \sin(2\omega s_i)}{\sum_i \cos(2\omega s_i)}. \tag{6.14}$$

If there are significant peaks in a periodogram, indicating strong spatial structures in the region's intensity distribution, this is assumed to be a potential hazard. Repeated occurrence of that structure in successive scans should be indicated by the presence of a periodic signal in the periodogram. To evaluate the significance of a peak in the spectrum $P_N(\omega)$ it is assumed that the pixels in the field of view have independent Gaussian random values, so that $P_N(\omega)$ has an exponential probability distribution with unit mean. Letting $X = P_N(\omega)$, the probability distribution p_Q that Q lies between $\tilde{q}$ and $(\tilde{q} + d\tilde{q})$ is given by

$$p_Q(\tilde{q})d\tilde{q} = e^{-\tilde{q}}d\tilde{q}. \tag{6.15}$$

Then the cumulative distribution F_z is represented as

$$F_Q(\tilde{q}) = P(Q < \tilde{q}) = 1 - e^{-\tilde{q}}. \tag{6.16}$$

The false-alarm probability P is the significance level of any peak in $P_N(\omega)$. If we have the largest value

$$Q_m = \max_M P_N(\omega) \tag{6.17}$$

in the spectrum over M independent frequencies, the probability that q is smaller than Q_m, i.e. that there is a significant periodic signal, is

$$P(Q_m > \tilde{q}) \equiv 1 - (1 - e^{-\tilde{q}})^M. \tag{6.18}$$

A relatively small P demonstrates the appearance of a periodic signal.

Figure 6.9 illustrates two triangles formed between matched feature points in successive frames. The associated spatial data and periodograms are shown in Figs. 6.10 and 6.11 respectively. In Fig. 6.10(A), there are 290 samples at sampling intervals from 1.7 to 2.0 pixels. This leads to a periodogram of $M = 290$ values, at a maximum frequency of 0.13 pixel^{-1} (see Fig. 6.11(A)). As shown in Table 6.1, we have a very significant peak,

$$P = 1 - (1 - e^{-44.5})^{290} \approx 290e^{-44.5} = 1.3 \times 10^{-17}, \tag{6.19}$$

which is expected from the clear periodic structure in the intensity data of Figs. 6.10(A) and 6.10(B) (caused by the lamp-post). There is a clear harmonic structure as shown by the frequency ratio $\frac{|F_{a_i} - F_{b_i}|}{|F_{a_i}|}$ and height of the second peak, also recorded in Table 6.1. In contrast, considering Fig. 6.11(C), there is no dominant peak or harmonic structure, and the significance level of the highest peak is

$$P = 1 - (1 - e^{-5.4})^{291} = 7.3 \times 10^{-1}. \tag{6.20}$$

Comparing the rows A and B in Table 6.1, we also note that the relative peak heights and frequency differences of the second harmonic suggest firstly that the same structure is observed, and secondly, that the object is not in the immediate path of the

Fig. 6.9 Two pairs of triangles used as examples. A and B refer to the upper triangles, C and D to the lower triangles in the images (a) and (b) respectively

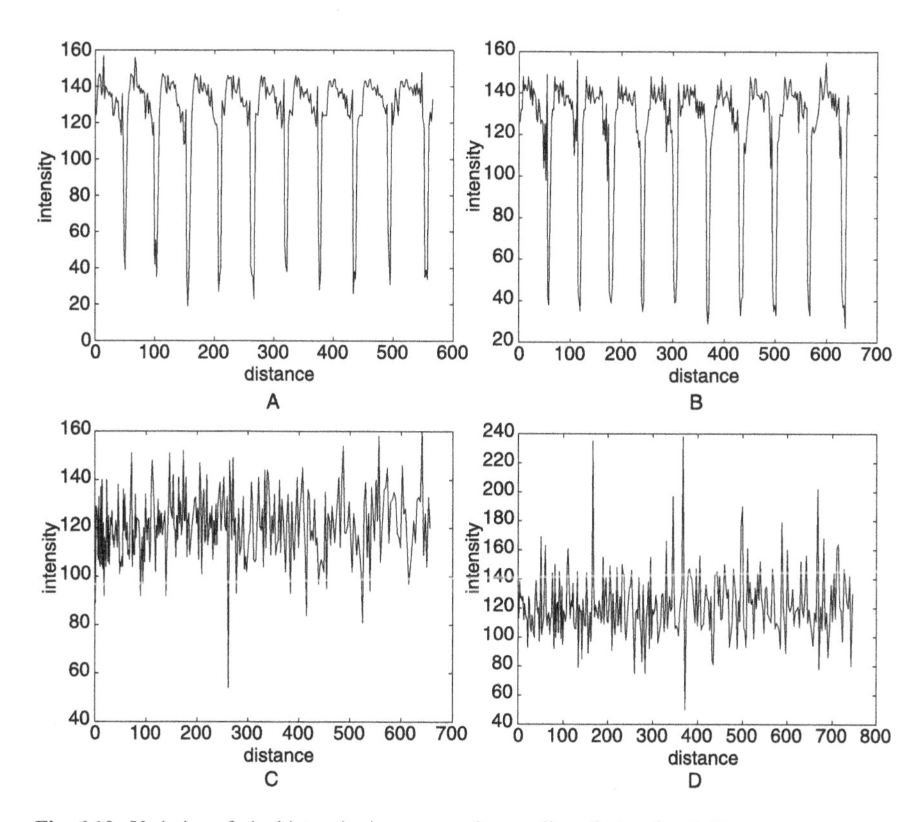

Fig. 6.10 Variation of pixel intensity in a composite scanline of triangles A–D

pedestrian. Given the perspective projection, near objects will result in much larger periodic differences between successive frames, as the corresponding triangle in the

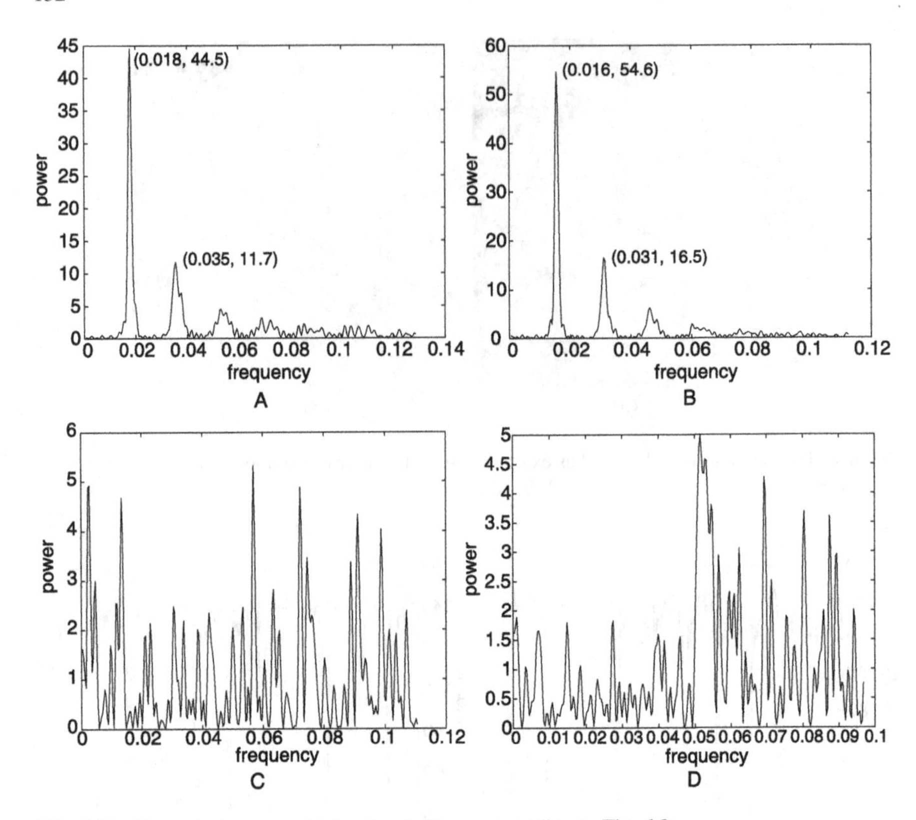

Fig. 6.11 The periodograms of triangles A–D corresponding to Fig. 6.9

Table 6.1 Periodogram characteristics for triangles A–D, where F—frequency; E—power

	Number of frequencies	Dominant peak $(a_i/b_i)(F_{a_i/b_i}, E_{a_i/b_i})$	$\frac{\lvert F_{a_i} - F_{b_i}\rvert}{\lvert F_{a_i}\rvert}$	$\frac{E_{b_i}}{E_{a_i}}$
A	290	$a_1(0.018, 44.5)$; $b_1(0.035, 11.7)$	0.944	26.3%
B	289	$a_2(0.016, 54.6)$; $b_2(0.031, 16.5)$	0.938	30.2%
C	291	no significant peak	n/a	n/a
D	289	no significant peak	n/a	n/a

mesh (e.g. the lower one in Fig. 6.9) becomes much larger while the fundamental frequency is reduced.

Figure 6.12 shows the complete mesh for the region in front of the pedestrian. The vertices in the picture are marked safe (clear) or unsafe (solid) by membership of set of ground plane points, the regions are safe (clear) or unsafe (solid) according to the periodic analysis. This figure illustrates the successful identification of the small (stones) and large (lamp-post) obstacles in front of the pedestrian, but also the mis-classification of additional points to the left and right of the central stone. These

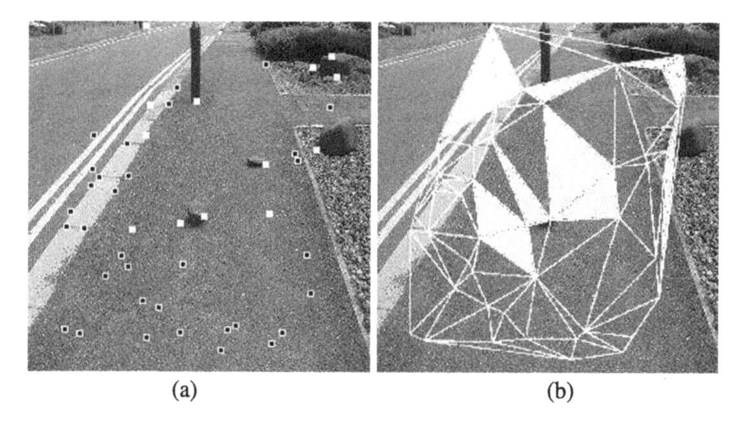

Fig. 6.12 (**a**) Labelled feature nodes, clear (co-planar) and filled (not co-planar); (**b**) Labelled regions, clear (unobstructed) and filled (obstructed)

points are not present in the original data, but are introduced by the incremental addition of points and co-planarity checking. This problem is not fully resolved but is caused essentially by error propagation from the tolerance in fitting the plane, computation of the fundamental matrix and tolerance on the coplanarity. Maintaining error consistency is not trivial and remains a significant goal. Figure 6.13 demonstrates the effect of the addition of extra points to make the mesh more dense at stage 2. It should be noted that these additional points, marked "+", are well within the error bound of the original points denoted by "*".

6.5 Further Evaluation

The intention of this section is to justify the gait-assisted strategy proposed in Sect. 6.3. In other words, once the feature points have been tracked over frames, we need to evaluate the performance of the new system in a sequence with lesser feature points on the floor. To do so, we utilise a new image sequence collected in Heriot-Watt University of United Kingdom to demonstrate the performance of the proposed strategy.

6.5.1 Estimation of Ground Floor

The determination of ground floor is based on the disparity velocities computed over a couple of images (see Sect. 6.3). Therefore, before tracking starts, two neighbouring images are collected. The short interval allows the gait velocity Z_{dis} to be ignored in the calculation of disparity velocity. The whole procedure of achievement in estimating ground floor is illustrated in Fig. 6.14. Figure 6.15 shows that there totally exist 11 features that are identified on the ground plane and labelled

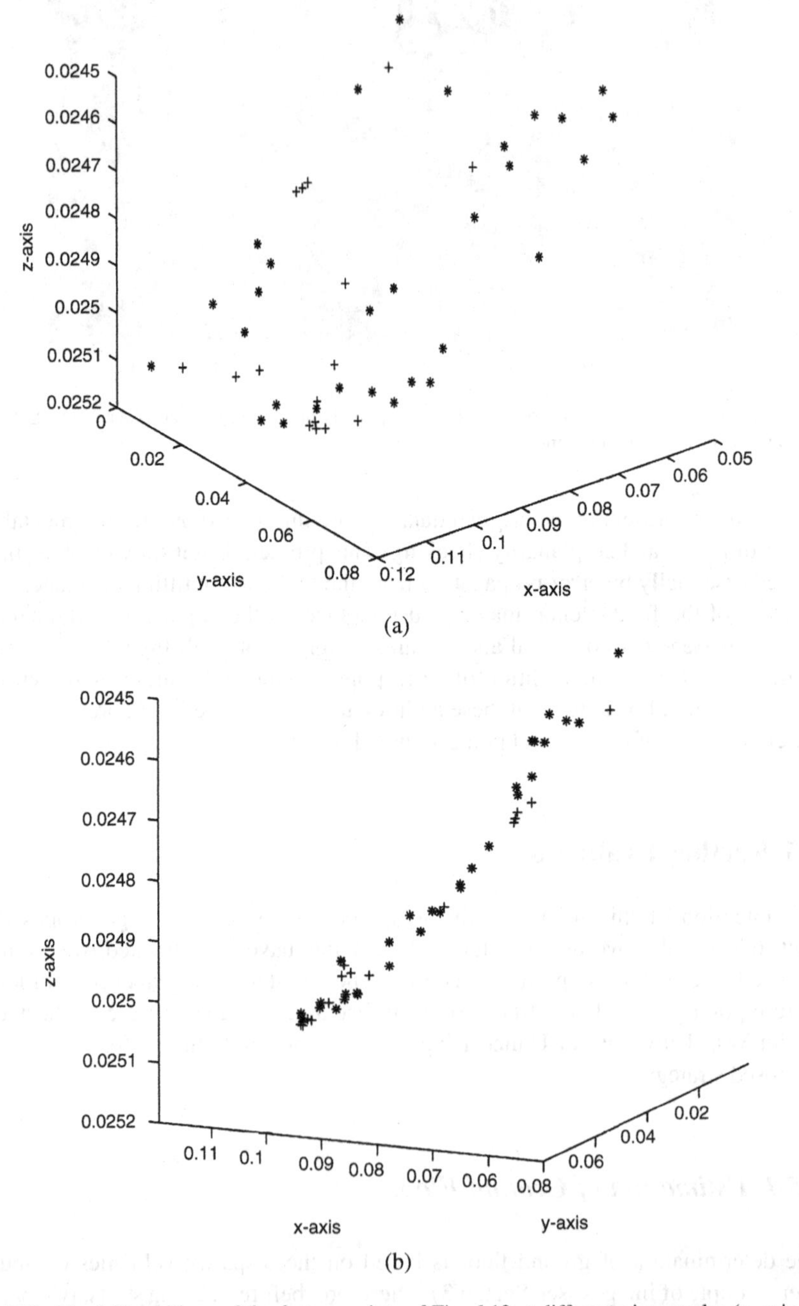

Fig. 6.13 The 3-D positions of the feature points of Fig. 6.12 at different view-angles (x-axis is vertical to the paper's plane, and y-axis is parallel to the direction of the sentences on the paper. Unit: meter). The matched features in stage 1 are marked by "*"; the additional points added at stage 2 by "+"

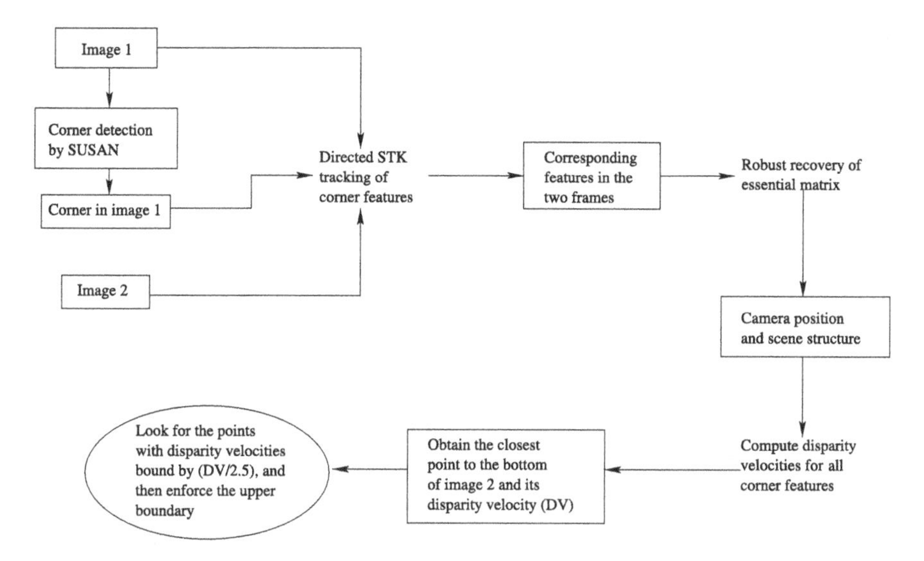

Fig. 6.14 The flow-chart of estimating ground floor

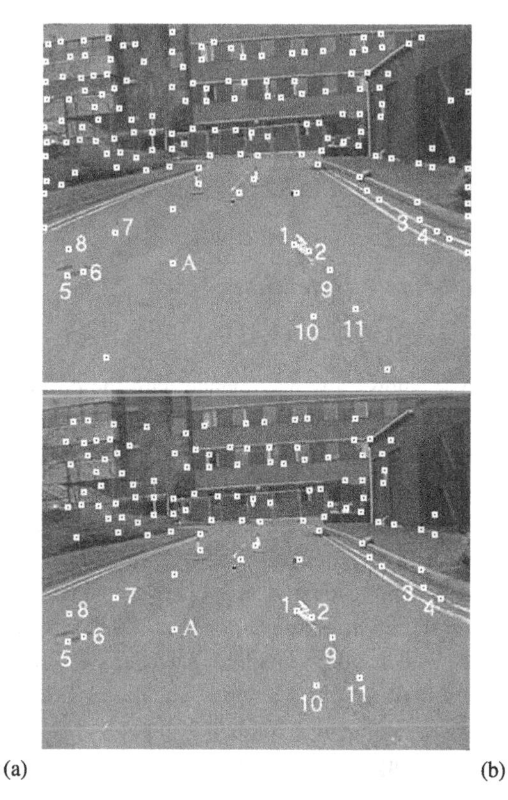

Fig. 6.15 Illustration of feature points on the ground floor, labeled as $1, 2, \ldots, 11$

Table 6.2 Disparity velocities against feature coordinates. F.C.—Feature coordinates [pixels]; D.V.—Disparity velocities [1/(ms)]

Feature no.	F.C.	D.V.
1	(213.7, 177.3)	0.001184
2	(225.8, 182.4)	0.001229
3	(316.5, 157.7)	0.000852
4	(330.8, 166.8)	0.000948
5	(26.7, 201.6)	0.001212
6	(39.7, 198.5)	0.001385
7	(66.4, 167.3)	0.000894
8	(28.0, 180.3)	0.001005
9	(243.1, 197.8)	0.001353
10	(230.1, 235.5)	0.001795
11	(264.8, 229.6)	0.001804
Plane parameters	c_1, c_2, c_3, c_4	0.030852, 0.694906, −0.056642, −0.716201

in the graph. Their disparity velocities are tabulated in Table 6.2, where the plane parameters in terms of the Cartesian equation are presented as well. Interestingly, the point marked as "A" has not been recognized as a ground point. By checking its disparity velocity, we find it is just 0.000251, which is much lower than the largest one, 0.001804. This is due to the mismatch of the feature points across frames.

6.5.2 Obstacle Detection

Figure 6.16 shows a real sequence with the corner features superimposed. As the feature points are tracked using the STK-based method, their number is gradually decreased due to leaving the view of the camera. The motion estimates conducted in 50 frames are illustrated in Fig. 6.17. Figure 6.18 shows the comparison of the 55th frame and its texture-map by using the gait-based scheme, which provides fast and consistent motion estimates [23]. In addition, all the tracked feature points in frame 55 can be recovered in 3-D space, which are used to study their properties on or over the floor. Figure 6.19 illustrates the tracked feature points, including five potential obstacle points labelled as a, b, ..., e. The 3-D positions of these five points are substituted into the Cartesian equation of the ground plane, whose computed residuals are tabulated in Table 6.3. It shows that alerts shall be triggered when the residuals are larger than 0.1 in this case.

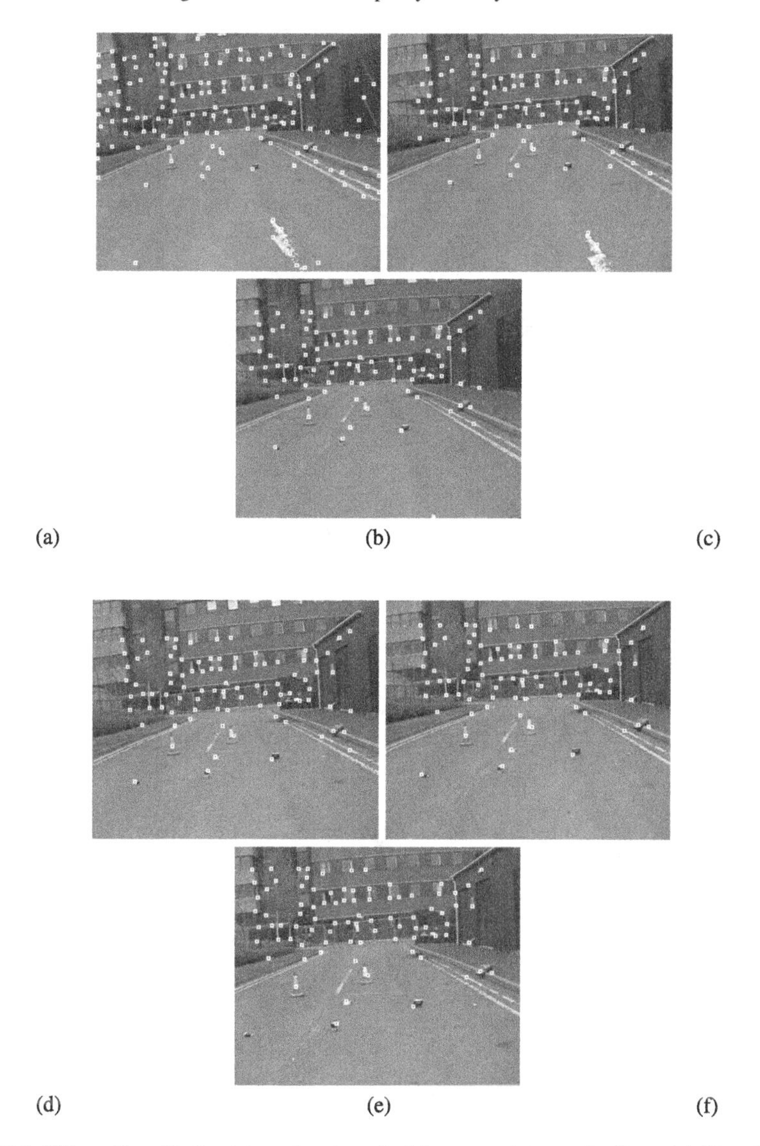

Fig. 6.16 Video clip with the detected and tracked feature points on the floor. (**a**) First frame. (**b**) 10th frame. (**c**) 20th frame. (**d**) 30th frame. (**e**) 40th frame. (**f**) 50th frame

6.6 Summary

We have described methods for detection of the ground plane and potential obstacles in an image sequence. Recovery of the 3-D geometry is dependent on significant displacement of features between frames, but this limitation is balanced by the observation that pedestrians concentrate on the area about 3 m ahead. The detection of structures by a periodic analysis does not discriminate between structural

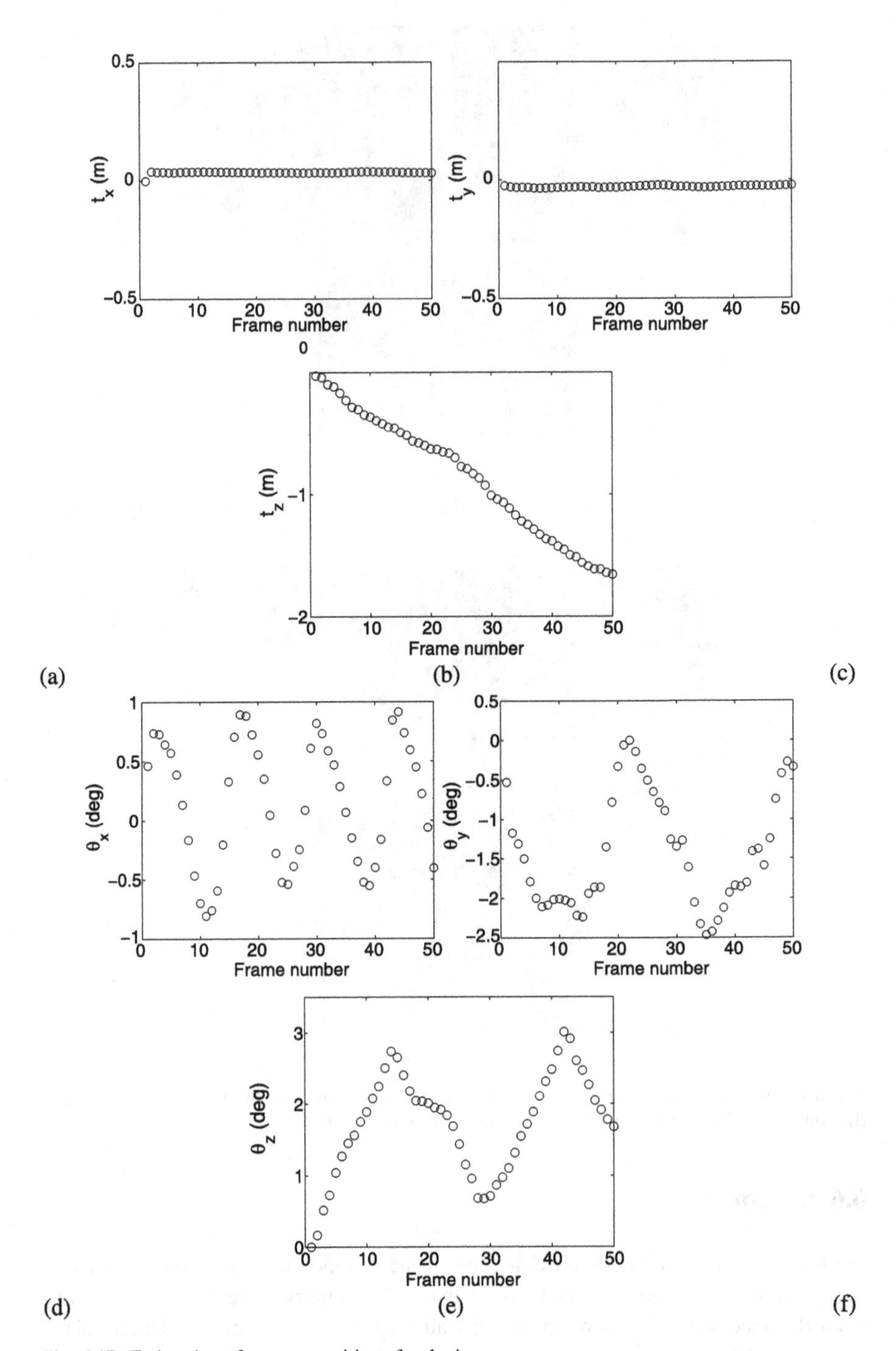

Fig. 6.17 Estimation of camera positions for the image sequence

(a) True 55th frame (b) Texture-mapped frame

Fig. 6.18 Comparison of a real frame and its texture-map in the image sequence

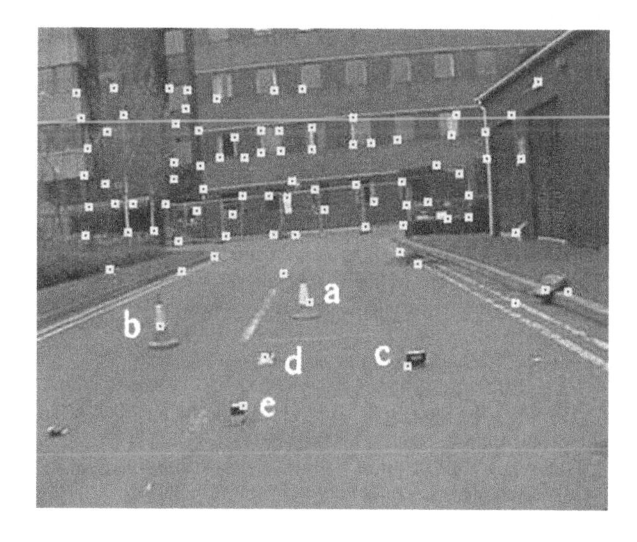

Fig. 6.19 Frame 55 and tracked feature points

Table 6.3 Residuals of estimating the ground plane

Features	Coordinates	Residuals (m)
a	(171.7, 163.7)	−2.5082
b	(81.2, 179.5)	−0.6047
c	(231.5, 202.2)	1.0351
d	(144.6, 196.9)	−0.5818
e	(131.6, 226.3)	0.0867

(i.e. hazards) and shadow or reflectance edges, so that greater dependence is placed on measurement of co-planarity and deviations from a ground plane. Indeed, the method assumes recovery of features from objects or textured data; if there is a featureless environment then some kind of active projection is necessary. In future work, we wish to extend the matching scheme to track feature points over several frames, using a model of gait to limit the search and 3-D recovery processes, and in due course to eliminate the need for the ground plane assumption. For effective deployment, it is also necessary to improve the efficiency, or to use parallel or hardware implementation of parts of the process, in order to achieve real time operation.

References

1. Kuan, D., Phipps, G., Hsueh, A.: Autonomous robotic vehicle road following. IEEE Trans. Pattern Anal. Mach. Intell., 648–658 (1988)
2. Hancock, J.: High-speed obstacle detection for automated highway applications. CMU-RI-TR-97-17, Pittsburgh, PA (1997)
3. Kunwar, F., Benhabib, B.: Rendezvous-guidance trajectory planning for robotic dynamic obstacle avoidance and interception. IEEE Trans. Syst. Man Cybern., Part B **36**(6), 1432–1441 (2006)
4. Zhu, Y., Comaniciu, D., Pellkofer, M., Koehler, T.: Reliable detection of overtaking vehicles using robust information fusion. IEEE Trans. Intell. Transp. Syst. **7**(4), 401–414 (2006)
5. Hatsopoulos, N., Gabbiani, F., Laurent, G.: Elementary computation of object approach by a wide-field visual neuron. Science **270**, 1000–1003 (1995)
6. Gehrig, S., Stein, F.: Collision avoidance for vehicle-following systems. IEEE Trans. Intell. Transp. Syst. **8**(2), 233–244 (2007)
7. Maylor, E., Wing, A.: Age differences in postual stability are increased by additional cognitive demands. J. Gerontol., 43–54 (1996)
8. Lacey, G., Dawson-Howe, K.: The application of robotics to a mobility aid for the elderly blind. Robot. Auton. Syst., 245–252 (1998)
9. Baker, T.: Automatic mesh generation for complex three-dimensional regions using a constrained Delaunay triangulation. In: Engineering in Computers, pp. 161–175. Springer, Berlin (1989)
10. Lei, G.: Recognition of planar objects in 3-d space from single perspective views using cross ratio. IEEE Trans. Robot. Autom., 432–437 (1990)
11. Delaunay, B.: Sur la sphere vide. Izv. Akad. Nauk SSSR, VII Seria, Otdelenie Matematicheskii i Estestvennyka Nauk, 793–800 (1934)
12. Shewchuk, J.: Lecture notes on Delaunay mesh generation. citeseer.nj.nec.com/shewchuk99lecture.html (1999)

13. Li, J., Brady, F., Reid, I., Hu, H.: Parallel image processing for object tracking using disparity information. In: Second Asian Conference on Computer Vision (ACCV'95), pp. 762–766 (1995)
14. Molton, N., Se, S., Brady, J., Lee, D., Probert, P.: A stereo vision-based aid for the visually impaired. Image Vis. Comput. **16**, 251–263 (1998)
15. Zhou, H., Wallace, A.M., Green, P.R.: A multistage filtering technique to detect hazards on the ground plane. Pattern Recognit. Lett. **24**, 1453–1461 (2003)
16. Horn, B.: Robot Vision. MIT Press, Cambridge (1986)
17. Sutherlan, D., Olshen, R., Biden, E., Wyatt, M.: The Development of Mature Walking. Blackwell Scientific Publications, Oxford (1988)
18. Shi, J., Tomasi, C.: Good features to track. In: IEEE Conf. on Comput. Vis. Pattern Recogn., pp. 593–600 (1994)
19. Zhou, H., Green, P., Wallace, A.: Efficient motion tracking using gait analysis. In: Proc. of International Conference on Acoustics, Speech, and Signal Processing (2004)
20. Zhang, Z., Deriche, R., Faugeras, O., Luong, Q.: A robust technique for matching two uncalibrated images through the recovery of the unknown epipolar geometry. Artif. Intell. **78**, 87–119 (1995)
21. Duda, R., Hart, P.: Pattern Classification and Scene Analysis. Wiley, New York (1973)
22. Scargle, J.: Studies in astronomical time series analysis, II: statistical aspects of spectral analysis of unevenly spaced data. Astrophys. J. **263**, 835–853 (1982)
23. Zhou, H., Wallace, A., Green, P.: Efficient tracking and ego-motion recovery using gait analysis. Signal Process. **89**(12), 2367–2384 (2009)

13. ... Brady, ... [illegible] ... information ... [illegible] ... 26 (1993)
14. ... N., ... [illegible] ... (1998)
15. ... Xu, W. ... [illegible] ... A multi... [illegible] ... obstacle ... the ground plane. Pattern Recognit. ... 24, 1451– ... (2003)
16. ... R., ... [illegible] ... Cambridge ... [illegible]
17. Sutherland, ... [illegible] ... (ed.) ... SPIE Scientific Publications ... (1985)
18. ... J., Tomasi, C.: Good features to track. In: ... Conf. on Computer Vision and Pattern Recognition, pp. 593– ... (1994)
19. Zhang, ... [illegible] ... analysis ... fundamental ... Signal Processing (200...)
20. Zhang, Z., Deriche, R., Faugeras, O., Luong, Q.: A robust technique ... two uncalibrated images through the recovery of the unknown epipolar geometry. ... 78, ... 119 (1995)

Duda, R.O., Hart, P.E.: ... [illegible] ... Wiley, New York ...
Scharstein, D., Szeliski, R.: ... [illegible] ... stereo correspondence algorithms. Int. J. ... 47(1), 7– ... (2002)
... Wang, ... [illegible] ... Effect of ... [illegible] ... using ... analysis ... (12), 23, 7... (2009)

Chapter 7
Learning and Vision-Based Obstacle Avoidance and Navigation

Jiandong Tian and Yandong Tang

Abstract A novel algorithm for camera calibration and correction is proposed in this chapter. A model of camera distortion is built without any prior knowledge. The calibration parameters are obtained by optimizing an objective function about the sum of the back projection errors using the LM algorithm. Also the distorted images are corrected by using the LM algorithm. A comparative study based on both synthetic data and real images corrupted by noise shows that the proposed algorithm successfully calibrated and corrected the distorted image.

7.1 Introduction

Robotic vehicles can perform exploration, search, rescue, and data collection in difficult or hazardous areas such as nuclear pollution, war, or extreme environmental conditions, *e.g.*, environments beyond our planet. Some particular purpose robots are shown in Fig. 7.1. Obstacle avoidance and autonomous navigation are fundamental abilities for mobile robots. They are also challenging issues, especially in unstructured outdoor environments. Mobile robots are dependent on sensory information for their operation. Popular sensors for range-based obstacle detection systems include laser rangefinders [28], ultrasonic sensors [8], and radar [2]. When it comes to the vision-based methods, the majority of them use binocular vision (using two cameras working together [18, 22], and optical flow [4]. However, none of these sensors is perfect. Ultrasonic sensors are cheap but suffer from cross talk and usually from limited sampling rate. Laser rangefinders and radars provide better resolution

---

J. Tian (✉) and Y. Tang
State Key Laboratory of Robotics, Shenyang Institute of Automation, Chinese Academy of Sciences, Shenyang, China
e-mail: tianjd@sia.cn; ytang@sia.cn

J. Tian
Graduate School of the Chinese Academy of Sciences, Beijing, China

H. Liu et al. (eds.), *Robot Intelligence*,
Advanced Information and Knowledge Processing,
DOI 10.1007/978-1-84996-329-9_7,

Fig. 7.1 Robotics in State Key Laboratory of Robotics, Shenyang Institute of Automation, Chinese Academy of Sciences, Shenyang, P.R. China

but are more expensive. They may bring dangers to human and animals and are of high power-consumption. While camera sensors have obvious advantages such as lightness, power saving, as well as provide rich information that make cameras suitable to be embedded in any robot, they cannot directly provide depth information. This chapter focuses on vision-based methods for obstacle avoidance and navigation. The main motivation comes from the fact that human and animals use their vision systems for walk in their world. The recent DARPA challenges (Learning Applied to Ground Robots 2004–2008; Grand Challenges 2005, 2006; Urban Challenge 2007) have emphasized the need for utilizing visual information in real-world applications.

For a robot to operate reliably in real world, it should have the ability to first perceive obstacles. For vision-based methods, they are classified into binocular vision and monocular vision. Binocular stereo vision is widely used in the most applications. It requires matching corresponding points in multiple images to reconstruct 3D information according to David Marr's theory [13]. However, looking for corresponding points is a challenging issue that has not been solved well until now. Furthermore, this technique assumes that camera parameters do not change. However, modern cameras tend to adjust their parameters automatically in order to acquire clearer images. In this situation, calibration has to be repeated every time the scene or the camera state changes. Due to the above-mentioned difficulties in binocular vision, some researchers start to pay attention to monocular vision recently. Y. Lila *et al.* [11] use multi-neural network to obtain object's depth ordering in monocular image. Jeff Michels *et al.* [14] proposed an algorithm that learns relative depth and can drive a car using only monocular visual cues on single images of outdoor environments. After supervised learning, they model the car control problem as a Markov decision process. Ashutosh Saxena *et al.* use a supervised learning and Markov Random Fields (MRFs) approach to recover depth in outdoor environments [19], and employ supervised learning and MRFs to learn the relationship between several monocular cues and depth and further to estimate 3D structure [20].

For the vision-based autonomous navigation, visual odometry and simultaneous localization and mapping are two popular techniques. Visual odometry [21] is the process of determining the position and orientation of a robot by analyzing the associated camera images. It has been used in a wide variety of robot applications, such as on the Mars Exploration Rovers [12]. Simultaneous localization and mapping (SLAM) [10, 24] is a technique for robots and autonomous vehicles to build up a map within an unknown environment that is used to calculate their current position. The SLAM is usually applied when the robot lacks a global positioning sensor. However, it has some problems that has been not solved effectively, such as dead reckoning, noisy sensory measurements, failure data association. These problems cause uncertainties in SLAM that could be solved in high-dimensional space, and require complex computation. Human being cannot perceive the accurate depth using their eyes, but they can do well in obstacle avoidance and path planning, even though with one eye. We believe that, with monocular vision and only with relative depth, it is feasible to find the passable path. In viewpoint of depth, robot can choose a path whose depth is deepest to pass through. In this chapter, our focus is on learning depth from a single camera inputs. We specifically focus on two topics: (a) Finding passable regions from a single still color image. (b) Making the robot vision less sensitive to illumination changes.

7.2 Depth Perception

Depth perception is the ability to perceive distance in three dimensions. It arises from a variety of depth cues. These cues are typically classified into binocular cues that provide depth information when viewing a scene with both eyes, and monocular cues that allow us to perceive depth with just one eye.

7.2.1 Absolute Depth and Binocular Vision

Absolute Depth Absolute depth is the accurate distance from objects to observer. In computer vision, absolute depth is calculated from binocular cues. As shown by the left illustration in Fig. 7.2, our eyes are spaced about 7 cm apart. The left and

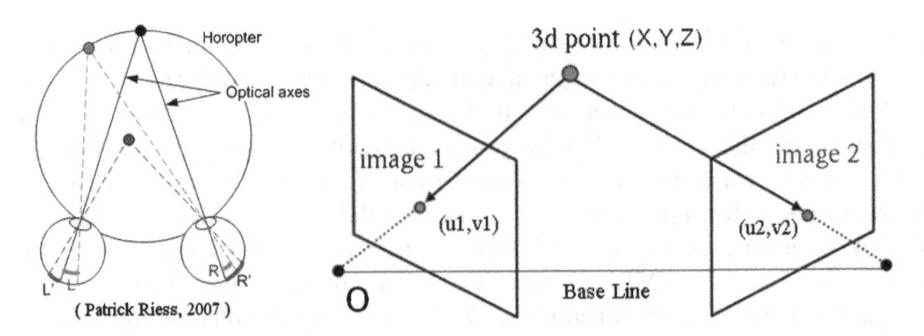

Fig. 7.2 Binocular disparity and 3D perception of human beings and cameras

right retinas receive slightly different images. This difference in the left and right images is called binocular disparity. Human brain integrates these two images into a single three-dimensional image, allowing us to perceive depth and distance.

As shown by the right illustration in Fig. 7.2, by using two images of the same scene obtained by two cameras from slightly different angles, the distance to an object can be triangulated with a high accuracy. A three dimensional point (X, Y, Z) in real world is projected to image 1 at pixel $(u1, v1)$ and to image 2 at pixel $(u2, v2)$ respectively. The projection matrix is $M1$ and $M2$, respectively (both $M1$ and $M2$ are 3×4 matrix, and can be gotten by camera calibration methods) [38]. The relationship between homogeneous coordinates of projected pixels and 3D point can be denoted as:

$$s1\begin{bmatrix} u1 \\ v1 \\ 1 \end{bmatrix} = M1\begin{bmatrix} X \\ Y \\ Z \\ 1 \end{bmatrix}, \tag{7.1}$$

$$s2\begin{bmatrix} u2 \\ v2 \\ 1 \end{bmatrix} = M2\begin{bmatrix} X \\ Y \\ Z \\ 1 \end{bmatrix} \tag{7.2}$$

where $s1$ and $s2$ are two scale factors and can be removed by row dividing. There are four equations by combing (7.1) and (7.2) with three unknowns. So, (X, Y, Z) can be solved. This process is called as binocular vision or stereo vision. Many binocular vision-based algorithms on depth computation and terrain 3D reconstruction for robot automatic navigation have been proposed in the past two decades. D. Murragy *et al.* [15] and G.N. Desouza *et al.* [5] presented several real-time stereo vision approaches for mobile robot navigation; C.F. Olson *et al.* [16, 17] presented a feature matching algorithm using a probabilistic formulation and applied it on the wide base-line stereo system of mars rover.

In Fig. 7.3, we show the reconstructed 3D terrain which can be used in obstacle avoidance and navigation. However, binocular disparity is only effective over a fairly short range. In Fig. 7.4, we show that as distance increases, the accuracy of 3D reconstruction decreases sharply. The denotations in Fig. 7.4 are listed as follows:

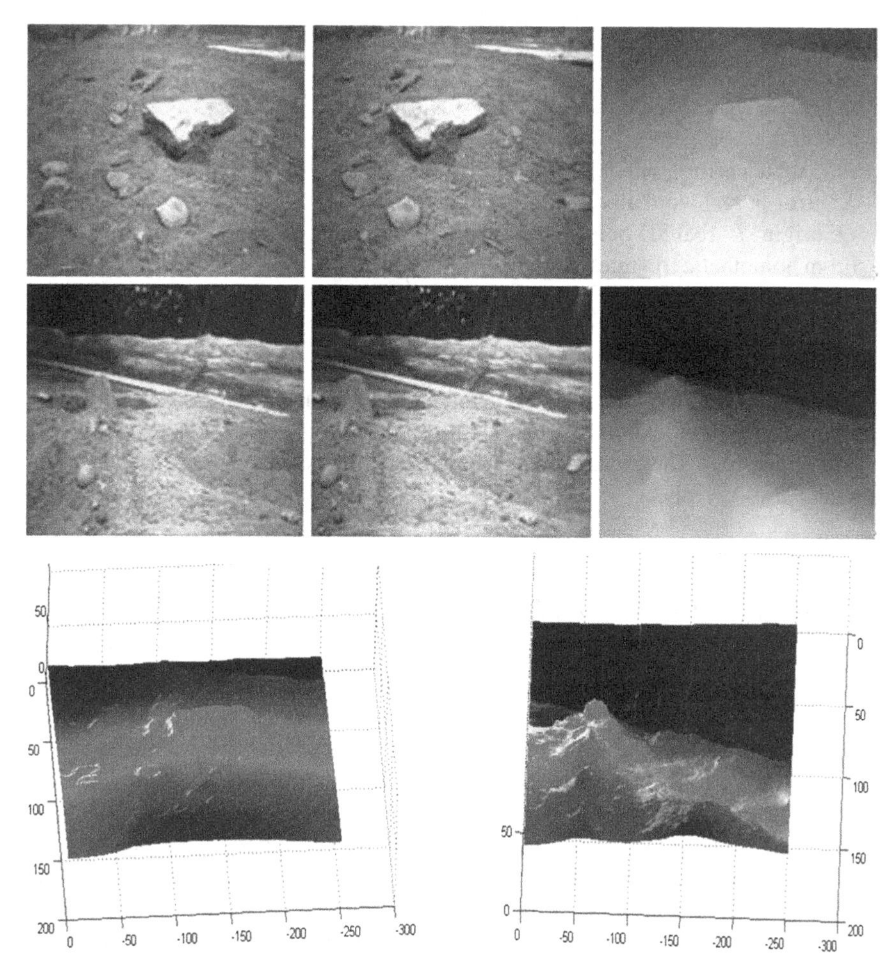

Fig. 7.3 In the *first row*, the first two images are captured by panoramic camera pair and the last is the disparity result. The images at the *second row* are captured by risk avoidance camera and the disparity result. The figures in the *third row* are reconstructed 3D terrain of the *first row*

Fig. 7.4 Relationship between 3D reconstruction accuracy and depth

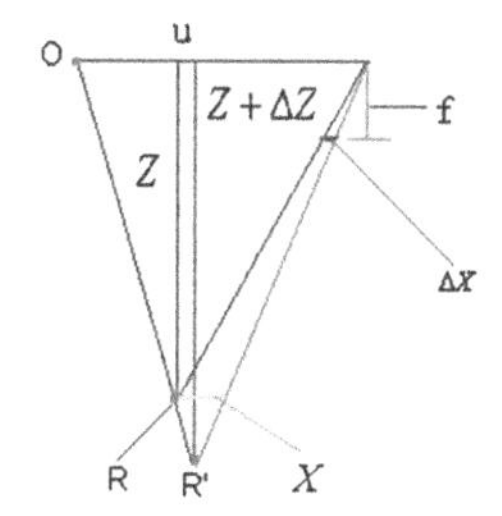

f: focus of the camera;
Z: real depth;
ΔZ: error of recovered depth;
u: length of base line;
ΔX: match error along base line direction;
X: error in real word along base line direction;
O: origin; R: real 3D point; R': recovered 3D point.
From homothetic triangles, we have $X = \frac{Z*\Delta X}{f}$, $\frac{\Delta Z}{\Delta Z+Z} = \frac{X}{u}$, and further yielding

$$\Delta Z = \frac{Z^2 * \Delta X}{uf - Z * \Delta X}. \tag{7.3}$$

Matching pixel without any error is impossible. From (7.3), we can find that if a pixel matching error ΔX exists, the depth recover error ΔZ is proportional to the square of depth Z. Therefore, as distance increases, the accuracy of 3D reconstruction decreases sharply.

7.2.2 Relative Depth and Monocular Vision

Relative depth is the relative distance of observed object depth position compared to other objects in a scene. It allows an observer to create a "ranking" of relative nearness (*e.g.*, the blue car is between my car and the red car). We can perceive the relative distance of the objects within our sight, even with one eye. Some cues on the scene in sight, such as object size, perspective, occlusion and motion cues, help us to perceive the object relative distance from us. These depth cues can be also used in monocular vision for depth perception. That is, these depth cues or features within an image can be used to perceive the object relative depth. The important depth cues within an image are interposition, edge direction and linear perspective, texture gradient, size cues, height cues, and motion parallax.

Edge Direction and Perspective Some edges in some directions on an image can tell the depth growing trend as illustrated in Fig. 7.5. Perspective is the special

Fig. 7.5 The edges labeled with red lines can tell the depth growing direction

Fig. 7.6 Closer objects show more detail, articulation than those farther away

Fig. 7.7 Relationship between pixel position and depth. (**a**) a real image; (**b**) pin-hole model

case of this situation (two parallel edges converging at a vanishing point). The most widely known example of this phenomenon is the illusion of a pair of railroad tracks receding into the distance-the two rails appear to grow smaller and closer together and finally to converge at infinity.

Clarity of Detail and Texture Gradient Near scenes show more details than farther ones. As shown in Fig. 7.6, near flowers can be clearly seen in terms of shape, size and color. As your vision shifts towards the distant road the texture cannot be clearly differentiated. We can also notice that the apparent texture of the narrow road changes over distance. The texture of the road near observer appears more detailed than that farther away. Texture change is a good cue for depth perception. When objects are placed at different locations along a texture gradient, judging their distance from you becomes fairly easy.

Pixel Position: As shown in Fig. 7.7, the pixel position in 2D image has a strong relation to depth in 3D structure. If ground is planar, then the distance is a monoton-

Fig. 7.8 As objects become smaller, they appear to recede into the distance or move farther away; objects that appear to be getting larger seem to be coming closer

Fig. 7.9 Closer objects appear to move faster than those farther away

ically increasing function of the pixel height in an image. This phenomenon can be explained by the image formation principle (perspective projection of the pin-hole model) of a camera. As shown in Fig. 7.7(b), an object nearer to observers occupies lower position in images.

Size Cues Another visual cue to apparent depth is closely related to size constancy. Through experience, we become familiar with the standard size of certain objects, such as, people, animals, cars, and houses. Knowing the size of these objects helps us judge our distance from them. As shown in Fig. 7.8, objects that appear to be getting smaller when moving farther away.

Motion Depth Cues This effect can be seen clearly on a moving car: nearby things pass quickly, while far objects appear stationary. Some animals that lack binocular vision due to wide placement of their eyes employ such motion depth

Fig. 7.10 Interposition: Objects that are nearer should occlude objects that are farther away

cues more explicitly than human for depth perception (*e.g.* some birds bob their heads to achieve motion depth perception) as shown in Fig. 7.9.

Occlusion (Also Referred to as Interposition) It is probably the most important monocular cue that provides information about relative distance. As shown in Fig. 7.10, when one object is overlapped or partly blocked by another object from our view, we judge the covered object as the farther one from us. This depth cue is all around us-look around you and notice how many objects are partly obscured by other objects.

7.3 Why Learning and How to Learn for Monocular Visions

7.3.1 The Role of Experience

Experience in interacting with the real world is vital to perception. Without visual experience, our visual system does not develop properly. Looking at these two upside-down pictures of Mona Lisa in Fig. 7.11, man could not easily find out the difference between the two images, they seems similar. However, in Fig. 7.12, man can find their difference very easily from the right side up. The Role of context visual experience is useful because it creates memories of past stimuli that can later serve as a context for perceiving new stimuli. Thus, we can think of experience as a form of context that we carry around with us.

7.3.2 Learning Methods

Existing study in neural science has provided much knowledge about biological multilayer networks. The popular applied artificial networks include Feed-Forward

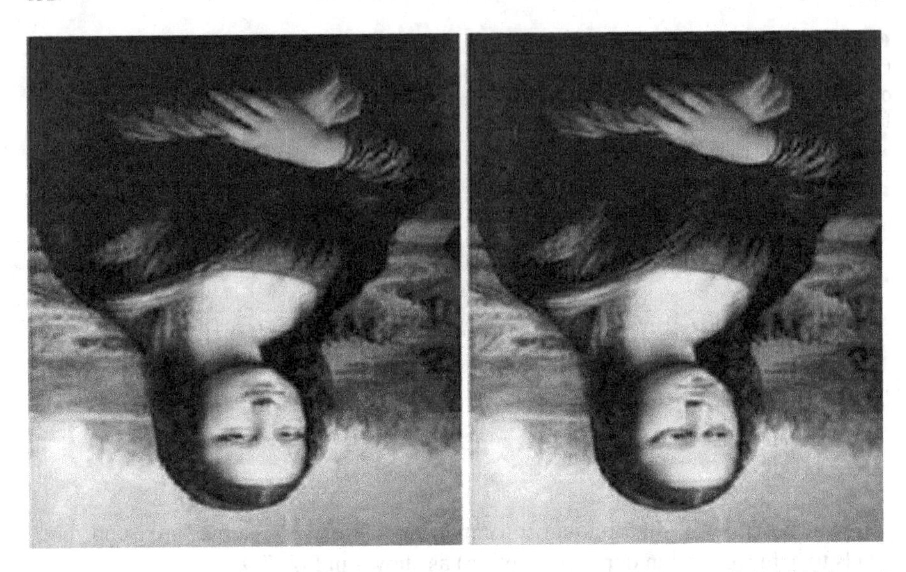

Fig. 7.11 Could you find the difference quickly? (http://www.exploratorium.edu/exhibits/mona/mona.html)

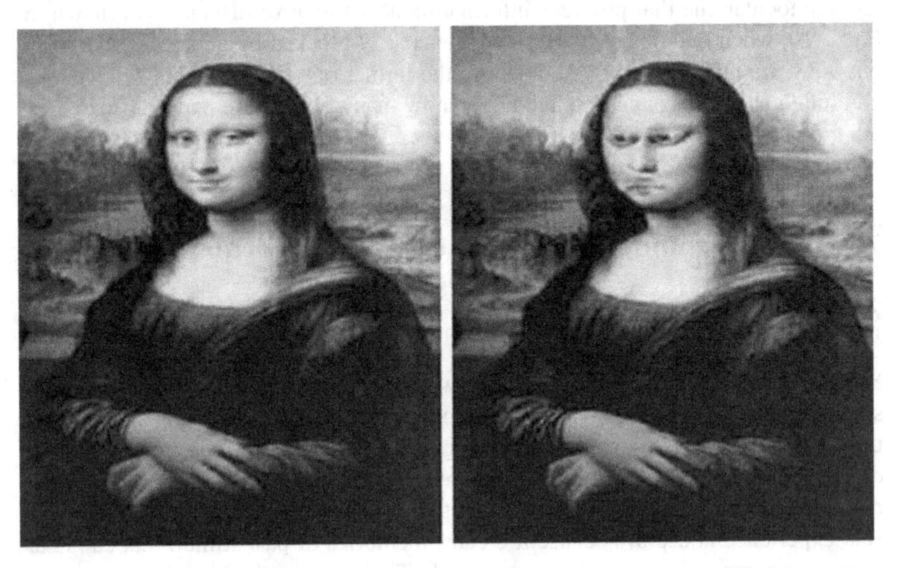

Fig. 7.12 You can find the difference very easily and detailed this time. (http://www.exploratorium.edu/exhibits/mona/mona.html)

Networks (FFN) with back-propagation learning, Radial Basis Functions (RBF), Error Back-Propagation, Linear Discriminant Analysis (LDA), Self-Organization Maps (SOM), Support Vector Machines (SVM), Cascade-Correlation Learning Architecture (CCLA) [6], and Incremental Hierarchical Discriminant Regression (IHDR) [9, 32]. Much research already exists in the field of learning using super-

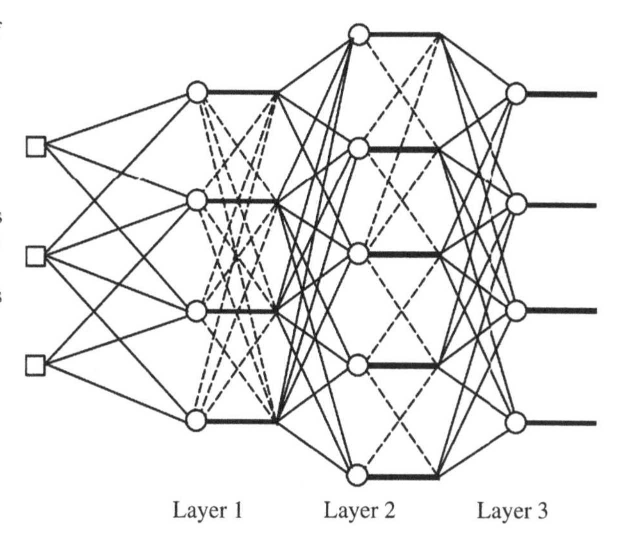

Fig. 7.13 The architecture of the Multi-layer In-place Learning Networks. A *circle* indicates a cell (neuron). The *thick segment* from each cell indicates its axon. The connection between a *solid* signal *line* and a cell indicates an excitatory connection. The connection between a *dashed* signal *line* and a cell indicates an inhibitory connection

vised, reinforcement, and unsupervised networks. For an autonomous vehicle operating in unstructured outdoor terrain, unsupervised learning is limited in use since in an unsupervised learning task, the system receives no feedback about desired outputs. The reinforcement learning in this domain consists of a reward or punishment, but the vision system receives no feedback on its outputs. Most of learning based navigation systems are supervised learning, which work well but generally need large amounts of training data.

In 2007, Juyang Weng *et al.* [33] proposed a biologically inspired network: Multilayer In-Place Learning Network (MILN). MILN is a biologically inspired network which is designed for autonomous mental development. Compared to popular artificial networks such as RBF, SOM, *etc.*, this network have some significant advantages such as simplicity, low computational complexity, free of local minima problem, enabling both unsupervised and supervised learning to occur concurrently, and in-place learning and so on. Gradient-based methods used by FFN, RBF, and SOM update the network along the greedy gradient direction computed by the last input data, without properly taking into account the observations along the past nonlinear search trajectory. In contrast, MILN takes into account all the observations along the nonlinear search path of every neuron. This can deal with the local minima problem (free from minima).

The architecture of MILN is shown in Fig. 7.13. The network takes a vector as input (*e.g.*, feature vector of image). The output of the network corresponds to control signals or classifications. It is a recurrent network. The output from each layer is not only used as input for the next layer, but is also feed back into other neurons in the same layer through lateral inhibition (dashed lines in the figure). For each neuron i, at layer l, there are three types of weights:

1. bottom-up (excitatory) weight vector ω_b that links input lines from the previous layer $l-1$ to this neuron;

2. lateral (inhibitory) weight ω_h that links other neurons in the same layer to this neuron;
3. top-down (excitatory or inhibitory) weight ω_p. It consists of two parts: (a) the part that links the output from the neurons in the next layer $l+1$ to this neuron. (b) The part that links the output of other processing areas (*e.g.*, other sensing modality) or layers (*e.g.*, the motor layer) to this neuron.

All the inputs to a neuron can be divided into three parts: bottom-up input from the previous layer y which is weighted by the neuron's bottom-up weight vector ω_b, lateral inhibition h from other neurons of the same layer which is weighted by the neuron's lateral weight vector ω_h, and the top-down input vector a which is weighted by the neuron's top-down weight vector ω_p. So, the response z from this neuron can be written as:

$$z = g(\omega_b \cdot y - \omega_h \cdot h + \omega_p \cdot a) \tag{7.4}$$

where g is nonlinear sigmoid function.

MILN Learning Initialize the time $t = 0$.

1. Grab the current input $x(t)$. Let $y_0 = x(t)$.
2. If the current desired output is given, set the output at layer l, $y_l \longleftarrow z(t)$.
3. For $j = 1, 3, \ldots, l$, run the lobe components analysis (LCA) algorithm (for detail, please refer to [33]) on layer j, $y_j = \mathrm{LCA}(y_{j-1})$, where layer j is also updated.
4. Produce output $z(t) = y_l$; $t \longleftarrow t+1$.

The importance of MILN is indicated by its conjunctive consideration of 8 challenging properties that are motivated by biological development and engineering applications:

1. high-dimensional (HD) input;
2. incremental learning;
3. without a significant local minima problem for regression (No Local Extrema, NLE);
4. without a significant loss of memory problem (Long Term Memory, LTM);
5. in-place learning;
6. enable supervised and unsupervised learning in any order suited for development;
7. local-to-global invariance from early to later processing (Soft invariant);
8. a rich completeness of response at the scale corresponding to a layer and the given layer resource (*e.g.*, the number of cells).

FFN suffers problems in (1), (3), (4), (5), (7), and (8). RBF has major problems with (3), (4), (7), and (8). SOM does not meet (7) and (8). CCLA is problematic in (1), (3), (5), (7) and (8). SVM is significantly limited in (1), (2) and does not meet (5), (6), (7) and (8). IHDR does well in (1) through (6) but does not satisfy (7) and (8). MILN introduced here is the only computational network designed for all Property (1) through (8). A summary of these comparison is provided in Table 7.1 [34].

Table 7.1 A comparison of major network models

Network	HD	Inc	NLE	LTM	In-place	Sup	Soft-Inv	Complete
FFN	N	Y	N	N	N	Y	N	N
RBF	N	Y	N	N	N	Y	N	N
SOM	Y	Y	Y	Y	Y	N	N	Y
CCLA	N	Y	N	Y	N	Y	N	N
SVM	N	N	Y	Y	N	Y	N	N
IHDR	Y	Y	Y	Y	N	Y	N	N
MILN	Y	Y	Y	Y	Y	Y	Y	Y

7.4 Special Problem: Illumination Changes in Outdoor Scenes

In vision (especially monocular vision) based depth perception and obstacle detection, lighting variance often brings great trouble to visual algorithm. For example, shadows may be detected as obstacles. In the recent DARPA challenges several approaches to illumination invariance on real-world mobile robot navigation [29] are used to improve the performance of robot vision systems. More recently, Finlayson *et al.* proposed projected method to derive an intrinsic image that not sensitive to illumination variance [7]. It relies upon finding a special direction in a 2D chromaticity feature space. When this "invariant direction" projected into 1D, produce a grayscale image which is approximately invariant to intensity and color of scene illumination. Since components for image formation involve uncertainty and nonlinearity, probabilistic and learning methods are becoming popular for shadow detection and illumination invariance. Tsin *et al.* [31] present a Bayesian MAP (maximum a posteriori) approach to achieve color constancy. Anzani *et al.* [1] describe another Bayesian approach for illumination invariance on wheeled robots. They use a Mixture of Gaussians (MoG) to model the distribution of each color. The labeling of color classes and association with mixture components is done by human supervision, and the Bayesian decision is used to determine the color label. Wu *et al.* employ the Bayesian learning approach [36] to extract shadows in a single image, but their method requires user interaction as input. In their work on color learning on legged robots, Sridharan and Stone [25, 27] model illuminations using autonomously collected image statistics. Mohan Sridharan and Peter Stone [26] gave a good survey of color learning and illumination invariance on mobile robots.

7.5 Finding Passable Regions for Obstacle Avoidance from Single Image Using MILN [30]

7.5.1 Feature Vector

We have shown in Sect. 7.2.2 that many image features can be used to predict relative depth. However, not all of them are easily to be applied. Taking size cues as

Fig. 7.14 Original image and its corresponding edge image

an example, object should be detected and recognized before measuring the size. In the following experiments for finding passable regions, we employ the edge, texture gradient, and pixel position as depth cues.

Edge Canny edge detector [3] is utilized to extract the edge features. The gotten edge image marked as binary image (1:white and 0:black) is as shown in Fig. 7.14.

Clarity of Detail and Texture Gradient Given a pixel (x, y) of image f, the extraction method of the clarity of detail and texture gradient is presented as following:

$$D(x, y) = f(x+1, y) + f(x-1, y) + f(x, y+1) + f(x, y-1) - 4 * f(x, y). \tag{7.5}$$

The result of D is binarized by the threshold T:

$$T = \frac{1}{M * N} \int\int D(x, y)dxdy \tag{7.6}$$

where $M * N$ is the size of image D. The output binary image is marked as image I_2 (1:white and 0:black) as shown in Fig. 7.15.

Color Similarity with Lighting Invariance Pixels with similar color may have the similar depth and passable property in an image. For example, in an images, the pixels of one upright tree or one vertical wall that have same depth from top to bottom in real world usually possess similar color values. Therefore, color similarity is a useful feature in relative depth measurement. Color similarity is sensitive to

Fig. 7.15 Original image and its corresponding clarity and detail image

lighting variance such as shadows or reflections. By far, there is not any method that can totally eliminate the lighting effect. Although there are many researches on illumination invariance, they are hard to be used practically because of the limitation of preconditions and the complexity of the algorithm. Here, we use popular normalized RGB to weaken the affection of shadows, based on observation that a shadow mainly change pixel intensity but seldom change the chromaticity [35].

$$\begin{bmatrix} r \\ g \\ b \end{bmatrix} = \begin{bmatrix} R/(R+G+B) \\ R/(R+G+B) \\ R/(R+G+B) \end{bmatrix} \tag{7.7}$$

where $[R, G, B]$is original color and $[r, g, b]$ is transformed color. The color similarity is calculated in normalized color space.

To measure the similarity, we select an initial region both in image I_1 and image I_2 as a reference. In the edge image I_1, we get line segment L_j for each column as follows: the line starts vertically from a point in the bottom of the image and ends at the point where it meets the first edge. The initial region R_1 in image I_1 is defined as:

$$R_1 = \left\{(x, y) \mid (x, y) \in \cup L_j \ (1 \le j \le c)\right\} \tag{7.8}$$

where c equals to the column number of image I_1. Defined δ as a scale parameter and T as the threshold, the initial region R_2 in image I_2 is composed of the points whose values are lower than δT in image

$$R_2 = \{(x, y) \mid D(x, y) < \delta T\}. \tag{7.9}$$

Here, δ is empirically set to 0.8; T equals to the threshold in (7.6). The initial region in the original RGB image is $R_1 \cap R_2$. The mean value of the initial region is calculated in each color channel as:

$$M(k) = \frac{1}{Area(R_1 \cap R_2)} \iint\limits_{R_1 \cap R_2} f(x, y, k)dxdy \tag{7.10}$$

where f is the original RGB image, $k = 1, 2, 3$ denotes the three color channels, and $Area(R_1 \cap R_2)$ the initial region in f:

$$Area(R_1 \cap R_2) = \iint\limits_{R_1 \cap R_2} dxdy. \tag{7.11}$$

The similarity between the initial region $Area(R_1 \cap R_2)$ and the rest parts in image f is described as:

$$I_3(x, y) = \begin{cases} 1 & \text{if } f(x, y, k) \in [T_1 M(k), T_2 M(k)], \\ 0 & \text{others.} \end{cases} \tag{7.12}$$

Here, 1 defined as similar and 0 as dissimilar. T_1 and T_2 are the thresholds to measure the degrees of similarity. We set T_1 and T_1 to be 0.9 and 1.1 respectively. In image I_3, all the pixels in the initial region $Area(R_1 \cap R_2)$ are set to 1 while others are determined by (7.12).

Pixel Position and Region Connection Because robots cannot stride across impassable regions or big obstacles, the connection between passable regions is taken into account in our method. For simplicity, we calculate the connection between one passable region and the below one next to it.

The edge, image clarity and details as well as color similarity with lighting invariance are the image content properties while pixel position and connection are the image geometry properties. We divided each image into equal 40×40 windows indexed by row in m and by column in n. 5 features are extracted in each window. Defined $H(m, n)$ as the feature vector, the three content properties are computed as following:

$$H(i, m, n) = \sum_{(x,y) \in window(m,n)} I_i(x, y), \quad i = 1, 2, 3. \tag{7.13}$$

For the two geometry properties:

$$\begin{aligned} H(4, m, n) &= m * n, \\ H(5, m, n) &= \sum_{i=1}^{3} H(i, m-1, n). \end{aligned} \tag{7.14}$$

After calculation, we get totally 1600 five-dimensional feature vectors for each image.

7.5.2 Training Data Generation and Experiment

We collect five hundred images downloaded from the internet, and most of them are different (including indoor and outdoor scenes, such as scene with sidewalks, building, grass, trees, even water, also including different lighting illumination as well as different weather conditions). Among these images, four hundred are used for training and the remaining one hundred for testing. Each training image is divided into 40×40 windows. For each window, five features described in Sect. 7.5.1 are extracted. 1600 five-dimensional feature vectors are obtained totally per image. The windows in each training image are marked as passable and non-passable regions manually. MILN combines the feature vectors as inputs and the corresponding passable regions as outputs for supervised learning. The trained MILN is to predict the passable regions of the testing images, which also takes the five-dimensional feature vectors as inputs and the passable regions as its outputs.

In the experiment, the MILN is of three layers. The number of neurons for each layer is $40 \times 40 = 1600$, $5 \times 5 = 25$, and 2, respectively. Every input neuron deals with one window, taking the 5-dimensional feature vector of the window as its input. The outputs are 0 and 1, where 0 denotes impassable regions and 1 denotes passable regions (shown in Fig. 7.16). In each window, if more than 50% pixels of the window belong to 0 or 1, the whole window is set to 0 or 1.

Figure 7.17 shows some results of passable regions found by our method. In order to be easily viewed, the output of MILN is superimposed on the original image and the passable regions are labeled with red color.

Fig. 7.16 Original image and its corresponding output from MILN

Fig. 7.17 (Color online) Some results of the experiments. The original images are on the *left*; the result images labeled with the passable regions by the *red zone* are on the *right*. Results (**a**), (**b**), (**c**), (**d**) show that our method can identify passable regions in different scenes, and can exclude obstacles from the passable regions. Result (**e**) shows that our method is not sensitive to shadow. Result (**f**) gives an example of outdoor scenes with complex texture ground

7.5.3 Performance Evaluation

We employ evaluation metrics introduced in [23, 37] for performance evaluation of our method. The evaluation metrics are defined as follows:

True positive (TP): the number of passable windows which are detected as passable windows.
True negative (TN): the number of non-passable windows which are detected as non-passable windows.
False positive (FP): the number of non-passable windows which are detected as passable windows.
False negative (FN): the number of passable windows which are detected as non-passable windows.
Correctness: $\frac{TP}{TP+FN}$.
Accuracy: $\frac{TP}{TP+FN+FP}$.
Branching: $\frac{TP}{TP}$.

The correctness metric is a measure of the correctly detected passable windows among all the passable windows. The accuracy metric reports the total accuracy of the method, which takes both *FP* and *FN* into account. The branching factor is a measure of the degree to which the system detects non-passable windows as passable windows. For a good vision system, correctness and accuracy should be high and branching factor should be low.

Just taking Fig. 7.18 as an example, in which, $TP = 491$, $TN = 1100$, $FP = 5$, $FN = 4$. So, it is calculated that correctness = 99.2%, accuracy = 98.2%, branching factor = 1%. Evaluation metrics of the results in Fig. 7.17 and Fig. 7.19 are tabulated in Table 7.2.

Experiment results showed that more than 90% of the tested images have fairly good response to the method (here, 'good' means correctness > 85%, ac-

Fig. 7.18 Passable property of each window

Fig. 7.19 Two examples of not good results

Table 7.2 Correctness, accuracy, and branching factor of results

	a	b	c	d	e	f	g	h
correctness	99.2%	98.8%	96%	96%	89.2%	97.1%	99.5%	75.6%
accuracy	98.2%	98.3%	93.2%	96%	88.3%	96%	94%	75.6%
branching factor	1%	0.6%	3%	0%	1.2%	1.1%	6%	0%

curacy > 85%, and branching factor < 5%), and two not good results were given in Fig. 7.19.

Figure 7.19 shows two results that are not good in the far field of the images. (g) has some false detect and (h) misses some passable regions. There are two reasons for the problem. On the one hand, in the false parts, the image quality is not high enough so that the features cannot be detected well; on the other hand, the features that we used may not be suitable for these two images, that is, the features do not have close relation to the depth perception of the scenes.

7.6 Control Law and Navigation

Once passable way or obstacles are detected, a mobile robot will take actions to avoid running into obstacles during navigation. The found passable regions can be used by the robot obstacle avoidance system to choose a sequence of actions to change the trajectory of the robot.

7.6.1 From Obstacle Boundaries to Motor Commands

To convert the output of the vision system into actual steering commands, a control policy must be developed. The robot should turn away from obstacles, and the degree of the turning depends on the depth information.

In Fig. 7.20, W is the width of image, and H is the height of image. A is the location of robot (for big robot or accurate control, the size of robot and the arrangement of camera on the robot should be considered). B is the center of the most top line of the image. Therefore, vector $\overline{AB}$ denotes the current direction of the robots. C is the largest depth in the passable regions, such that:

$$C(x, y) = \arg\max_{y}(\{(x, y) \mid (x, y) \in P\}) \tag{7.15}$$

where P denotes the passable regions.

The turning angle is determined by:

$$Turning\_angle = k \cdot \cos^{-1}\left(\frac{\overline{AB} * \overline{AC}}{|\overline{AB}| \cdot |\overline{AC}|}\right) \tag{7.16}$$

where k is a constant dependent on motor and $*$ denotes dot product. A trajectory of navigation is shown in Fig. 7.21.

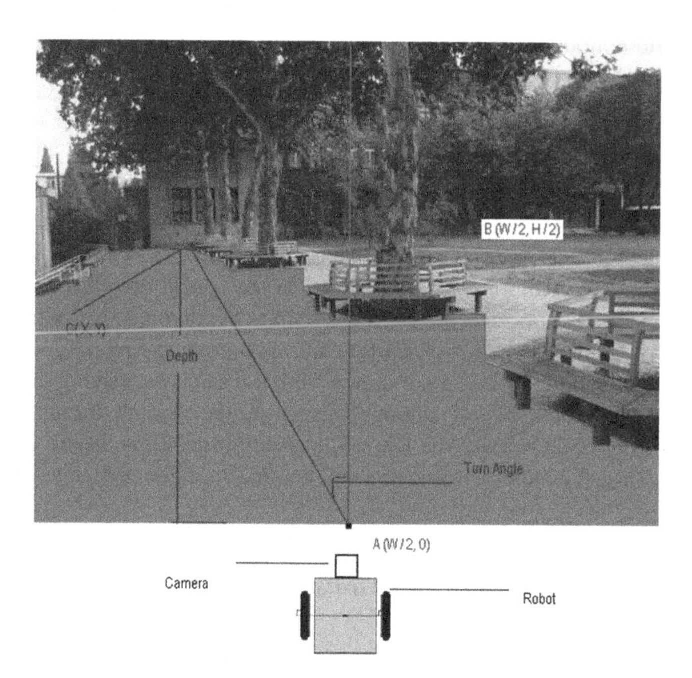

Fig. 7.20 From passable regions to steering commands

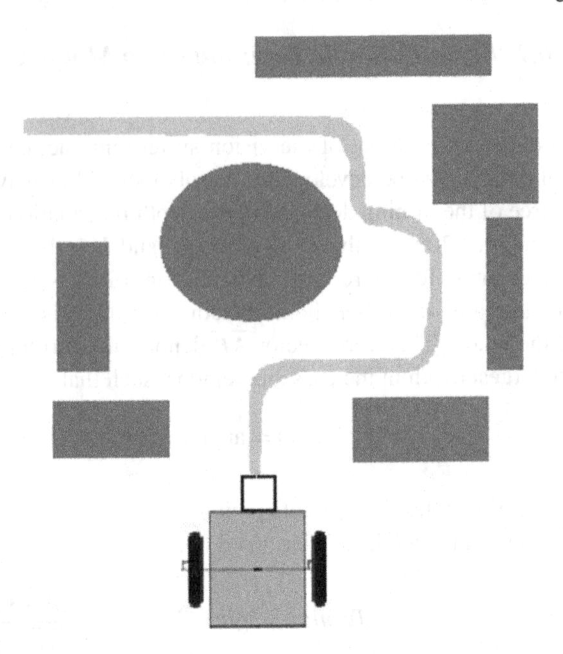

Fig. 7.21 Trajectory of navigation

7.7 Discussion

A mobile robot should be able to autonomously learn environmental features, to recognize environmental changes, and to adapt the learned models in response to such changes. Autonomous learning and adaptation on mobile robots is a challenging problem. Some basic issues need further study.

7.7.1 Learning Ability

The major challenge to mobile robots is the ability to learn, to adapt and to operate autonomously. A mobile robot deployed in real-world domains is frequently required to operate in new environments where the surroundings, including objects, may change dynamically. However, current learning systems generally assume that the training data and the test data are drawn from the same underlying distribution. When the new environments are sharply different to the train environments, learning methods may trend to fail. The theory on ability to learn, ability to generalize, network structure, *e.g.*, number of layers and neurons, remain challenge issues.

7.7.2 Changing Lighting Conditions

The sensitivity to illumination is a major challenge to the use of color on robot vision. As seen in Sect. 7.4, illumination invariance (*i.e.*, color constancy) has been a major research focus for years, but the problem is still far from being solved. Algorithms often make unrealistic assumptions, such as being able to measure the properties of the operating environment or knowing all possible illuminations ahead of time. Robot vision algorithms for on-board robot do not have robust performance under a range of illuminations. In complex environments, how to make mathematical model of the effects of illuminations whose parameters can be learned autonomously by the robot remains still a difficult problem.

7.7.3 Learning from Experience

For one application, we program our designed algorithm and our priori knowledge into robots. It is similar with that we open a robot's 'brain' and put our knowledge and intelligence into it. That is, we impose our intelligence into robots. Hence, machine intelligence today, actually, is human's intelligence rather than machine's. Robots should learn their knowledge by communicating with their environments, should promote its intelligence through experience. Unfortunately, we have little knowledge on how human learn from experience and communication. Just taking size constancy of an object for an example: When an object is near to us, its image on the retina is large. When the object is far away, its image on the retina is small. In spite of the image size on the retinal changes, we perceive the real object as the same size. For example, when you see a person at a great distance from you, you do not perceive that person as very small. Instead, you think that the person is of normal size and far away. People learn the general size of objects through experience and use this knowledge to judge object size and depth. Learning this experience seems very simple for human, while it is extremely difficult for computer vision. Learning from experience is still an open problem in artificial intelligence.

Acknowledgements The work in this chapter was supported in part by Natural Science Foundation of China (Grant Number: 60871078 and 60835004).

References

1. Anzani, F., Bosisio, D., Matteucci, M., Sorrenti, D.: On-line color calibration in non-stationary environments. Lect. Notes Comput. Sci. **4020**, 396 (2006)
2. Bertozzi, M., Bombini, L., Cerri, P., Medici, P., Antonello, P., Miglietta, M.: Obstacle detection and classification fusing radar and vision. In: IEEE Intelligent Vehicles Symposium, 608–613 (2008)
3. Canny, J.: A computational approach to edge detection. IEEE Trans. Pattern Anal. Mach. Intell. **8**(6), 679–698 (1986)
4. Coombs, D., Herman, M., Hong, T., Nashman, M.: Real-time obstacle avoidance using central flow divergence, and peripheral flow. IEEE Trans. Robot. Autom. **14**(1), 49–59 (1998)
5. DeSouza, G., Kak, A.: Vision for mobile robot navigation: a survey. IEEE Trans. Pattern Anal. Mach. Intell. **24**, 237–267 (2002)
6. Fahlman, S., Lebiere, C.: The cascade-correlation learning architecture. Technical report (1990)

7. Finlayson, G., Drew, M., Lu, C.: Entropy minimization for shadow removal. Int. J. Comput. Vis. **85**(1), 35–57 (2009)
8. Gutierrez-Osuna, R., Janet, J., Luo, R.: Modeling of ultrasonic range sensors for localization of autonomous mobile robots. IEEE Trans. Ind. Electron. **45**(4), 654–662 (1998)
9. Hwang, W., Weng, J.: Hierarchical discriminant regression. IEEE Trans. Pattern Anal. Mach. Intell. **22**(11), 1277–1293 (2000)
10. Lemaire, T., Berger, C., Jung, I., Lacroix, S.: Vision-based slam: stereo and monocular approaches. Int. J. Comput. Vis. **74**(3), 343–364 (2007)
11. Lila, Y., Lursinsap, C., Lipikorn, R.: Object's depth ordering in monocular image by using multi-neural network classification. In: SICE-ICASE International Joint Conference, pp. 587–592 (2006)
12. Maimone, M., Cheng, Y., Matthies, L.: Two years of visual odometry on the mars exploration rovers. J. Field Robot. **24**(3), 169–186 (2007)
13. Marr, D.: Vision. Freeman, New York (1982)
14. Michels, J., Saxena, A., Ng, A.: High speed obstacle avoidance using monocular vision and reinforcement learning. Proceedings of the 22nd International Conference on Machine Learning, pp. 593–600 (2005)
15. Murray, D., Little, J.: Using real-time stereo vision for mobile robot navigation. Auton. Robots **8**, 161–171 (2000)
16. Olson, C.: Maximum-likelihood image matching. IEEE Trans. Pattern Anal. Mach. Intell. **24**, 853–857 (2002)
17. Olson, C., Abi-Rached, H., Ye, M., Hendrich, J.: Wide-baseline stereo vision for Mars rovers. In: Proceedings of the IEEE/RSJ International Conference on Intelligent Robots and Systems, pp. 1302–1307 (2003)
18. Rosen, D.B., Rosen, A.: A robotic neural net based visual-sensory motor control system that reverse engineers the motor control functions of the human brains. In: Proceedings of International Joint Conference on Neural Networks, pp. 11–17 (2007)
19. Saxena, A., Chung, S., Ng, A.: 3-d depth reconstruction from a single still image. Int. J. Comput. Vis. **76**(1), 53–69 (2008)
20. Saxena, A., Sun, M., Ng, A.: Make3D: Learning 3D scene structure from a single still image. IEEE Trans. Pattern Anal. Mach. Intell. **31**(5), 824–840 (2009)
21. Scaramuzza, D., Siegwart, R.: Appearance-guided monocular omnidirectional visual odometry for outdoor ground vehicles. IEEE Trans. Robot. **24**(5), 1015–1026 (2008)
22. Scharstein, D., Szeliski, R.: A taxonomy and evaluation of dense two-frame stereo correspondence algorithms. Int. J. Comput. Vis. **47**(1), 7–42 (2002)
23. Shufelt, J.: Performance evaluation and analysis of monocular buildingextraction from aerial imagery. IEEE Trans. Pattern Anal. Mach. Intell. **21**(4), 311–326 (1999)
24. Sola, J., Monin, A., Devy, M., Vidal-Calleja, T.: Fusing monocular information in multi-camera SLAM. IEEE Trans. Robot. **24**(5), 958–968 (2008)
25. Sridharan, M.: Robust structure-based autonomous color learning on a mobile robot. Ph.D. thesis (2007)
26. Sridharan, M., Stone, P.: Color learning and illumination invariance on mobile robots: a survey. Robot. Auton. Syst. **57**(6), 629–844 (2009)
27. Stone, M.: Structure-based color learning on a mobile robot under changing illumination. Autonomous Robots **23**(3) (2007)
28. Swain-Oropeza, R., Devy, M., Hutchinson, S.: Sensor-based navigation in cluttered environments. In: IEEE International Conference on Intelligent Robots and Systems, vol. 3, pp. 1662–1669 (2001)
29. Thrun, S., Montemerlo, M., Dahlkamp, H., Stavens, D., Aron, A., Diebel, J., Fong, P., Gale, J., Halpenny, M., Hoffmann, G., et al.: Stanley: the robot that won the DARPA grand challenge. J. Field Robot. **23**(9), 661–692 (2006)
30. Tian, J., Dong, W., Tang., Y.: Finding passable regions from a single still image. International journal of computer visionrobotics and automation **14**(3) (2009)

31. Tsin, Y., Collins, R., Ramesh, V., Kanade, T.: Bayesian color constancy for outdoor object recognition. In: IEEE Computer Society Conference on Computer Vision and Pattern Recognition, p. 1 (2001)
32. Weng, J., Hwang, W.: Online image classification using IHDR. Int. J. Doc. Anal. Recognit. **5**(2), 118–125 (2003)
33. Weng, J., Luwang, T., Lu, H., Xue, X.: A multilayer in-place learning network for development of general invariances. Int. J. Humanoid Robot. **4**(2), 281–320 (2007)
34. Weng, J., Luwang, T., Lu, H., Xue, X.: Multilayer in-place learning networks for modeling functional layers in the laminar cortex. Neural Netw. **21**, 661–692 (2008)
35. Wren, C., Azarbayejani, A., Darrell, T., Pentland, A.: Pfinder: Real-time tracking of the human body. IEEE Trans. Pattern Anal. Mach. Intell. **19**(7), 780–785 (1997)
36. Wu, T., Tang, C.: A Bayesian approach for shadow extraction from a single image. In: IEEE International Conference on Computer Vision, pp. 480–487 (2005)
37. Yao, J., Zhang, Z.: Hierarchical shadow detection for color aerial images. Comput. Vis. Image Underst. **102**(1), 60–69 (2006)
38. Zhang, Z., et al.: A flexible new technique for camera calibration. IEEE Trans. Pattern Anal. Mach. Intell. **22**(11), 1330–1334 (2000)

Chapter 8
A Fraction Distortion Model for Accurate Camera Calibration and Correction

Yonghuai Liu, Ala Al-Obaidi, Anthony Jakas, and Junjie Liu

Abstract While camera calibration is a fundamental yet challenging problem for 3D measurement, it has attracted intensive attention from 3D vision community. In this paper, we propose a new model to characterise camera distortion in the process of the camera calibration. This model attempts to blindly characterise the overall camera distortion without taking the specific radial, decentring, or thin prism distortion into account. To estimate the parameters of interest, the well-known Levernburg-Marquardt algorithm is applied. To initialise the Levernburg-Marquardt algorithm, the results from the classical Tsai algorithm are estimated. After both the camera intrinsic and distortion parameters have been estimated, the distorted image points are corrected using again the Levernburg-Marquardt algorithm initialised by these distorted image points themselves. The performance of algorithms is measured as absolute and relative correction errors and collinear fitting errors. The experimental results based on both synthetic data and real images show that the proposed algorithm often successfully characterises the camera overall distortion, producing encouraging results for camera calibration and correction.

8.1 Introduction

The acquisition and analysis of 3D data attract more and more attention from the robot vision community for the representation of the objects of interest in the process of object modelling, classification and recognition [1]. Accurate camera calibration is a pre-requisite step for 3D metric measurement using either the latest laser scanning techniques or stereo vision systems. Without correction, the captured

Y. Liu (✉), A. Jakas, and J. Liu
Department of Computer Science, Aberystwyth University, Ceredigion SY23 3DB, UK
e-mail: yyl@aber.ac.uk; ajj08@aber.ac.uk; jul06@aber.ac.uk

A. Al-Obaidi
Smart Light Devices, Ltd., Aberdeen AB24 2YN, UK
e-mail: ala@sldltd.com

H. Liu et al. (eds.), *Robot Intelligence*,
Advanced Information and Knowledge Processing,
DOI 10.1007/978-1-84996-329-9_8,

data are distorted, which may give an illusion to the shape of the objects of interest. Subsequent analysis of such data is not accurate or even meaningless. As a matter of fact, there are a number of parameters (intrinsic, extrinsic and distortion parameters) to be calibrated. These parameters often interact with each other to achieve some objectives (e.g., minimisation of the average squared back-projection errors). Various models (Table 8.1) have been proposed to model the camera distortions, and calibration usually involves non-linear optimisation which is usually sensitive to both the initialisation and the parameters inside the non-linear optimisation algorithm such as the Levernburg-Marquardt algorithm. Consequently, accurate and stable camera calibration is challenging and still remains open.

8.1.1 Previous Work

The existing camera calibration techniques can be classified broadly into the following four main categories: calibration rig based, self calibration, autocalibration, and parameter free correction:

- The calibration rig based approaches require the knowledge of correspondences between the 3D calibration rig and its projective image. The rig contains either non-coplanar points [12, 13, 20] or coplanar points [7, 22]. While the approach in [22] is based on the rigidity constraint on the rigid rotation matrix, that in [16, 20] is based on the radial alignment constraint. The calibration usually involves two-steps. In the first step, a crude estimate of the parameters of interest is obtained often with closed form solutions. In the second step, all parameters of interest are globally and iteratively optimised through minimising some error functions. One of the most widely used objective functions is the sum of the squared back-projection errors;
- The self calibration techniques requires just a single projective image. The distortion parameters can be estimated using either the projective geometry [5, 6, 9] or in the frequency domain [8]. The former applies the property that the projections of lines are still lines in the corrected image. Thus, the curvatures of line segments in the distorted image are due to the camera distortion. In this case, various objective functions can be constructed: minimising distances from the points to best fit line segments [5], slope variation of the line segments [9], etc. In contrast, the latter observes that the Fourier transform of the signal before and after distortion is correlated which can be defined as the bicoherence. The first order radial distortion parameter can be estimated as one of the candidates sampled from an interval that minimises the bicoherence. The experimental results show that the frequency domain method is by no means comparable to that based on calibration rigs for camera calibration and correction;
- Autocalibration approaches require at least two projective images of the same scene and are often based on epipolar geometry [14, 18]. To calibrate the intrinsic camera parameters, the constraints on the plane at infinity and the quadrics in that plane are often constructed. Usually, autocalibration approaches do not consider

the camera distortion. Even so, it was observed in [3] that it is often difficult to estimate the focal length, the estimate of the principal point is often unstable, and there is an ambiguity in calibrating the focal length, the principal point, and the camera position; and

- Parameter free approaches [11] take only the radial distortion into account and estimate the distortion factor for each point. To estimate the distortion factors, a planar calibration grid is required and a constraint on the center of distortion and the homography **H** transforming calibration grid points in the Euclidean space into the undistorted image points in pixel coordinates is derived. Interestingly enough, the constraint is very similar to that on the fundamental matrix which results in the estimation of the first two rows of **H**. To estimate the third row of **H**, an assumption of monotonicity of distortion with regard to the radial distance is made. This assumption implies that the neighbouring undistorted points should not differ significantly whose squared difference can be minimised, resulting in the relative parameters being estimated in the least squares sense.

8.1.2 The Proposed Work

In this chapter, we propose a novel camera distortion model which attempts to blindly model the overall camera distortion. In this case, no knowledge is required about what the camera distortion is: radial, decentring, or thin prism. This attempt is practical, since in reality, we probably have little idea about what distortion the captured image is subject to. On the other hand, the novel model attempts to combat the imaging noise. This is very important to the subsequent data analysis, since all imaging devices introduce some amount of noise caused by point sampling, quantisation of measurements, reflective properties, etc.

In order to estimate the parameters in the novel camera distortion model, we employ the Levernburg-Marquardt (LM) algorithm to globally optimise all the parameters of interest through minimising the sum of squared differences between the transformed distorted projected 3D world points and their given distorted image points. Four parameters are used to model the image formation: focal length f, aspect ratio s, and the principal point (u_0, v_0). Seven parameters are used to model the camera orientation and position: a quaternion q is used to describe the camera orientation and a 3D vector $\mathbf{t}$ is used to describe the camera position in the world coordinate system. Five parameters are used to describe the camera distortions.

We implement the classical Tsai algorithm [20]. The resulting estimates of both the intrinsic and extrinsic parameters are then used to initialise the LM algorithm with the camera distortion parameters all set to zero. After both the camera intrinsic and distortion parameters have been obtained, the distorted image points are finally corrected using again the LM algorithm that minimises the squared difference between the distorted image points and the transformed corrected points. The LM algorithm is initialised by the distorted image points themselves. Since the distortion is generally small, the distorted and undistorted image points thus should not be too far apart from each other, implying that the distorted points often provide a good initialisation for their optimised correction.

For the sake of fair testing, we did not use the results from the Tsai algorithm directly for a comparative study. Instead, we employ again the LM algorithm to globally optimise its estimation results through minimising the sum of squared differences between the transformed distorted projected 3D world points and their given distorted image points with the camera distortion explicitly modelled as both radial and decentring ones in 4 parameters altogether. The improved Tsai algorithm is called the Tsai+LM algorithm in this chapter. Such a comparative study is valuable, since it can reveal whether the explicit knowledge of the camera distortion is beneficial to successful image point correction. The performance of camera calibration algorithms are measured from two aspects: (1) absolute and relative correction errors. This is in contrast with the commonly used average correction error which does not take into account the fact that the distortion in the middle area of images is little, while the distortion in the peripheral area is more pronounced. As such, the average correction error may not be informative to the performance of algorithms; and (2) the collinearity constraint on points. While the collinear points lie on curves in the distorted image, they should lie on the collinear line segments after correction. To get a further idea about the extent to which the proposed camera calibration and correction algorithm reduces the collinear fitting errors, we also implement for a comparative study the self-calibration algorithm [6] that operates for the calibration and correction of the distorted points in the projective image on the collinearity constraint alone. The collinear fitting errors are measured as: maximum fitting error (MFE), average fitting error (AFE), and the standard deviation of fitting errors (SDFE) of points on different line segments. The experiments based on both synthetic data and real images show that the proposed algorithm produces encouraging camera calibration and correction results.

The rest of this chapter is structured as follows: Sect. 8.2 proposes a novel camera distortion model, Sect. 8.3 proposes a novel camera calibration and correction algorithm, while Sect. 8.4 presents the experimental results based on both synthetic data and real images. Finally, Sect. 8.5 draws some conclusions and indicates future research directions.

8.2 A New Distortion Model

The following notations are used throughout this chapter: capital letters denote vectors or matrices, lower case letters denote scalars, $|\cdot|$ denotes the absolute value of a scalar, $\|\cdot\|$ denotes the Euclidean norm of a vector, $\mathbf{I}$ is an identify matrix, $\mathbf{a} \times \mathbf{b}$ denotes the cross product of vectors $\mathbf{a}$ and $\mathbf{b}$, $\det(\mathbf{A})$ denotes the determinant of the square matrix $\mathbf{A}$, the subscripts w and c denote the 3D points described in the world and camera centred coordinate systems, the subscripts f, u, and d denote the points in the frame buffer, undistorted, and distorted image points, respectively, variables with $\hat{}$ signs denote the corrected or estimated ones, and superscript T denotes the transpose of a vector or a matrix.

Table 8.1 Commonly used camera distortion models

Type	Model	Applicability
Radial polynomial model	$r_d = r_u(1 + \kappa_1 r_u^2 + \kappa_2 r_u^4 + \cdots + \kappa_p r_u^{2p})$	Projected/Image points
Radial model [17]	$r_d = r_u(1 + \kappa_1 r_u + \kappa_2 r_u^2)$	Projected points
Radial model [20]	$r_d = r_u(1 + \kappa_1 r_u^2)$	Projected points
Radial division model [10]	$\hat{P}_f = \frac{1}{1+\kappa_1 \lVert P_f \rVert^2} P_f$	Image points
Radial model [19]	$\hat{P}_f - C = (P_f - C)\sum_{i=0}^{d} \kappa_i \lVert P_f - C \rVert^i$	Image points
Radial rational model [4]	$\hat{P}_f = \left(\frac{A_1^T \chi(i,j)}{A_3^T \chi(i,j)}, \frac{A_2^T \chi(i,j)}{A_3^T \chi(i,j)}\right)$	Image points
Decentring model [21]	$x_d = p_1(r_u^2 + 2x_u^2) + 2p_2 x_u y_u$	Projected points
	$y_d = 2p_1 x_u y_u + p_2(r_u^2 + 2y_u^2)$	
Thin Prism model [21]	$x_d = s_1 r_u^2$	Projected points
	$y_d = s_2 r_u^2$	

Inspired by the various camera distortion models summarised in Table 8.1 proposed in the literature and with an attempt to blindly model the overall distortion the camera is subject to and combat the imaging noise for accurate camera calibration and correction, we propose the following camera distortion model:

$$\begin{aligned} x_{di} &= x_{ui}\frac{1 + \kappa_2 x_{ui} + \kappa_3 y_{ui}}{1 - \kappa_1 r_{ui}^2}, \\ y_{di} &= y_{ui}\frac{1 + \kappa_4 x_{ui} + \kappa_5 y_{ui}}{1 - \kappa_1 r_{ui}^2} \end{aligned} \tag{8.1}$$

where κ_1, κ_2, κ_3, κ_4, and κ_5 are unknown distortion parameters to be calibrated, $\mathbf{P}_{di} = (x_{di}, y_{di})^T$ is distorted image points, $\mathbf{P}_{ui} = (x_{ui}, y_{ui})^T$ is undistorted image points: $x_{ui} = f\frac{\mathbf{R}_1\mathbf{p}_{wi}+t_x}{\mathbf{R}_3\mathbf{p}_{wi}+t_z}$, $y_{ui} = f\frac{\mathbf{R}_2\mathbf{p}_{wi}+t_y}{\mathbf{R}_3\mathbf{p}_{wi}+t_z}$, $\mathbf{R}_1$, $\mathbf{R}_2$, and $\mathbf{R}_3$ are the three rows of the camera orientation matrix $\mathbf{R}$ in the world coordinate system, t_x, t_y and t_z are three components of the camera position $\mathbf{t}$, and $r_{ui}^2 = x_{ui}^2 + y_{ui}^2$ is the squared radial distance from the principal point.

This model which is called a fraction model, is applied to the projected 3D world points, and clearly has the property that when $r_u = 0$, the point has no distortion at all. With the increase of r_u, the distortion in the points also increases. This property conforms to the normal observation that the points in the middle of the image are subject to little distortion, while the distortion of the points in the peripheral area is more pronounced, as illustrated in Figs. 8.1 through 8.4.

8.3 A Novel Calibration Algorithm

In this section, we estimate the unknown parameters in the proposed camera distortion model (see (8.1)) described in the last section and then summarise the main steps in the proposed algorithm for camera calibration and correction. To this end,

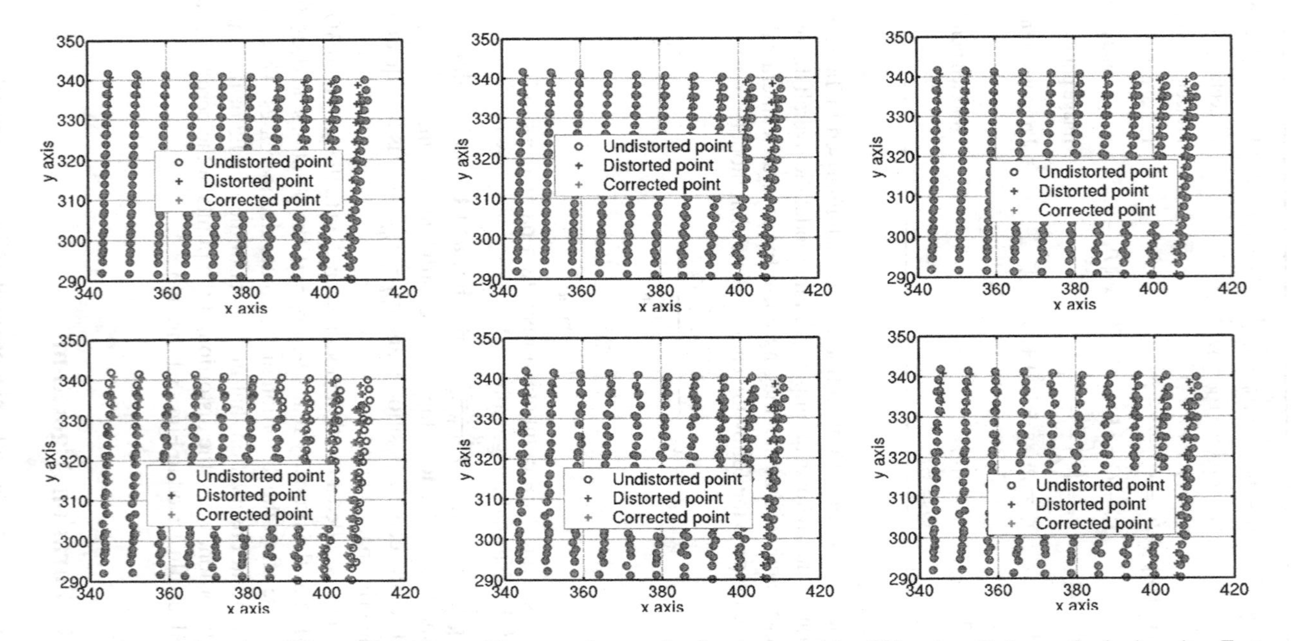

Fig. 8.1 The correction results using different algorithms with non-coplanar point data corrupted by different multiples τ of a basic noise. *Top row*: $\tau = 1$; *Middle row*: $\tau = 10$; *Bottom row*: $\tau = 20$. *Left column*: Self-calibration; *Middle column*: The Tsai+LM algorithm; *Right column*: FMC

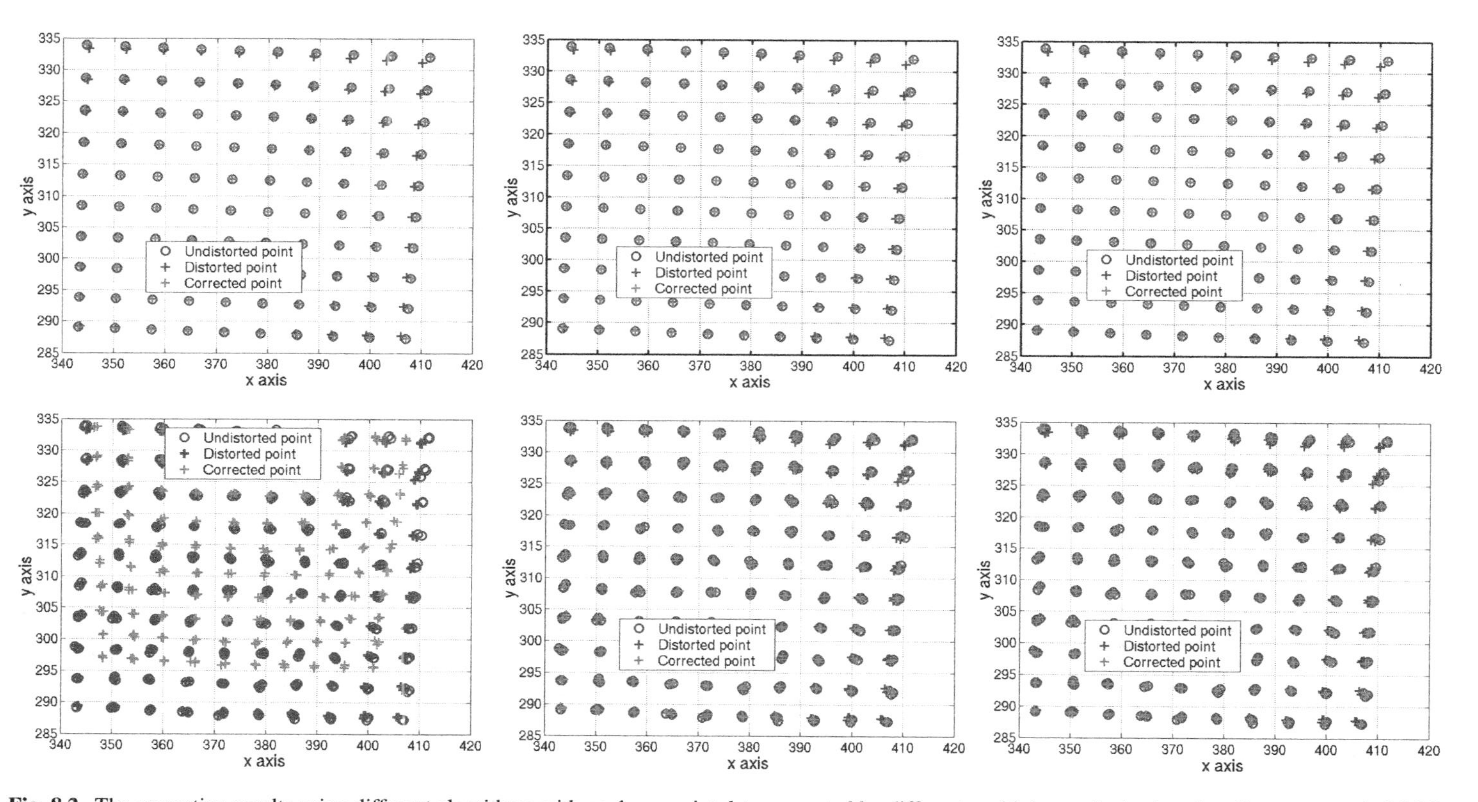

Fig. 8.2 The correction results using different algorithms with coplanar point data corrupted by different multiples τ of a basic noise. *Top row*: $\tau = 1$; *Middle row*: $\tau = 10$; *Bottom row*: $\tau = 20$. *Left column*: Self-calibration; *Middle column*: The Tsai+LM algorithm; *Right column*: FMC

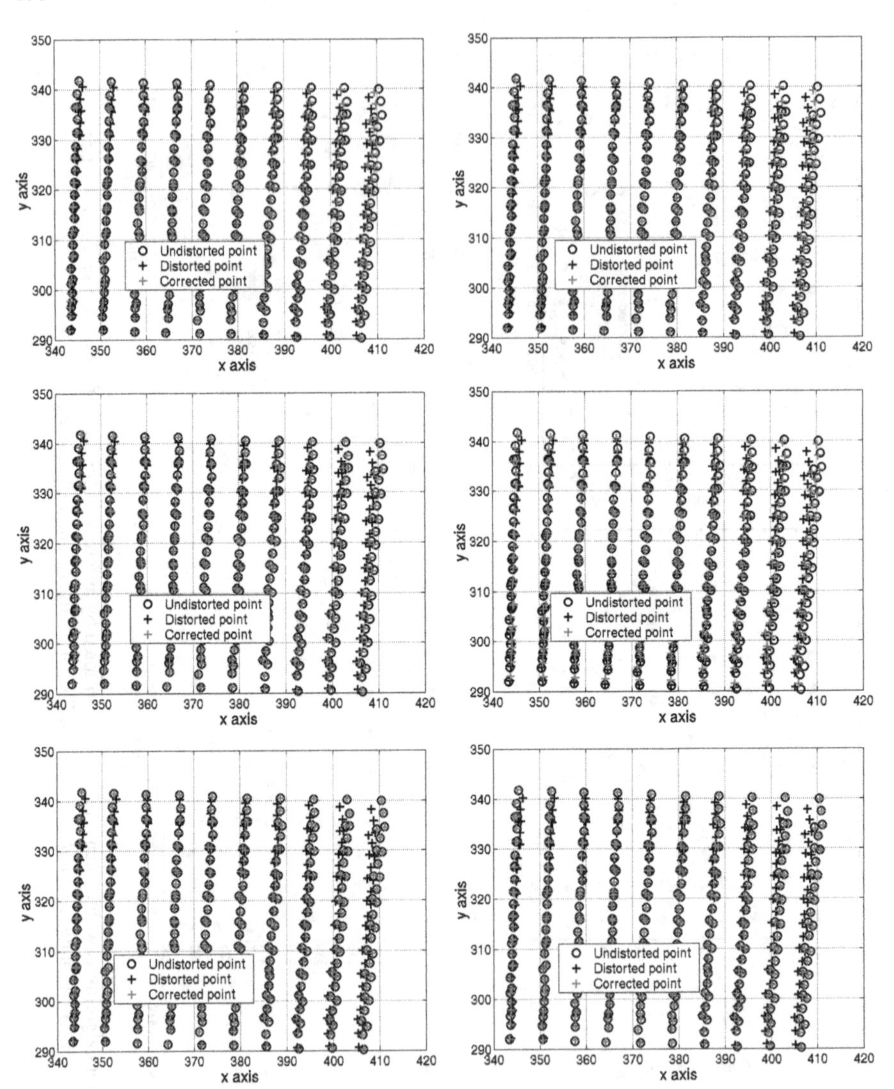

Fig. 8.3 The correction results using different algorithms with non-coplanar point data subject to more severe radial distortion as well as decentring distortion and corrupted by a multiple 10 of a basic noise. *Left column*: $\kappa_1 = -0.1$; *Right column*: $\kappa_1 = -0.12$. *Top row*: Self-calibration; *Middle row*: The Tsai+LM algorithm; *Bottom row*: FMC

we minimise the sum of the squared differences between the transformed distorted projected 3D world points $\mathbf{p}_{wi} = (x_{wi}, y_{wi}, z_{wi})^T$ and their given corresponding distorted projective image points $\mathbf{P}_{fi} = (x_{fi}, y_{fi})^T$ $(i = 1, 2, \ldots, n)$. The details of optimisation are given as follows.

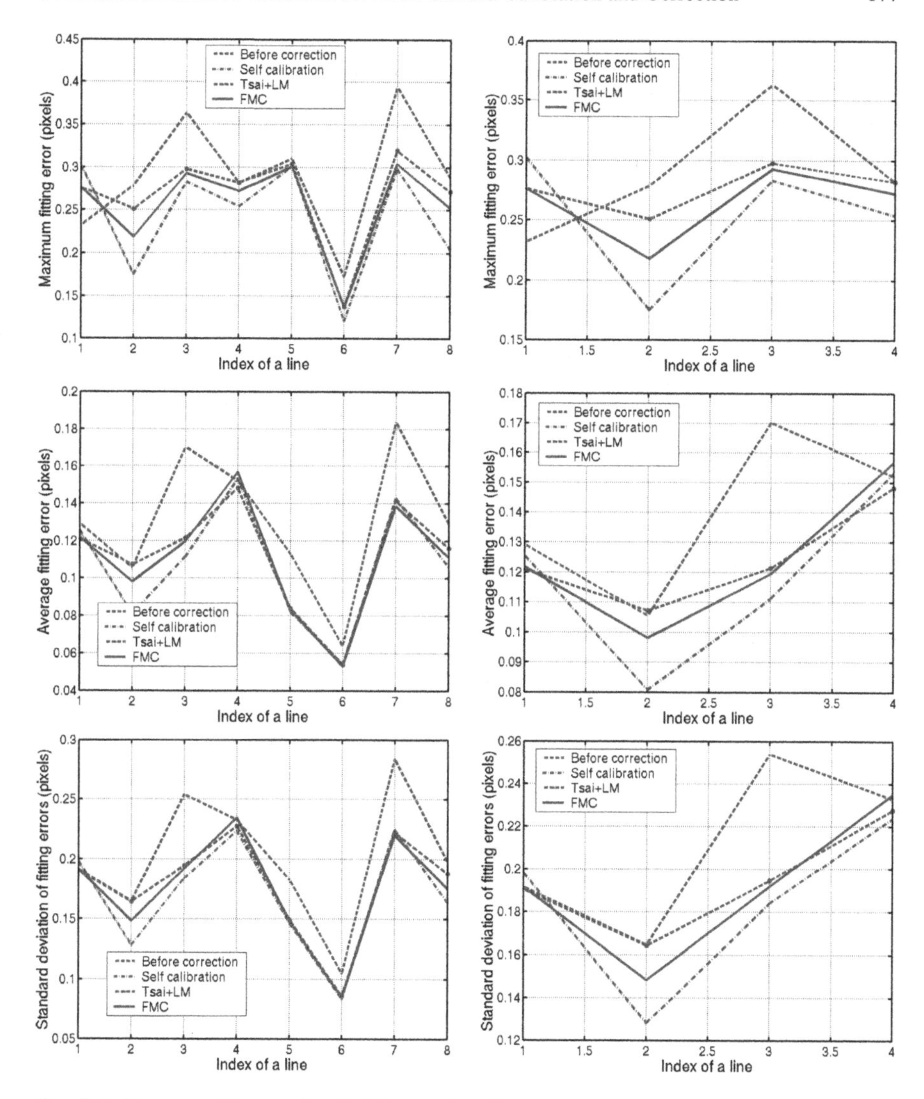

Fig. 8.4 The correction results of different algorithms measured by the collinearity constraint based on synthetic data corrupted by a multiple 10 of a basic noise. *Left column*: non-coplanar points; *Right column*: coplanar points. *Top row*: maximum fitting error (MFE); *Middle row*: average fitting error (AFE); *Bottom row*: standard deviation of fitting errors (SDFE)

8.3.1 Pin-Hole Camera Model

The relationship between a 3D world point $\mathbf{p}_{wi} = (x_{wi}, y_{wi}, z_{wi})^T$ and its image point $\mathbf{P}_{fi} = (x_{fi}, y_{fi})^T$ in the image plane without distortion can be represented as:

$$Z_{ci}\begin{pmatrix} x_{fi} \\ y_{fi} \\ 1 \end{pmatrix} = \mathbf{A}(\mathbf{R}\ \mathbf{t})\begin{pmatrix} \mathbf{p}_{wi} \\ 1 \end{pmatrix} = \mathbf{H}\mathbf{p}_{wi} + \mathbf{T} \tag{8.2}$$

where $\mathbf{H} = \mathbf{AR}$, $\mathbf{T} = \mathbf{At}$, matrix

$$\mathbf{A} = \begin{pmatrix} sf/dx & 0 & u_0 \\ 0 & f/dy & v_0 \\ 0 & 0 & 1 \end{pmatrix}$$

encodes the camera intrinsic parameters, s is the aspect ratio of the pixel, f is the focus length of the camera, dx and dy are the actual width and height of the pixel, (u_0, v_0) is the principal point, and Z_{ci} is the depth of 3D world point $\mathbf{p}_{wi}$ in the camera centred coordinate system. dx and dy are assumed to be known which can usually be found in the specifications of the cameras.

8.3.2 Optimisation of All Parameters

Taking the proposed camera distortion model equation (8.1) and the pin-hole camera model equation (8.2) into account, the distorted image points are modelled as:

$$\begin{aligned}
x_{ui} &= f\frac{\mathbf{R}_1\mathbf{p}_{wi} + t_x}{\mathbf{R}_3\mathbf{p}_{wi} + t_z}, \\
y_{ui} &= f\frac{\mathbf{R}_2\mathbf{p}_{wi} + t_y}{\mathbf{R}_3\mathbf{p}_{wi} + t_z}, \\
x_{di} &= x_{ui}\frac{1 + \kappa_2 x_{ui} + \kappa_3 y_{ui}}{1 - \kappa_1 r_{ui}^2}, \\
y_{di} &= y_{ui}\frac{1 + \kappa_4 x_{ui} + \kappa_5 y_{ui}}{1 - \kappa_1 r_{ui}^2}, \\
x_{fi} &= s x_{di}/dx + u_0, \\
y_{fi} &= y_{di}/dy + v_0.
\end{aligned} \tag{8.3}$$

Then altogether 16 parameters need to be optimised: 4 intrinsic parameters $\mathbf{IP} = (f, s, u_0, v_0)$, 7 extrinsic parameters $\mathbf{EP} = (\mathbf{q}, \mathbf{t})$ where a quaternion $\mathbf{q}$ is used to represent the camera orientation matrix $\mathbf{R}$ and three parameters are used to represent the camera position $\mathbf{t}$ in the world coordinate system, and 5 distortion parameters $\mathbf{DP} = (\kappa_1, \kappa_2, \kappa_3, \kappa_4, \kappa_5)$. For more accurate and efficient optimisation, the normalisation of the quaternion $\mathbf{q}$ is ignored: $\mathbf{q} \leftarrow \mathbf{q}/\|\mathbf{q}\|$. To optimise these 16 parameters, the following objective function is built:

$$J_1 = \min_{\mathbf{IP},\mathbf{EP},\mathbf{DP}} \sum_{i=1}^{n} ((\hat{x}_{fi} - x_{fi})^2 + (\hat{y}_{fi} - y_{fi})^2) \tag{8.4}$$

which minimises the differences between the re-projected 3D points $\hat{\mathbf{P}}_{fi} = (\hat{x}_{fi}, \hat{y}_{fi})^T$ in the image plane and given image points $\mathbf{P}_{fi} = (x_{fi}, y_{fi})^T$. The optimisation is performed using the Levernburg-Marquardt algorithm. The initialisation was provided using the intrinsic and extrinsic parameters estimated by the classical Tsai algorithm [20] with the distortion parameters all set to 0.

8.3.3 The Correction of the Distorted Image Points

From (8.1), it can be seen that given the distorted image points $\mathbf{P}_{fi} = (x_{fi}, y_{fi})^T$, generally, there is no closed form solution to the undistorted points $\mathbf{P}_{ui} = (x_{ui}, y_{ui})^T$. While some researchers turn to the approximation method [12, 13], in this chapter, we propose using the LM algorithm to correct the distorted points. To this end, the following objective function is built:

$$J_2 = \min_{\hat{x}_{ui}, \hat{y}_{ui}} (\hat{x}_{di} - x_{di})^2 + (\hat{y}_{di} - y_{di})^2 \tag{8.5}$$

where $\hat{x}_{di} = \hat{x}_{ui} \frac{1+\hat{\kappa_2}\hat{x}_{ui}+\hat{\kappa_3}\hat{y}_{ui}}{1-\hat{\kappa_1}\hat{r}^2_{ui}}$, $\hat{y}_{di} = \hat{y}_{ui} \frac{1+\hat{\kappa_4}\hat{x}_{ui}+\hat{\kappa_5}\hat{y}_{ui}}{1-\hat{\kappa_1}\hat{r}^2_{ui}}$, $\hat{r}^2_{ui} = \hat{x}^2_{ui} + \hat{y}^2_{ui}$, $x_{di} = (x_{fi} - \hat{u}_0)dx/\hat{s}$, and $y_{di} = (y_{fi} - \hat{v}_0)dy$, $\hat{\kappa_1}$, $\hat{\kappa_2}$, $\hat{\kappa_3}$, $\hat{\kappa_4}$, $\hat{\kappa_5}$, $\hat{s}$, $\hat{u}_0$ and $\hat{v}_0$ are the distortion and intrinsic parameters calibrated in the last section. The LM algorithm was initialised by the distorted image points. Since the undistorted points around the principal point are always very close to the distorted points and the undistorted points should not be far away from the distorted points, the distorted points thus often provide a good initialisation for their optimised correction. As long as both the intrinsic and distortion parameters have been calibrated with reasonable accuracy, then the objective function J_2 usually well poses the image correction problem, leading the distorted points to be accurately corrected.

8.3.4 Summary of the Novel Camera Calibration and Correction Algorithm

From (8.4), it is known that 15 camera intrinsic, extrinsic and distortion parameters need to be calibrated. From (8.2) and (8.3), it can be seen that a single 3D-2D correspondence provides two constraints. Consequently, at least 8 such 3D-2D correspondences are required for the calibration of these 15 parameters. Given a set of 3D-2D correspondences $(\mathbf{p}_{wi}, \mathbf{P}_{fi})$ $(i = 1, 2, \ldots, n | n \geq 8)$, the main steps in the proposed camera calibration and correction algorithm can be summarised as follows:

1. Use (8.1) to model the camera distortion;
2. Use the traditional Tsai algorithm [20] to estimate both the intrinsic and extrinsic parameters;

3. Apply the LM algorithm to minimise the objective function equation (8.4) for an optimal estimation of the camera intrinsic, extrinsic and distortion parameters. The LM algorithm is initialised by the intrinsic and extrinsic parameters that are estimated above by the Tsai algorithm and the distortion parameters that are all set to zero;
4. Apply again the LM algorithm to minimise the objective function equation (8.5) to correct the distorted points with the LM algorithm initialised by the distorted points themselves.

Since the proposed algorithm is based on a fraction distortion model for camera calibration, it is called the FMC algorithm in the rest of this chapter. Since it operates directly on 3D-2D correspondences, it has a computational complexity of $O(n)$ where n is the number of 3D-2D correspondences used for camera calibration and correction.

8.4 Experimental Results

In this section, we experimentally validate the proposed camera calibration algorithm described in Sect. 8.3 using both synthetic data and real images. For a comparative study, we also implemented the Tsai+LM algorithm, which is an improved version of the traditional two-step Tsai algorithm [20]. The improvement lies in two aspects: (1) the camera distortion was explicitly modelled as the second order radial and decentring ones, instead of just the first-order radial one:

$$\begin{aligned} x_{di} &= x_{ui}(1+\kappa_1 r_{ui}^2+\kappa_2 r_{ui}^4)+2\nu_1 x_{ui} y_{ui}+\nu_2(r_{ui}^2+2x_{ui}^2), \\ y_{di} &= y_{ui}(1+\kappa_1 r_{ui}^2+\kappa_2 r_{ui}^4)+\nu_1(r_{ui}^2+2y_{ui}^2)+2\nu_2 x_{ui} y_{ui}, \end{aligned} \tag{8.6}$$

and (2) the LM algorithm was used to globally optimise the calibration results of the Tsai algorithm.

8.4.1 Synthetic Data

In this section, we use the synthetic data to validate the proposed FMC algorithm for camera calibration and correction. The following parameters were used to generate the synthetic data: $Ncx = 649$, $Nfx = 512$, $dx = dy = 0.015$, $u_0 = 367.6093$, $v_0 = 305.8503$, and aspect ratio: $s = Ncx/Nfx$. 3D calibration patterns were created as: $x_m = i * \delta_x + O_x$ $(i = 1, 2, \ldots, n_x)$, $y_m = j * \delta_y + O_y$ $(j = 1, 2, \ldots, n_y)$, and $z_m = k * \delta_z + O_z$ $(k = 1, 2, \ldots, n_z)$, where $m = ((k-1) * n_x + i - 1) * ny + j$, δ_x, δ_y, and δ_z are the distance between neighbouring points along x, y and z axes respectively, O_x, O_y, and O_z are the offsets of the starting points of the calibration pattern in the world coordinate system. Then these points were subject to a rotation, consisting of a rotation angle of 30° around the x axis, 1° around the y axis, and 2° around the z axis followed by a translation vector $\mathbf{t} = (-100, -85, 2000)^T$. The transformed points

Table 8.2 Absolute and relative calibration errors of different algorithms based on non-coplanar and coplanar points subject to different levels of noise defined as a multiple τ of a basic noise

Points	τ	Self-calibration		Tsai+LM		FMC	
		ACE (pixels)	RCE (%)	ACE (pixels)	RCE (%)	ACE (pixels)	RCE (%)
Non-coplanar	1	0.0434	2.50	0.1516	8.74	0.2529	14.58
	10	0.8963	51.32	0.2441	13.98	0.1726	9.88
	20	2.0735	114.73	0.4008	22.62	0.3361	19.15
Coplanar	1	0.1819	12.18	0.1004	6.77	0.1482	10.01
	10	0.9064	59.30	0.3478	22.37	0.2209	14.21
	20	8.3068	507.66	0.4249	25.97	0.4216	25.76

were described in the camera centred coordinate system and projected onto a plane with a focal length equal to $f = 8.3431$. Finally, the projected image points were distorted using the model described in (8.6) above. In the experiments, the following parameters, unless otherwise stated, were used: $n_x = 10$, $n_y = 10$, $n_z = 3$, $O_x = 10$, $O_y = 10$, $O_z = 0$, $\delta_x = 20$, $\delta_y = 20$, $\delta_z = 20$, $\kappa_1 = -0.08186$, $\kappa_2 = 0.008755$, $\nu_1 = -0.0003275$, and $\nu_2 = -0.0001565$. For coplanar points, we simply set $z_m = 0$.

In order to simulate the real world imaging noise due to quantisation, different reflectance properties of the object surface, mechanical errors, etc, the Gaussian white noise with standard deviation $\sigma = 0.04\tau$ was added to the coordinates of 3D world points $\mathbf{p}_{wi}$ $(i = 1, 2, \ldots, 300)$ and $\sigma = 0.005\tau$ was added to image points $\mathbf{P}_{fi}$ where τ varied from 0 to 16 at intervals of 2, simulating different levels of noise.

Due to the fact that the image points around the principal point are subject to little distortion and only the peripheral pixels are subject to heavier distortion, the average and standard deviation of the differences between the corrected image points and the given undistorted image points may not be informative to the performance of algorithms. In this case, two new performance measures were defined: absolute correction error (ACE) and relative correction error (RCE). ACE was defined as the maximum coordinate difference in pixels between the corrected image points and the given undistorted image points. RCE was defined as: ACE/MCD*100% where MCD is the maximum coordinate difference in pixels between the distorted and undistorted image points. ACE and RCE measure the extent to which the most heavily distorted points have been corrected. Once such points have been accurately corrected, then the other points will also be accurately corrected. The experimental results are presented in Tables 8.2, 8.3, 8.4, and 8.5 and Figs. 8.1, 8.2, 8.3, 8.4, and 8.5.

8.4.1.1 Calibration and Correction

From Figs. 8.1 and 8.2 and Table 8.2, it can be seen that (1) both the proposed FMC and the improved Tsai+LM algorithms successfully calibrate the camera parameters and superimpose the corrected points accurately on the undistorted points, no matter

Table 8.3 Absolute and relative calibration errors of different algorithms based on non-coplanar points subject to different radial distortions κ_1 as well as decentring distortion corrupted by a multiple 10 of a basic noise

Points	Self-calibration		Tsai+LM		FMC	
	ACE (pixels)	RCE (%)	ACE (pixels)	RCE (%)	ACE (pixels)	RCE (%)
$\kappa_1 = -0.10$	0.9502	44.23	0.4352	20.26	0.1757	8.18
$\kappa_1 = -0.12$	0.9988	38.54	1.1645	44.93	0.2509	9.68

Table 8.4 The average μ and standard deviation δ of the maximum fitting error (MFE), average fitting error (AFE) and standard deviation of fitting errors (SDFE) of points on different line segments using different algorithms based on synthetic data corrupted by a multiple 10 of a basic noise

Points	Algorithm	μ (pixels)			δ (pixels)		
		MFE	AFE	SDFE	MFE	AFE	SDFE
Non-coplanar	BeforeCorrection	0.2902	0.1309	0.2016	0.0652	0.0359	0.0522
	Self-calib.	0.2423	0.1070	0.1691	0.0641	0.0314	0.0458
	Tsai+LM	0.2675	0.1116	0.1778	0.0530	0.0286	0.0425
	FMC	0.2567	0.1102	0.1743	0.0527	0.0303	0.0444
Coplanar	BeforeCorrection	0.2751	0.1231	0.1913	0.0308	0.0204	0.0286
	Self-calib.	0.2235	0.0890	0.1559	0.0471	0.0151	0.0220
	Tsai+LM	0.2335	0.1003	0.1595	0.0226	0.0163	0.0221
	FMC	0.2467	0.1008	0.1605	0.0491	0.0203	0.0309

Table 8.5 The average μ and standard deviation δ of calibration errors of the parameters of interest using different algorithms based on the synthetic data corrupted by different multiples of a basic noise

Para.	Tsai+LM		FMC	
	μ	δ	μ	δ
ACE (pixels)	0.35	0.15	0.32	0.11
RCE (%)	19.20	7.73	17.46	5.67
f (%)	−2.20	3.51	−5.86	4.12
s (%)	−0.11	0.07	−0.03	0.07
u_0 (%)	1.35	0.64	0.29	0.84
v_0 (%)	0.65	0.67	0.73	0.64
$\mathbf{q}$ (%)	0.38	0.19	0.24	0.17
$\mathbf{t}$ (%)	20.31	14.00	20.32	14.00

whether these points lie in the middle or peripheral area of the image. Such accurate superimposition indicates that the distorted points have been satisfactorily corrected with remaining errors as small as 0.40 pixels; (2) when the noise levels are too low, the Tsai+LM algorithm is more accurate than the proposed FMC algorithm. This is because the Tsai+LM algorithm makes full use of the prior knowledge of the cam-

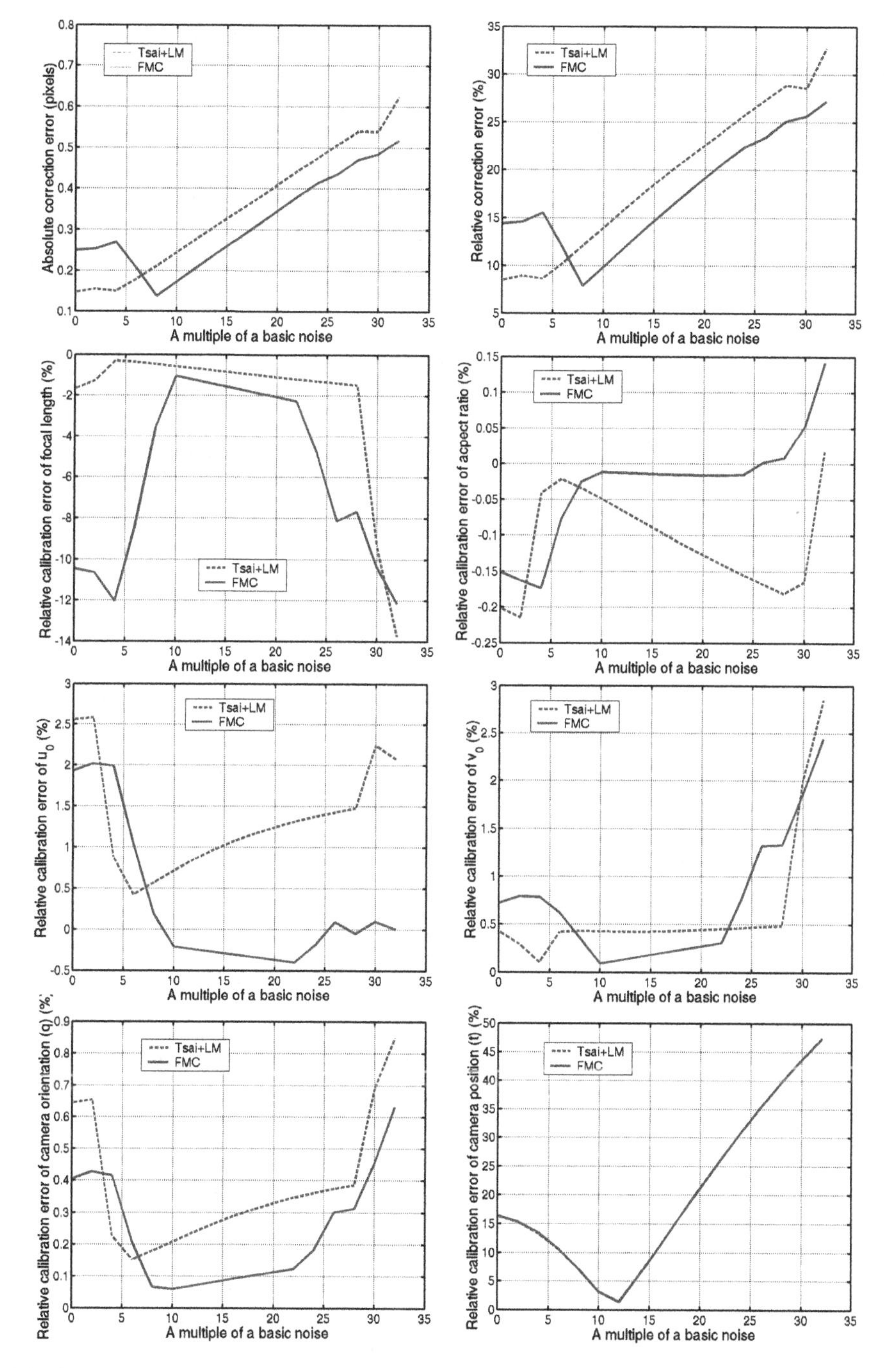

Fig. 8.5 The relative calibration errors of different parameters using different algorithms based on synthetic non-coplanar point data with various multiples of a basic noise. *Top left*: ACE; *Top right*: RCE. *Second row left*: focal length; *Second row right*: aspect ratio. *Third row left*: u_0; *Third row right*: v_0. *Bottom left*: rotation; *Bottom right*: translation

era distortion models. However, with the noise level increasing, the noise weakens the regularities existing in the data about the distortions the camera was subject to. In this case, the Tsai+LM algorithm failed to take advantages of the prior knowledge of the camera distortions for the camera calibration and correction. This shows that the knowledge of radial and decentring distortions does not necessarily always bring benefits to the Tsai+LM algorithm for accurate camera calibration and correction. The proposed FMC algorithm clearly outperforms the Tsai+LM algorithm for the correction of the distorted points with the RCE being reduced up to 8%; (3) when the coplanar points are corrupted by higher levels of noise, both the proposed FMC and Tsai+LM algorithms produced worse results. This shows that the non-coplanar points can better pose the calibration problem than the coplanar points; and (4) heavy noise often renders all the algorithms to degrade their performance, as expected.

From Fig. 8.3 and Table 8.3, it can be observed that after the radial distortion coefficient κ_1 was increased from -0.08186 to -0.12, both the self-calibration and the Tsai+LM algorithms displace the corrected points with regard to the undistorted points, while the proposed FMC algorithm superimposes the corrected points perfectly on the undistorted points. This observation shows that severe radial and decentring distortions render both the self-calibration and the Tsai+LM algorithms to degrade considerably, they do not really impose a significant impact on the proposed FMC algorithm for accurate camera calibration and correction. In this case, the latter decreases the RCE by 35% compared with the Tsai+LM algorithm and by 28% compared with the self-calibration algorithm despite the fact that the proposed FMC algorithm knew nothing about that the camera was subject to severe radial distortion as well as decentring distortion. Such a significant performance improvement clearly shows that the proposed fraction camera distortion model successfully described both the radial and decentring distortions and effectively combated imaging noise for accurate camera distortion calibration and correction.

8.4.1.2 Collinearity Constraint

To get a further idea about the extent to which the proposed FMC algorithm is accurate, in this section, we compare it against the self-calibration algorithm proposed in [6]. This self-calibration algorithm does not require a calibration rig and attempts to straighten the collinear points which appear as a curve in the distorted image. To this end, we manually extracted the non-coplanar points which should lie on 8 line segments in this chapter with the largest curvatures in the distorted image and then estimated the coordinates of the principal point and the coefficients κ_1 and κ_2 for the first two orders of the radial distortion. After these parameters of interest had been estimated, all points (including both the selected and non-selected points for camera distortion calibration) were corrected using these parameters. It is interesting to note that we also applied Equation 8.6 to model both the radial and decentring distortions. However, no better results were obtained. Thus, we finally modelled the radial distortion alone. The performance of algorithms is measured as: maximum fitting error (MFE), average fitting error (AFE) and the standard deviation of fitting errors (SDFE) of points on different line segments. The experimental results are presented in Fig. 8.4 and Table 8.4.

From Fig. 8.4 and Table 8.4, it can be seen that (1) All the self-calibration, Tsai+LM and FMC algorithms successfully reduce the collinear fitting errors. This shows that all these three algorithms somewhat successfully correct the distorted points; (2) While the self-calibration algorithm manipulates the collinearity constraint, it usually produces the smallest collinear fitting errors. Even though the proposed FMC algorithm did not take into account the collinearity constraint, it usually produces smaller collinear fitting errors than the Tsai+LM algorithm. This shows that the proposed FMC algorithm is more accurate than the Tsai+LM algorithm for camera calibration and correction despite the fact that it does not require the prior knowledge of the camera distortion which is usually difficult to obtain; and (3) As demonstrated in Figs. 8.1 and 8.2, when the noise corrupting the points is light, the self-calibration algorithm is sufficient for accurate camera distortion correction. As demonstrated in Fig. 8.1, when the noise was heavy and used to corrupt the non-coplanar points, it failed completely to correct the distorted points, since the corrected points are very close to the distorted points, instead of the undistorted points. As demonstrated in Fig. 8.2, when the noise was heavy and used to corrupt the coplanar points, it produced completely wrong correction results, since the corrected points have been shifted a very large distance from their corresponding undistorted points. This shows that the self-calibration algorithm is very sensitive to imaging noise. In contrast, both the proposed FMC and Tsai+LM algorithms however still successfully correct the distorted points, superimposing the corrected points accurately on the undistorted points.

8.4.1.3 Different Levels of Noise

In this section, we present experimental results based on non-coplanar points corrupted by different levels of noise defined as various multiples of a basic noise. The experimental results are presented in Fig. 8.5 and Table 8.5. From Fig. 8.5, it can be clearly observed that (1) the proposed FMC algorithm is always more accurate than the Tsai+LM algorithm in the sense of correcting the distorted image points when the imaging noise is heavy enough ($\tau > 5$, for example) and in the sense of calibrating the intrinsic and camera rotational parameters when the data were corrupted by a medium level of noise ($\tau \in [10, 25]$, for example). It is interesting to note that even when the data were free from noise, both the proposed FMC and the Tsai+LM algorithms did not correct the camera distortion completely. This is because the objective function (see (8.4)) is a highly non-linear function of 16 parameters. Even though the data are free from noise, the global minimum of (8.4) is often difficult for the LM algorithm to find. In this case, the correction error still exists; (2) even though all parameters have not been perfectly calibrated, the correction error can still be small. This is because the intrinsic and distortion parameters that eventually determine the amount of correction errors interact with each other and their errors counteract with each other. When images have been successfully corrected, the estimation of the camera position still can have a relative error of up to 48%; (3) both the intrinsic and distortion parameters are crucial to successful image correction. If they have a large error, then image correction will generally fail; (4) while the aspect ratio is the most reliably calibrated, the camera position is consistently the most difficult to calibrate. This explains why the depth from the surface of the object of interest to the camera is often difficult to measure, since it is a function of the camera position; and (5) heavy noise usually leads to less accurate estimation of the parameters of interest, which is often within expectation.

8.4.2 Real Images

In this section, we use real image data to validate the proposed FMC algorithm for camera calibration and correction. The real image data were downloaded from [15] and used in [13] and from the website of the Calibrated Imaging Laboratory at CMU [2]. The control points [15] are circles which were extracted using the moment and curvature preserving ellipse detection technique and the renormalisation conic fitting. 491 out of 512 points were finally used for camera calibration. The experimental results are presented in Tables 8.6, 8.7, 8.8, and 8.9 and Fig. 8.6.

From Tables 8.7, 8.8, and 8.9, it can be seen that both the proposed FMC and Tsai+LM algorithms produce similar results for the calibration of intrinsic and extrinsic parameters, as expected, while the distortion parameters are different and have different interpretations. It seems that both the intrinsic and distortion parameters are critical for the correction of distorted images, the extrinsic parameters impose little impact on the correction of the images, even though they do affect the optimisation of the intrinsic and distortion parameters. This conclusion has been confirmed by Fig. 8.6. Both the proposed and Tsai+LM algorithms successfully correct distorted images as expected with curved line segments having been

Table 8.6 The correction made in pixels by different algorithms based on different real images

Algorithm	Grid	Castle	WallandTower
Tsai+LM	13.34	8.02	4.85
FMC	11.10	8.08	4.90

Table 8.7 The calibration results of intrinsic parameters using different algorithms based on different real images

Para.	Grid		Castle		WallandTower	
	Tsai+LM	FMC	Tsai+LM	FMC	Tsai+LM	FMC
f	8.35	8.32	57.45	57.42	148.39	147.54
s	1.0034	1.0038	1.0003	1.0009	1.0014	1.0015
u_0	367.31	371.23	258.59	253.36	255.34	257.53
v_0	306.63	288.76	197.24	202.20	207.05	203.05

Table 8.8 The calibration results of the extrinsic parameters using different algorithms based on different real images

Para.	Grid		Castle		WallandTower	
	Tsai+LM	FMC	Tsai+LM	FMC	Tsai+LM	FMC
q_0	0.9066	0.9079	0.9997	0.9999	0.9999	0.9999
q_1	−0.1174	−0.1093	0.0004	−0.0006	0.0056	0.0059
q_2	0.3669	0.3696	−0.0038	−0.0049	−0.0011	−0.0009
q_3	0.0646	0.0611	0.0056	0.0057	0.0032	0.0032
t_x	−0.6154	−1.7976	−558.0944	−554.6412	−318.8117	−320.8479
t_y	−97.8920	−92.7628	−506.0734	−509.5744	−301.0551	−301.9105
t_z	308.8881	370.4704	1741.5957	1741.7843	1133.8702	1133.3641

Table 8.9 The calibration results of distortion parameters using different algorithms based on different real images

Para.	Grid		Castle		WallandTower	
	Tsai+LM	FMC	Tsai+LM	FMC	Tsai+LM	FMC
κ_1	−0.003130	0.000059	0.000538	−0.000024	0.000519	0.000053
κ_2	0.000044	−0.001355	0.000001	0.000136	0.000001	0.000004
κ_3	n/a	−0.002594	n/a	0.000595	n/a	0.000558
ν_1	0.000009	0.000248	−0.000066	−0.000033	0.000051	0.000018
ν_2	−0.000035	−0.000991	0.000071	0.000035	−0.000040	0.000034

straightened. The points in the middle areas of the images before and after correction perfectly superimpose, while those in the peripheral area stay apart due to the correction made by different algorithms. Even though we have no knowledge about the ground truth of the actual distortion of different points, we believe that the correction of the distorted points made is reasonable, since the calibrated intrinsic and extrinsic parameters are very close to the values provided on the website [2].

8.5 Conclusion

In this chapter, we have proposed a novel algorithm for camera calibration and correction. This algorithm is based on a novel camera distortion model which attempts to blindly model the overall distortion of the camera and combat the imaging noise for more accurate camera calibration and correction. To estimate the parameters of interest, we globally optimise an objective function about the sum of the back-projection errors using the LM algorithm. For the initialisation of the LM algorithm, the classical Tsai algorithm was implemented. The initialisation must be sufficiently accurate. Otherwise, the LM algorithm can easily fail to optimise the parameters

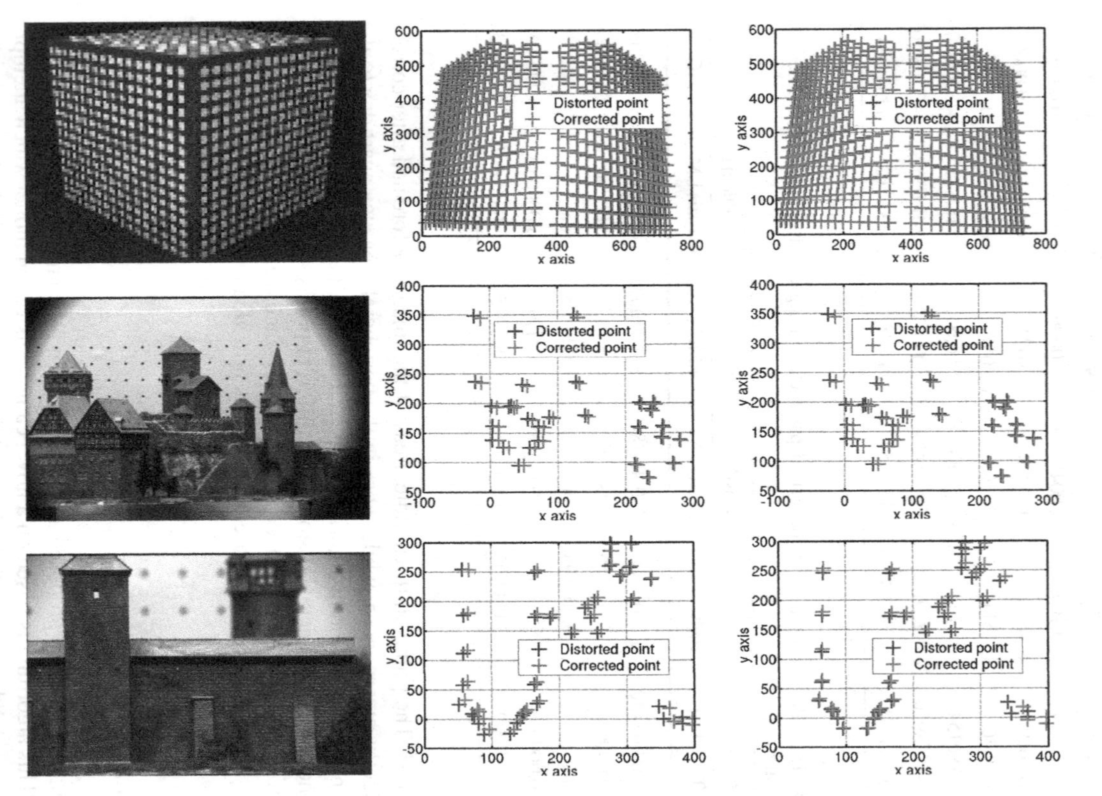

Fig. 8.6 Camera calibration and correction results of different algorithms based on different real images. *Left column*: real image. *Middle column*: Tsai-LM; *Right column*: FMC. *Top row*: grid; *Middle row*: castle; *Bottom row*: wall and tower

of interest. After both the intrinsic and distortion parameters have been calibrated, the distorted image points have been corrected using again the LM algorithm initialised by the distorted image points themselves. This LM algorithm minimises the squared difference between the distorted points and the transformed corrected points. As long as both the intrinsic and distortion parameters have been calibrated with reasonable accuracy, then the minimisation is usually successful. The proposed algorithm took fewer than 2 seconds on a Pentium IV, 2.80 MHz computer for any experiment reported in this chapter.

In summary, we have made the following contributions in this chapter:

- we have proposed a fraction formula to blindly model the overall camera distortion without any prior knowledge about what distortion the camera is subject to;
- we have proposed using the LM algorithm to correct the distorted images initialised by the distorted points themselves;
- two novel parameters have been defined to measure the performance of camera calibration and correction algorithms; and
- we experimentally demonstrate that the prior knowledge about the camera distortion does not necessarily always bring benefits to algorithms for accurate camera calibration and correction. What the accurate camera calibration and correction algorithm requires is a powerful model to blindly characterise the overall camera distortion and effectively resist the imaging noise in data.

A comparative study based on both synthetic data and real images corrupted by reasonable levels of noise shows that the proposed algorithm successfully calibrated and corrected the distorted image points, while the self-calibration method often produced good results when the noise level was very low. Further research is to investigate the reason why while the distorted images can be satisfactorily corrected, the calibration of the camera position parameters is often poor, and how the depth of points in the camera coordinate system can be accurately estimated. Research is underway and results will be reported in the future.

References

1. Andreetto, M., Brusco, N., Cortelazzo, G.M.: Automatic 3D modelling of textured cultural heritage objects. IEEE Trans. Image Process. **13**, 354–369 (2004)
2. Bolles, R.C., Baker, H.H., Hannah, M.J.: The JISCT Stereo Evaluation. In: Proceedings of the ARPA Image Understanding Workshop, pp. 263–274 (1993)
3. Bougnoux, S.: From projective to Euclidean space under any practical situation, a criticism of self-calibration. In: Proceedings of the ICCV, pp. 790–796 (1998)
4. Claus, D., Fitzgibbon, A.W.: A rational function lens distortion model for general cameras. In: Proceedings of International Conference on Computer Vision and Pattern Recognition, pp. 213–219 (2005)
5. Devernay, F., Faaugeras, O.: Automatic calibration and removal of distortion from scenes of structured environments. Proc. SPIE **2567**, 62–72 (2006)
6. Devernay, F., Faugeras, O.: Straight lines have to be straight. Mach. Vis. Appl. **13**, 14–24 (2001)

7. Douxchamps, D., Chihara, K.: High-accuracy and robust localization of large control markers for geometric camera calibration. IEEE Trans. Pattern Anal. Mach. Intell. **31**, 376–383 (2009)
8. Farid, H., Popescu, A.C.: Blind removal of lens distortion. J. Opt. Soc. Am. A **18**, 2072–2078 (2001)
9. El-Melegy, M.T., Farag, A.A.: Nonmetric lens distortion calibration: closed-form solutions, robust estimation and model selection. In: Proceedings of International Conference on Computer Vision, pp. 554–559 (2003)
10. Fitzgibbon, A.W.: Simultaneous linear estimation of multiple view geometry and lens distortion. In: Proceedings of International Conference on Computer Vision and Pattern Recognition, pp. 125–132 (2001)
11. Hartley, R., Kang, S.B.: Parameter free radial distortion correction with center of distortion estimation. IEEE Trans. Pattern Anal. Mach. Intell. **29**, 1309–1321 (2007)
12. Heikkila, J., Silven, O.: A four-step camera calibration procedure with implicit image correction. In: Proceedings of International Conference on Computer Vision and Pattern Recognition, pp. 1106–1112 (1997)
13. Heikkila, J.: Geometric camera calibration using circular control points. IEEE Trans. Pattern Anal. Mach. Intell. **22**, 1066–1077 (2000)
14. Kim, J.-S., Kweon, I.S.: Camera calibration based on arbitrary parallelograms. Comput. Vis. Image Underst. **113**, 1–10 (2009)
15. http://www.ee.oulu.fi/jth/calibr/
16. Liu, Y., Al-Obaidi, A., Jakas, A., Li, L.: Accurate camera calibration and correction using rigidity and radial alignment constraints. In: Proceedings of International Symposium on 3D Data Processing, Visualization and Transmission, pp. 145–152 (2008)
17. Ma, L., Chen, Y., Moore, K.L.: Flexible camera calibration using a new analytical radial undistortion formula with application to mobile robot localization. In: Proceedings of 2003 IEEE International Symposium on Intelligent Control, pp. 799–804 (2003)
18. Pollefeys, M., Koch, R., Gool, L.V.: Self-calibration and metric reconstruction in spite of varying and unknown intrinsic camera parameters. Int. J. Comput. Vis. **32**, 7–25 (1999)
19. Tardif, J., Sturm, P., Roy, S.: Self-calibration of a general radially symmetric distortion model. In: Proceedings of the European Conference on Computer Vision, pp. 186–199 (2006)
20. Tsai, R.Y.: A versatile camera calibration technique for high-accuracy 3D machine vision metrology using the off-the-shelf TV cameras and lenses. IEEE J. Robot. Autom. **3**, 323–344 (1987)
21. Weng, J., Cohen, P., Herniou, M.: Camera calibration with distortion models and accuracy evaluation. IEEE Trans. Pattern Anal. Mach. Intell. **14**, 965–980 (1992)
22. Zhang, Z.: A flexible new technique for camera calibration. IEEE Trans. Pattern Anal. Mach. Intell. **22**, 1330–1334 (2000)

Chapter 9
A Leader-Follower Flocking System Based on Estimated Flocking Center

Zongyao Wang and Dongbing Gu

Abstract This chapter introduces a robot flocking system in which only minority members are the group leaders who have global trajectory knowledge, while majority members are the group followers who do not have global trajectory information, but can communicate with neighbors. The followers even do not know who the leaders are in the group. In order to keep the flocking group connected, all the group members estimate the position of flocking center by using a consensus algorithm via local communication. Based on the estimated positions of flocking center, a leader-follower flocking algorithm is proposed. A group of real robots, "wifibots", are used to test the feasibility of the flocking algorithm. The simulation is conducted for a large group to demonstrate its scalability. The results show that this leader-follower flocking system can track desired trajectories led by the leaders.

9.1 Introduction

For animals that forage or travel in groups, few individuals have global information, such as knowledge about the location of a food source or a migration route. The informed individuals play an important role in guiding those who are less experienced [2]. For example, the foraging behavior of fish shoals is known to be influenced by few informed individuals and such a group can navigate towards a target [11]. Honeybee swarms can be guided to a new site by very few individuals [12]. Such a leader-follower flocking strategy can be used for mobile robots, where few robots possess complicated sensory ability, such as GPS, and powerful computation ability to plan the global trajectory. The global trajectory is a desired robot trajectory to which the flocking system is designed to track. The followers simply use local sensors or limited communication to exchange information and flock without separation and collision.

Z. Wang (✉) and D. Gu
School of Computer Science and Electronic Engineering, University of Essex, Essex, UK
e-mail: zwangf@essex.ac.uk; dgu@essex.ac.uk

H. Liu et al. (eds.), *Robot Intelligence*,
Advanced Information and Knowledge Processing,
DOI 10.1007/978-1-84996-329-9_9,

In robotics, the leader-follower strategy has been applied in formation control for many years. The formation control needs to build dynamic models of relative distance and bearing between leaders and followers. The chain structure is used, in which a follower can also be a leader of another follower and every follower knows who the leaders are. Some formation systems can change the leader follower relationships. In [3], a framework for multi-robot coordination that allows robots to maintain or change formation while following a specified trajectory or performing cooperative manipulation tasks is introduced. In the system, a follower has to follow a specific leader and a switching protocol is required to allow robots to select the most appropriate formation depending on the environment. The work in [4] investigates the formation changing by using graph theory. A transition matrix is introduced to govern the addition and deletion of edges in the network and manage the changing of communication protocol.

In robotic flocking systems, neighbor robots exchange information via local wireless communication. A flocking network is formed according to the wireless communication connection. The topology of flocking network is varied with time due to the motion of robots. The flocking robots should have a cohesion control ability in order to keep the flocking network connected. Cohesion control is a challenging issue in the leader-follower robot flocking systems. In recent years, many distributed algorithms have been provided. However, the group split into multiple components may happen during manoeuvres under these distributed algorithms, which prevents a robot group from forming a flock. The system stabilization is always established by assuming that the topology of the flocking network is connected at all time. A flocking system is investigated in terms of controllability and optimal control given that the flocking network is connected [8]. In [14], a convergent condition is constructed by using the contraction theory. However, the convergent condition can only be applied in a flocking system with connected network topology.

If all the members know the global trajectory information, the cohesive force generated by this global trajectory information can keep the flocking group connected [9, 10]. Some researchers have attempted to find a distributed way to keep the connectivity of a flocking system. One approach is to define a measure of local connectedness to check if the connection is maintained and the measure can be used as a constraint for the motion control law [13]. Using a specific potential function can also maintain the connectivity [5, 7, 15] as the potential function generates an infinite force to keep them connected when the distance between two connected robots approaches a threshold.

In the leader-follower flocking system considered in this chapter, only leaders have the global or desired trajectory information, while followers do not have. Thus the followers have the potential to move away from the flocking group and lead to group split. To solve this problem, we propose to estimate the position of the flocking center by all the robots in the flocking system. All the members including followers can use this knowledge to keep their connections by moving to the position of the flocking center. The position of the flocking center is the global information. Its accurate calculation requires a centralized approach. In this chapter, we propose to use a consensus algorithm to estimate the position of the flocking center. It only requires the robots to communicate with their neighbors and therefore it is a distributed algorithm.

The experimental tests are conducted on a group of wifibots [1]. The wifibots are networked robots which can use wireless communication to exchange information with neighbors. In the proposed algorithm, followers do not need to know the global or desired trajectory, but all the members need to know their own positions. Our testing is carried out in an in-door environment (circular Robot Arena in University of Essex, around 100 m^2). There is a 3D tracking system (the Vicon tracking system) equipped in the lab with high accuracy. All wifibots can connect with the Vicon system via TCP/IP protocol and acquire their positions.

In the rest of this chapter, Sect. 9.2 introduces the structure of the flocking system and the estimation of flocking center. Section 9.3 presents the leader-follower flocking algorithm. The algorithm stability is analyzed in Sect. 9.4. Section 9.5 shows experimental results. Three real robots are used to test the algorithm. To show the system scalability, a simulation for thirty robots is conducted in Sect. 9.6. A brief conclusion is given in Sect. 9.7.

9.2 Flocking System

We consider a flocking system consisted of N robots. The state of these N robots can be represented by a vector $q = [q_1^T, q_2^T, \dots, q_N^T]^T \in \mathbf{R}^{2N}$, where $q_i = [x_i, y_i]^T$ is the position of robot i. The topology of flocking system can be represented by a graph $\mathcal{G} = (\mathcal{V}, \mathcal{E})$ where $\mathcal{V}$ is the set of vertices (robots) and $\mathcal{E}$ is the set of edges (communication channels). The communication range of robots is denoted as C. The topology of flocking system depends on the distance between robots; a link only exists between a robot pair when their distance is smaller than C. When robot j is the neighbor of robot i, we have $j \in \mathcal{N}_i = \{j \in \mathcal{V}, j \neq i : \|q_i - q_j\| \leq C\}$. We assume that the communication range is the same for all the robots, so the graph of flocking system is undirected. For a connected graph, Laplacian matrix $\mathcal{L}$ is symmetric and positive semi-definite [6].

Each robot can contact with the tracking system (Vicon system) to obtain its own state q_i. Each robot is required to send out its state q_i via wireless communication in order to estimate the position of flocking center. Due to the limited space used in the experiment, all the robots can receive states from all other robots. To simulate the limited range, a robot only receives states from neighbors who are located within the distance of C and discards states from other robots whose distances are larger than C.

Figure 9.1 is a diagram of the internal software structure of robot. The core component is the controller which takes state information as input and outputs forward velocity v_i and rotation velocity ω_i to move the robot. The robot model we used is kinematic model:

$$\begin{cases} \dot{x}_i^b = v_i \cos\theta_i, \\ \dot{y}_i^b = v_i \sin\theta_i, \\ \dot{\theta}_i = \omega_i \end{cases} \tag{9.1}$$

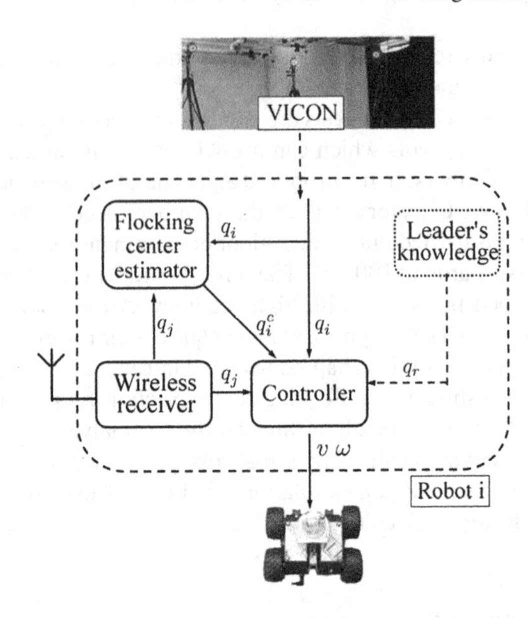

Fig. 9.1 Software structure

where θ_i is the heading of robot i and ($q_i^b = [x_i^b, y_i^b]^T$) represents the center point of robot. The robot hand position $q_i = [x_i, y_i]^T$ is a point located at the heading axis with distance L to the center of robot. In the following, we will use q_i as the robot state for our investigation. By using the hand position, the relationship between $[x_i, y_i]$ and $[v_i, \omega_i]$ can be written as:

$$\begin{cases} \dot{x}_i = v_i \cos\theta_i + L\omega_i \cos(\theta_i + \frac{\pi}{2}), \\ \dot{y}_i = v_i \sin\theta_i + L\omega_i \sin(\theta_i + \frac{\pi}{2}). \end{cases} \tag{9.2}$$

A coordination transformation can be made to obtain v_i, ω_i from $\dot{q}_i$.

$$\begin{cases} v_i = \dot{x}_i \cos\theta_i + \dot{y}_i \sin\theta_i, \\ \omega_i = \frac{1}{L}[\dot{x}_i \cos(\theta_i + \frac{\pi}{2}) + \dot{y}_i \sin(\theta_i + \frac{\pi}{2})]. \end{cases} \tag{9.3}$$

All other components in Fig. 9.1 are related to the state estimation, including acquiring neighbor's state q_j via wireless communication and estimating the position of flocking center q_i^c. The TCP/IP protocol is used to obtain the state q_i from the Vicon system. We assume a full state observation without noise can be obtained.

The estimation of flocking center is a key element in our algorithm. It can be used for each robot to move close towards each other and avoid flocking split. Flocking robots can reach agreement on their aggregate information via consensus between adjacent robots. The consensus algorithm of robot i to estimate the position of flocking center is as follows:

$$\dot{q}_i^c = -\sum_{j \in N_i} (q_i^c - q_j^c) - q_i^c + q_i \tag{9.4}$$

where q_j^c is the estimate made by neighbor j. The vector form

$$\dot{q}^c = -\mathcal{L}q^c + q - q^c. \tag{9.5}$$

For a connected graph, the symmetric and positive semi-definite $\mathcal{L}$ guarantees that q_i^c will asymptotically converge to the average of coordinates q_i:

$$q_i^c \to \frac{1}{N}\sum_{i=1}^{N} q_i. \tag{9.6}$$

9.3 Flocking Algorithms

The leader-follower flocking system in this chapter is composed of minority leaders and majority followers. The leaders have knowledge of a desired trajectory and need to track this trajectory. At the same time, they also need to avoid collision and move to the flocking center. Eventually, the leaders can lead the entire flocking group to track the desired trajectory. The followers do not know who the leaders are in the group and do not know the desired trajectory. They only need to avoid collision and move to the flocking center. Assume that there are L leader robots and F follower robots in a group and $L + F = N$. The leader states denoted as $q_l = [q_{l_1}^T, q_{l_2}^T, \ldots, q_{l_L}^T]^T$ and $q_f = [q_{f_1}^T, q_{f_2}^T, \ldots, q_{f_F}^T]^T$ are follower robot states.

The position estimation of flocking center is the same to both leaders and followers. Also all of them need to avoid collision with other members or external obstacles. The collision avoidance can be established by keeping neighbor robots a specific distance. If the distance between neighbor robots is too small, they attempt to separate. There are several approaches to design a separation potential function $H_s(\|q_i - q_j\|)$ where $\|q_i - q_j\|$ is the Euclidean distance between robots i and j. We have used fuzzy logic to design $H_s(d_{ij})$. It is continuous function and its gradient $\nabla H_s(\|q_i - q_j\|)$ owns the following properties:

- When the distance $(\|q_i - q_j\|)$ between robots i and j is smaller than a specific distance C, $\nabla H_s(\|q_i - q_j\|)$ is negative. Robot i moves away from j.
- When the distance $(\|q_i - q_j\|)$ between robots i and j is larger than the specific distance C, $\nabla H_s(\|q_i - q_j\|)$ is close to zero and equal to zero when $d_{ij} = C$.

Figure 9.2 illustrates an example of $H_s(\|q_i - q_j\|)$ and its corresponding $\nabla H_s(\|q_i - q_j\|)$ designed by using fuzzy logic.

The follower flocking algorithm is designed as follows:

$$\dot{q}_{f_i} = -\left[k_c(q_{f_i} - q_{f_i}^c) + \sum_{j \in N_i} \nabla_{q_{f_i}} H_s(\|q_i - q_j\|)\right] \tag{9.7}$$

where k_c is the cohesion control gain.

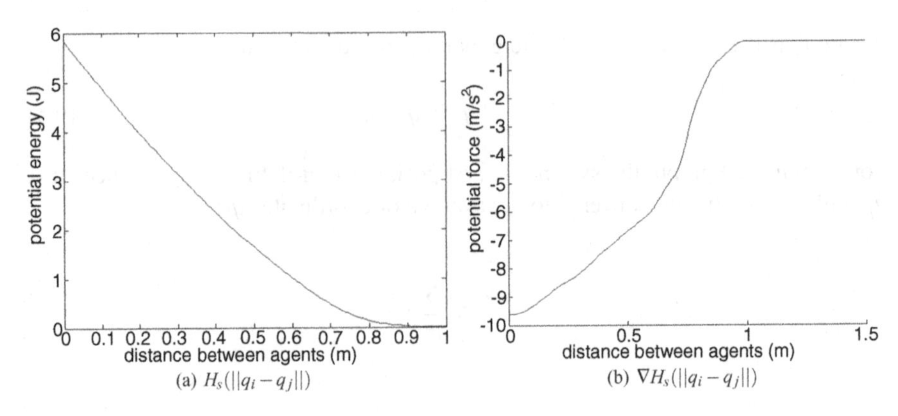

(a) $H_s(\|q_i - q_j\|)$ (b) $\nabla H_s(\|q_i - q_j\|)$

Fig. 9.2 Separation potential function and its force function

The leader controller is designed as follows:

$$\dot{q}_{l_i} = -\left[k_c(q_{l_i} - q^c_{l_i}) + \sum_{j \in N_i} \nabla_{q_{l_i}} H_s(\|q_{l_i} - q_j\|) + k_t(q_{l_i} - q_r)\right] \tag{9.8}$$

where k_t is the tracking control gain.

9.4 Algorithm Stability

The key to use stability theory to analyze the algorithm stability is to find a proper potential function for the entire flocking system. The system potential function should include several potential sub-functions, which reflect the system performance. Firstly all the robots have two common potential sub-functions. One is the separation potential sub-function $H_s(\|q_i - q_j\|)$, which is used for robots to move away from neighbor robots to avoid collisions. Another one is the cohesion potential sub-function $H_c(\|q_i - q^c_i\|) = \frac{1}{2}k_c\|q_i - q^c_i\|^2$, which is used for robots to keep the flocking system connected. Furthermore, the leader robots need to track the desired trajectory q_r and an additional potential sub-function for the leader tracking is required $H_t(\|q_{l_i} - q_r\|) = \frac{1}{2}k_t\|q_{l_i} - q_r\|^2$. In our analysis, another common potential sub-function for all the members is considered: $H_a(\|q^c_i - q^c_j\|) = \frac{1}{2}k_c\|q^c_i - q^c_j\|^2$. This potential sub-function is used for all robots to achieve the consensus on the position estimation of flocking center. Summarizing all potential sub-functions together, the potential function of the entire flocking system is defined as:

$$\begin{aligned} H = &\sum_{i=1}^{N}\sum_{j \in N_i} H_a(\|q^c_i - q^c_j\|) + \sum_{i=1}^{N} H_c(\|q_i - q^c_i\|) \\ &+ \sum_{i=1}^{N}\sum_{j \in N_i} H_s(\|q_i - q_j\|) + \sum_{i=1}^{L} H_t(\|q_{l_i} - q_r\|). \end{aligned} \tag{9.9}$$

The derivative of the potential function can be found according to the definitions of the potential sub-functions:

$$\begin{aligned}\dot{H} = &\sum_{i=1}^{N}\sum_{j\in N_i} k_c(q_i^c - q_j^c)^T \dot{q}_i^c + \sum_{i=1}^{N} k_c(q_i - q_i^c)^T \dot{q}_i - \sum_{i=1}^{N} k_c(q_i - q_i^c)^T \dot{q}_i^c \\ &+ \sum_{i=1}^{N}\sum_{j\in N_i} \nabla_{q_i} H_s^T \dot{q}_i + \sum_{i=1}^{L} k_t(q_{l_i} - q_r)^T \dot{q}_{l_i} - \sum_{i=1}^{L} k_t(q_{l_i} - q_r)^T \dot{q}_r \end{aligned} \quad (9.10)$$

or

$$\begin{aligned}\dot{H} = &\sum_{i=1}^{N} k_c\left[\sum_{j\in N_i}(q_i^c - q_j^c)^T - (q_i - q_i^c)^T\right]\dot{q}_i^c \\ &+ \sum_{i=1}^{L}\left[k_c(q_{l_i} - q_{l_i}^c)^T + \sum_{j\in N_i} \nabla_{q_{l_i}} H_s^T + k_t(q_{l_i} - q_r)^T\right]\dot{q}_{l_i} \\ &+ \sum_{i=1}^{F}\left[k_c(q_{f_i} - q_{f_i}^c)^T + \sum_{j\in N_i} \nabla_{q_{f_i}} H_s^T\right]\dot{q}_{f_i} - \sum_{i=1}^{L} k_t(q_{l_i} - q_r)^T \dot{q}_r. \end{aligned} \quad (9.11)$$

To simplify the presentation, define:

$$\begin{aligned}E = &\sum_{i=1}^{N} k_c\left\|\sum_{j\in N_i}(q_i^c - q_j^c) - (q_i - q_i^c)\right\|^2 \\ &+ \sum_{i=1}^{L}\left\|k_c(q_{l_i} - q_{l_i}^c) + \sum_{j\in N_i} \nabla_{q_{l_i}} H_s + k_t(q_{l_i} - q_r)\right\|^2 \\ &+ \sum_{i=1}^{F}\left\|k_c(q_{f_i} - q_{f_i}^c) + \sum_{j\in N_i} \nabla_{q_{f_i}} H_s\right\|^2. \end{aligned} \quad (9.12)$$

Substituting the controllers (9.7), (9.8) and the consensus algorithm (9.4) into the derivative of the potential function (9.11), we have

$$\dot{H} = -E - \sum_{i=1}^{L}(q_{l_i} - q_r)^T \dot{q}_r. \quad (9.13)$$

In the theoretical analysis, we can consider the tracking trajectory as a reference point at each time step. Namely, the velocity of the tracking trajectory can be considered as zero ($\dot{q}_r = 0$). Then,

$$\dot{H} = -E. \quad (9.14)$$

According to the definition E, we have:

$$E \geq 0. \tag{9.15}$$

From (9.14), we have:

$$\dot{H} = -E \leq 0. \tag{9.16}$$

According to the above results and LaSalle's invariance principle, H will decrease until $\dot{H} = 0$ where the $\Omega = \{q, q^c | E = 0\}$ is an invariance set. It also proves that the flocking system continuously consumes its energy until reaching a stable state. From $E = 0$, it is concluded that the equilibrium of follower states lies in $\|k_c(q_{f_i} - q_{f_i}^c) + \sum_{j \in N_i} \nabla_{q_{f_i}} H_s\| = 0$ and the equilibrium of follower states lies in $\|k_c(q_{l_i} - q_{l_i}^c) + \sum_{j \in N_i} \nabla_{q_{l_i}} H_s + k_t(q_{l_i} - q_r)\| = 0$.

9.5 Experiments

The experimental tests were conducted in the Robot Arena at University of Essex. The Vicon system was used to provide position information to each of wifibots. We used the TCP/IP protocol to implement the communication.

In the experiment, it was assumed that there existed only local neighbor-to-neighbor information exchange among the robots and the robots can only receive their own states from the Vicon system. The maximum velocity of robot was set to 2000 mm/s. The global or desired trajectory was a circle centered at $[0, 0]$ with a radius of 1700 mm and the leader robot knew this trajectory:

$$x_r = 1700 \cos 0.05t,$$
$$y_r = 1700 \sin 0.05t.$$

Firstly, we tested the flocking algorithm without external obstacles. Four robots were used and all the four robots were leaders. The main purpose of this experiment was to examine the performance of flocking center estimation and distributed cohesion control.

Figure 9.3 shows the video snapshots of the tracking process at time $t = 0$ s, $t = 47$ s, $t = 94$ s and $t = 142$ s. At the initial state, the four wifibots were placed randomly in the robot arena. The desired trajectory is plotted in Fig. 9.3(a) with the solid line. It can be seen that the flocking system kept cohesion and no collision happened during the whole process.

The trajectory of the flocking system with four leader robots is shown in Fig. 9.4. The solid line denotes the desired trajectory. Four dotted lines are the trajectories of four robots. The connections between robots at four time instants $t = 0$ s, 47 s, 94 s, 142 s are illustrated. The square mark on the solid trajectory represents the desired position at four time instants. Initially four robots were not very close to the desired position at $t = 0$ s. After $t = 142$ s, the four robots gradually moved close to the

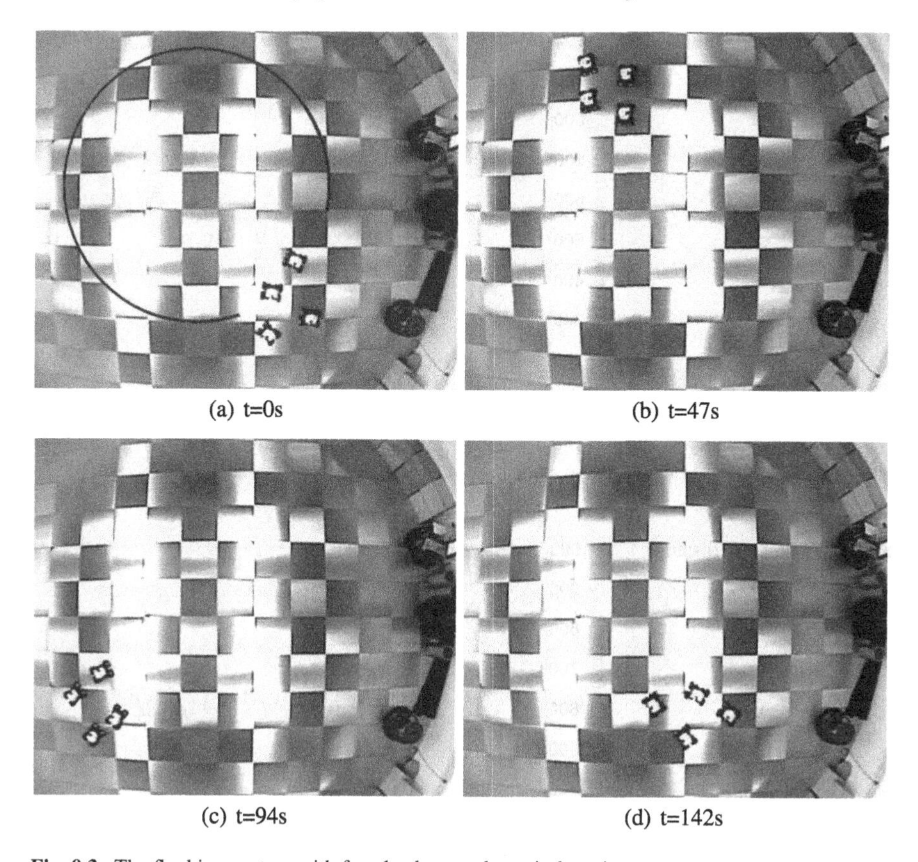

(a) t=0s (b) t=47s

(c) t=94s (d) t=142s

Fig. 9.3 The flocking system with four leaders tracks a circle trajectory

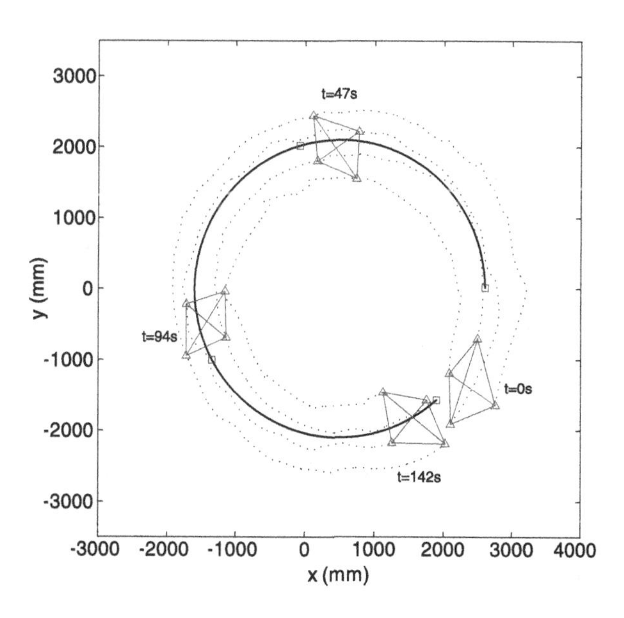

Fig. 9.4 Flocking system experiment with four leader robots

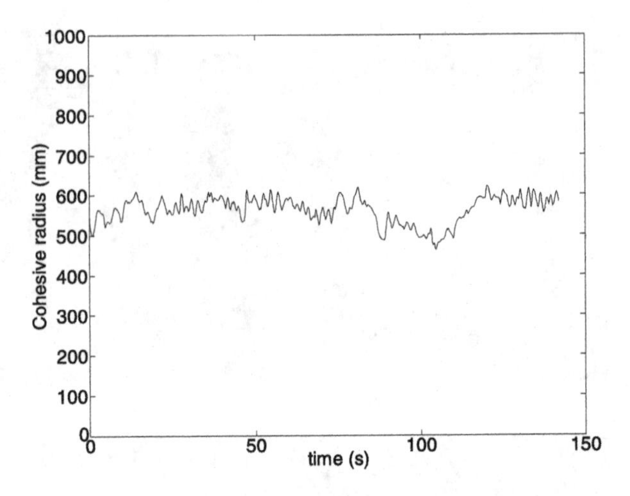

Fig. 9.5 Cohesion radius

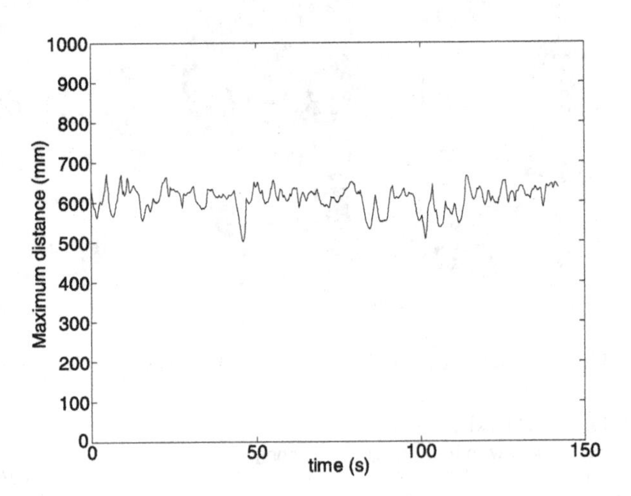

Fig. 9.6 Minimum distance between robots

desired position. Generally speaking, four robots can track the desired trajectory as a connected group during the whole process.

The minimum distance between any two robots was used to show if there was a collision in the flocking system. The cohesion radius and minimum distance are shown in Figs. 9.5 and 9.6. Both of them kept stable during the tracking process. They show that the flocking group kept the group connection without separation and there was no collision.

The results of flocking center estimation by four robots are shown with four dotted lines in Fig. 9.7. The solid line shows the real flocking center, i.e. the position of the flocking center calculated by averaging four robot positions. All estimated lines were very close to the real flocking center. Four triangle marks encircled the square mark at four time instants. This justifies the stability of the flocking center estimation algorithm.

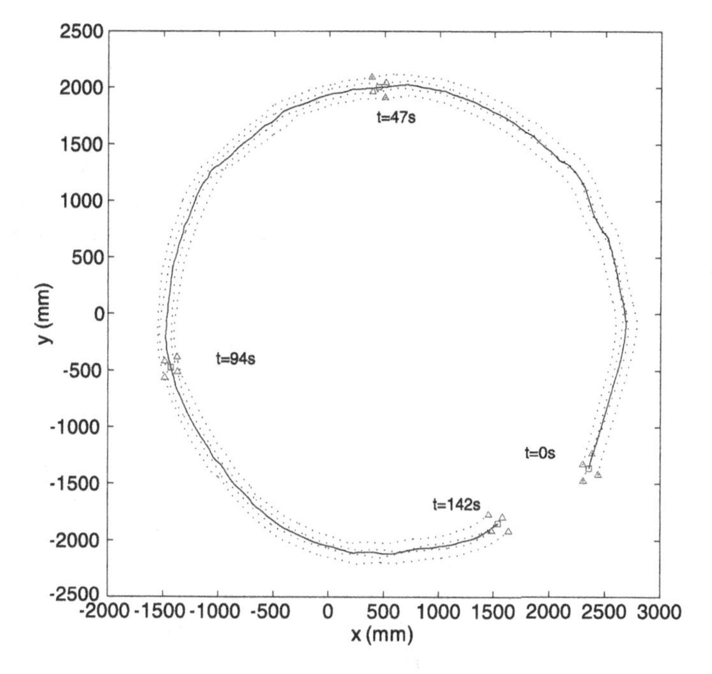

Fig. 9.7 Estimated flocking centers and true flocking center

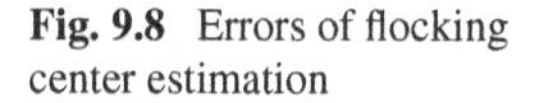
Fig. 9.8 Errors of flocking center estimation

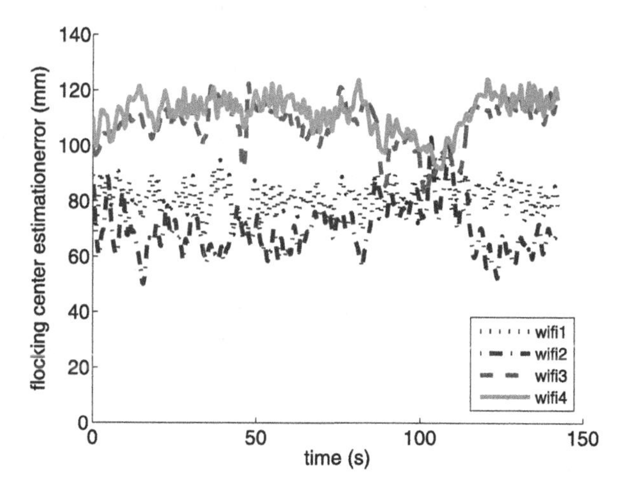

Figure 9.8 shows the errors of flocking center estimation. The four lines are the estimation errors of the four wifibots. It can be seen that all the errors of estimation were less than 120 mm. Compared with the cohesion radius of the flocking system (600 mm) shown in Fig. 9.5, the errors were much smaller and the accuracy was acceptable.

Secondly, we tested the obstacle avoidance of the flocking system by placing an obstacle on the desired trajectory. The obstacle position was provided to each robot by the Vicon system. No change was made to the flocking controller. Both leaders and followers treated the obstacle as a neighbor and avoided it through the separation function.

It has been mentioned that the advantage of the flocking system is the leader-follower relationship can be changed to adapt to the environment. Any robot controlled by the flocking controller can freely join or leave the flocking system at any time. To test the ability of the flocking system, we designed an experiment in which one single robot (Wifibot1) was placed in the front of the flocking system and it did not have a neighbor at the initial state. Wifibot2 was the leader of the flocking system and has one neighbor: Wifibot3. Wifibot3 was placed at the back of the flocking system and followed the leader (Wifibot2). Furthermore, an obstacle was placed on the trajectory as well. Our purpose was to make Wifibot1 join the flocking system, and the flocking system pass over the obstacle.

Figure 9.9 shows the video snapshots of the whole process. At the initial state (Fig. 9.9(a)), Wifibot1 was placed at a point where no neighbor robots can be connected. Wifibot2 led Wifibot3 running towards Wifibot1. At that moment, Wifibot1 did not sense the other robots because of the limited communication range. Figure 9.9(b) shows that Wifibot1 sensed Wifibot2 and started joining the flocking system. Under the flocking control, Wifibot1 was moving to the estimated position of the flocking center. In Figs. 9.9(c) and 9.9(d), the robots gradually reached a stable state and a new flocking network was formed. The flocking system was passing the obstacle in Fig. 9.9(e). It can be seen that the flocking network was changed into a line pattern. After passing the obstacle, the flocking system changed its network back to the triangular pattern (Fig. 9.9(f)).

Figure 9.10 illustrates the trajectories of the flocking system. The dashed lines denote the trajectories of the three robots. The connections between robots at four time instants $t = 22$ s, 45 s, 74 s, 105 s are illustrated. It shows that Wifibot1 moved back at the beginning as it wanted to move close to the estimated flocking center. The middle robot (marked by a triangle) was the leader. Gradually three robots moved from a line pattern to a triangle pattern ($t = 45$ s). The obstacle is marked with a big solid circle on the top of the figure. It can be seen that the flocking system smoothly avoided the obstacle and the flocking network changed to a line pattern ($t = 74$ s). Finally, the flocking system regrouped together after passing the obstacle ($t = 105$ s).

Figure 9.11 illustrates the cohesion radius of the flocking system. Initially, the cohesion radius was quite large. At $t = 22$ s, Wifibot1 joined the group and the cohesion radius was stabilized at about 600 mm. When the flocking system encountered the obstacle at about $t = 74$ s, the cohesion radius increased due to the manoeuvre of obstacle avoidance. However they were still connected as a group. After passing the obstacle, the cohesion radius decreased and the flocking system reached a stable state again.

The errors of flocking center estimation of the three robots are shown in Fig. 9.12. At the beginning of the tracking, all the errors were quite large because Wifibot1 cannot communicate with the flocking system and the flocking network was dis-

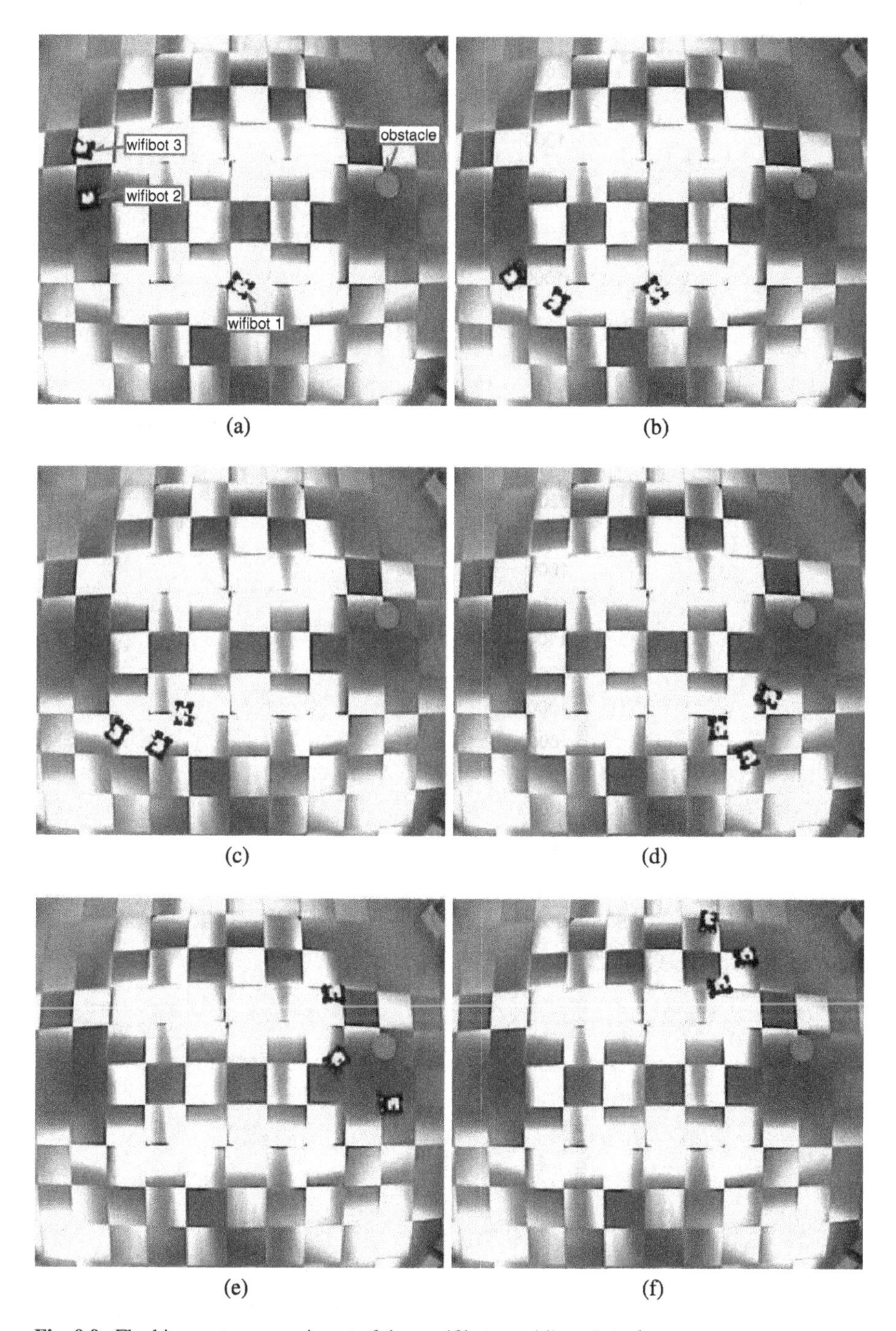

Fig. 9.9 Flocking system experiment of three wifibots avoiding obstacle

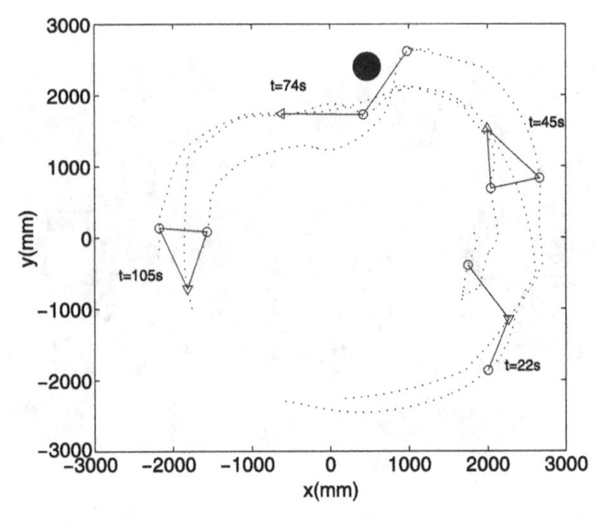

Fig. 9.10 Flocking system with three robots avoids obstacle

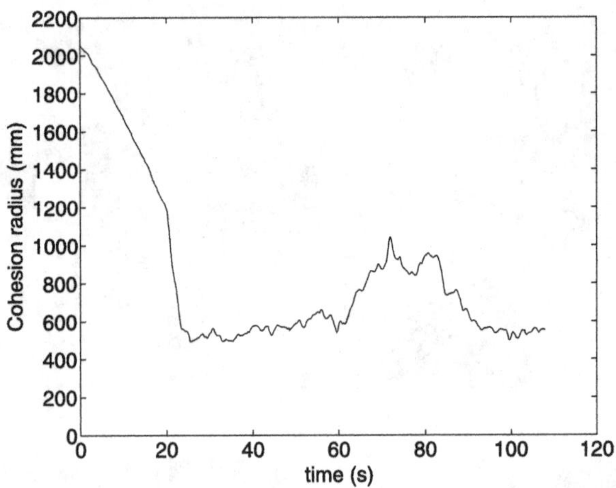

Fig. 9.11 Cohesion radius

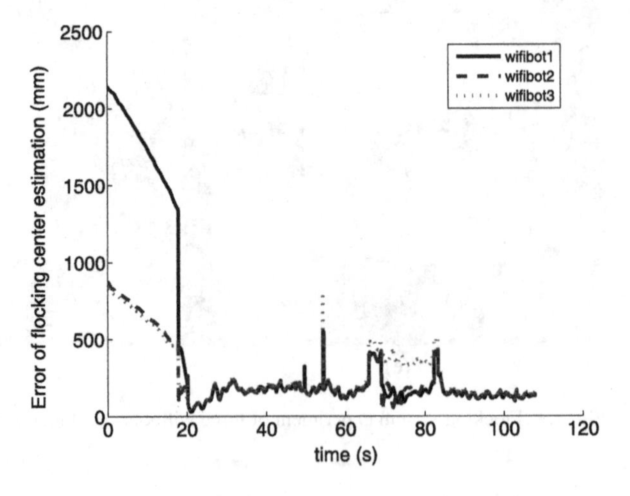

Fig. 9.12 Errors of flocking center estimation

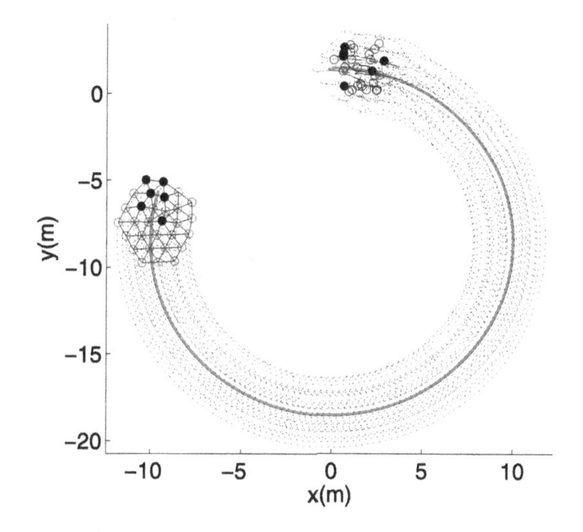

Fig. 9.13 Thirty robots track a circle

connected. After Wifibot1 joined the flocking system (at about $t = 22$ s), the errors gradually decreased. When the flocking network was formed and changed from the line pattern to the triangle pattern, the errors of the estimation results became smaller and were stabilized at around 150 mm. The fluctuation at about $t = 70$ s was caused by the behavior of obstacle avoidance.

9.6 Simulations

In the simulation, thirty robots were placed randomly in a $5 \times 5\ \mathrm{m}^2$ area. Their initial velocities were 0 m/s. The desired trajectory was a circle with center [0, −10 m] and radius 10 m. Figure 9.13 illustrates the flocking system with six leaders tracking the desired trajectory. The desired trajectory is represented by a solid line. The leaders are marked with solid circles. It clearly shows that the flocking system was formed and it tracked the circle. Furthermore, almost all of the individuals in the flocking system maintained the specific distance. The flocking center position was calculated during the tracking and is shown in Fig. 9.14 (the solid line). The dashed lines are the results of the flocking center estimation by robots. It can be seen that the estimates gradually reached a stable state and closely attached to the real flocking center.

The mean error of flocking center estimation was also calculated during the tracking. Figure 9.15 shows that the mean error slightly fluctuated at the beginning of the tracking. Finally, the mean error was stabilized around 0.5 m. Figure 9.16 shows that the cohesion radius of the flocking system gradually reached a reasonable value and maintained a relatively stable level (2.5 m). It proves that the flocking system kept cohesion during the tracking and the cohesion controller worked properly.

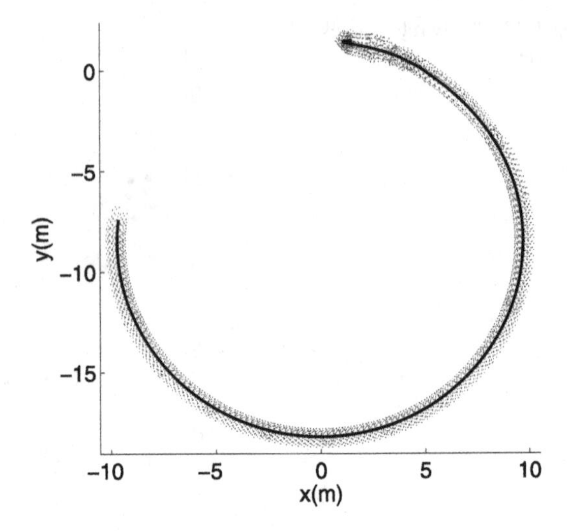

Fig. 9.14 Real flocking center position (*real line*) and flocking center estimates by robots (*dashed lines*)

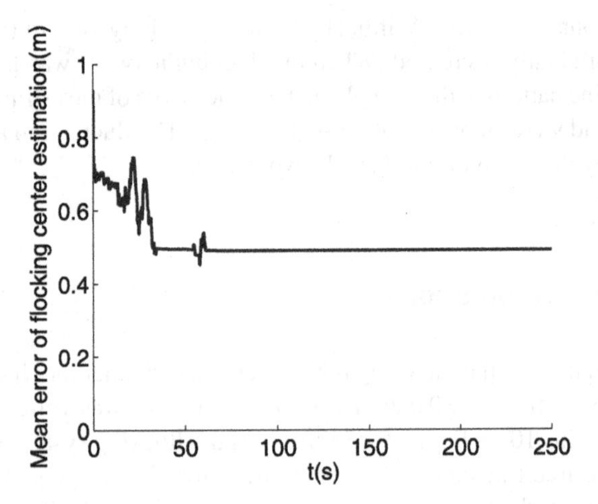

Fig. 9.15 Mean error of flocking center estimation

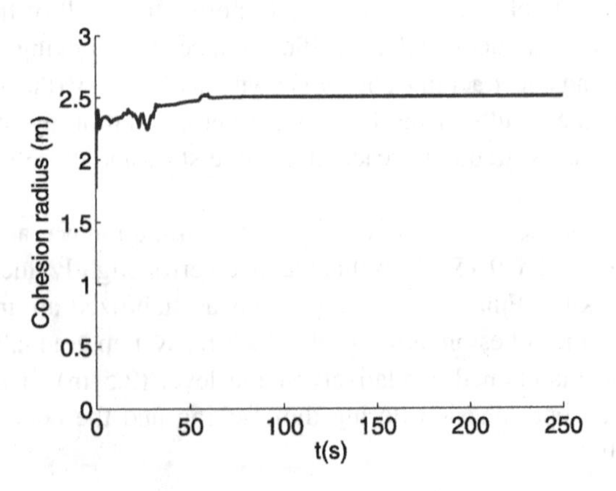

Fig. 9.16 Cohesion radius

9.7 Conclusions

Some biological flocking systems have shown that a leader-follower flocking group consists of few individuals who have global information and other members who even do not know which individuals, if any, have such information. The informed individuals play an important role in guiding those who are less experienced. This chapter proves such a leader-follower flocking system is stable and demonstrates the simulations and experiments of such a leader-follower flocking system.

To keep the cohesion of the flocking system, a consensus algorithm is used to estimate the position of the flocking center in a distributed way. The robots keep the flocking system cohesion by moving to the estimated position of the flocking center. The potential function approach is used to design a leader follower flocking controller. The fuzzy potential function is used to achieve the separation control and the flocking center estimation is used as a key element for the cohesion control.

Real robot experiments and simulations are conducted for testing the flocking controller. In both experiments and simulations, the robots only used local communication to exchange the information. The results experimentally proved that the flocking system works as a flocking group to track the desired trajectory without separation.

Our further work will focus on the scalability problem of this flocking algorithm. We are proving the group can keep the connectivity by using the flocking center information.

References

1. Information of wifibot [online]. http://www.wifibot.com
2. Couzin, I.D., Krause, J., Franks, N.R., Levin, S.A.: Effective leadership and decision-making in animal groups on the move. Nature **433**, 513–516 (2005)
3. Das, A., Fierro, R., Kumar, V., Ostrowski, J., Spletzer, J., Taylor, C.: A vision-based formation control framework. IEEE Trans. Robot. Autom. **18**, 813–825 (2002)
4. Desai, J.P., Ostrowski, J.P., Kumar, V.: Modelling and control of formations of nonholonomic mobile robots. IEEE Trans. Robot. Autom. **17**, 905–908 (2001)
5. Dimarogonas, D., Kyriakopoulos, K.J.: On the rendevzous probleme for multiple nonholonomic agents. IEEE Trans. Robot. Autom. **52**, 916–922 (2007)
6. Godsil, C., Royle, G.: Algebraic Graph Theory. Springer, Berlin (2001)
7. Ji, M., Eagnus, M.: Distributed coordination control of multiagent systems while preserving connectedness. IEEE Trans. Robot. **23**(4), 693–703 (2007)
8. Ji, M., Muhammad, A., Egerstedt, M.: Leader-based multi-agent coordination: controllability and optimal control. In: Proceedings of the American Control Conference (2006)
9. Olfati-Saber, R.: Flocking for multi-agent dynamic systems: algorithms and theory. IEEE Trans. Autom. Control **51**, 401–420 (2006)
10. Olfati-Saber, R.: Flocking for multi-agent dynamic systems: algorithms and theory. IEEE Trans. Autom. Control **51**, 401–420 (2006)
11. Reeds, S.G.: Can a minority of informed leaders determine the foraging movements of a fish school. Anim. Behav. **59**, 403–409 (2000)
12. Seeley, T.D.: Honeybee Ecology: a Study of Adaptation in Social Life. Princeton University Press, Princeton (1985)

13. Spanos, D., Murray, R.: Robust connectivity of networked vehicles. In: Proceedings of the 43rd IEEE Conference on Decision and Control, Atlantis, Paradise Island, Bahamas, December 2004
14. Wang, W., Slotine, J.E.: A theoretical study of different leader roles in networks. IEEE Trans. Autom. Control **51**(7), 1156–1161 (2006)
15. Zavlanos, M.M., Pappas, G.J.: Potential fields for maintaining connectivity of mobile networks. IEEE Trans. Robot. **23**(4), 812–816 (2007)

Chapter 10
A Behavior Based Control System for Surveillance *UAVs*

John Oyekan, Bowen Lu, Bo Li, Dongbing Gu, and Huosheng Hu

Abstract Unmanned Aerial Vehicles (*UAVs*) is required to carry out duties such as surveillance, reconnaissance, search and rescue and security patrol missions. Autonomous operation of *UAVs* is a key to the success of these missions. In this chapter, we propose to use a behavior based control architecture to implement autonomous operation for *UAV* surveillance missions. This control architecture consists of two layers: a low level control layer and a behavior layer. The low level control layer decomposes $3D$ motion of *UAVs* into several atomic actions, such as yaw, roll, pitch, altitude, and $2D$ position control. These atomic actions together serve as a basis for the behavior layer. The behavior layer consists of a number of necessary behaviors used for surveillance missions, including take-off, object tracking, hovering, landing, trajectory following, obstacle avoidance amongst other behaviors. These behaviors can be instantiated individually or collectively to fulfill the required missions issued by human operators. To evaluate the proposed control architecture, the commercially available DraganFlyer QuadRotor was used as the *UAV* platform. With the aid of an indoor positioning system, several atomic actions and a group of behaviors were developed for the DraganFlyer. Real testing experiments were conducted to demonstrate the feasibility and performance of the proposed system.

10.1 Introduction

Within the last decades, the use of Unmanned Aerial Vehicles (*UAVs*) in military and security operations has increased tremendously, including surveillance, reconnaissance and even search and destroy missions because of the significant advantages over manned aerial vehicles. For example, small size of an *UAV* enables it to penetrate an enemy's radar defences while maintaining a low radar signature. Its low

J. Oyekan (✉), B. Lu, B. Li, D. Gu, and H. Hu
School of Computer Science and Electronic Engineering, University of Essex, Wivenhoe Park, Colchester CO3 4SQ, UK
e-mail: jooyek@essex.ac.uk; blv@essex.ac.uk; dgu@essex.ac.uk; hhu@essex.ac.uk

H. Liu et al. (eds.), *Robot Intelligence*,
Advanced Information and Knowledge Processing,
DOI 10.1007/978-1-84996-329-9_10,

noise level gives it a low profile when stalking or collecting status data of an object of interest. This prevents it from being detected easily and hence shot down. An *UAV* also carries more payload than a manned aerial vehicle since the space that is occupied by a human pilot and life support systems can be filled up with equipment. In addition to these advantages, if an *UAV* were shot down, no human life would be lost as the operator would not be on-board [1].

However, most of *UAV*s require a human operator to remotely control it. This re-introduces the weakness of the human factor into the control loop as humans may lose concentration, get tired and bored after extended periods of time leading to mistakes. Furthermore, the probability of making mistakes increases with an increase with the degree of complexity required to fly the *UAV*. This was made more obvious with the amount of effort required to manually fly the platform used in our work.

Currently enormous effects on autonomous operations have been made to address this problem. The evidence can be found from many research projects that have carried out investigations into controlling *UAV*s autonomously. The Multiple Agent Intelligent Coordination and Control (*MAGICC*) lab at the Brigham Young University used fixed-wing *UAV*s during their outdoor investigation of autonomous tracking of way-points [2]. They used the force field vector method to fly *UAV*s to a desired way-point. The *OAT*s project of the University of Oxford implemented autonomous visual tracking of intelligent ground targets and way-points using a commercial airframe [3]. The Stanford Testbed of Autonomous Rotorcraft for Multi-Agent Control (*STARMAC*) project [4] used a QuadRotor platform to investigate autonomous multi-agent control of numerous QuadRotors in real-world scenarios. The platform was developed so that it carried sufficient computing power in order to carry out computations in real-time. The *UAV*s in this project followed a trajectory autonomously in order to reach a desired way-point [5]. The Vanderbilt Embedded Computing Platform for Autonomous Vehicle (*VECPAV*) project at Vanderbilt University aimed to develop an autonomous intelligent control system that replaces a human operator in the flight control process of an *UAV* [6]. The Raven project of the Massachusetts Institute of Technology investigated the development of a test bed for the rapid prototyping of *UAV* technologies. It used a multi-vehicle platform comprising of autonomous ground and aerial vehicles. It aimed to develop mission-level algorithms [7]. This project made use of a motion-capture system during its investigations. The Bear Aerobot team of the Berkeley University used a rotary winged platform to investigate obstacle avoidance and trajectory path planning in an urban environment [8]. The Carnegie Mellon Robotics Institute used a rotary winged helicopter to investigate autonomous flight and have achieved autonomous takeoff, trajectory following and landing [9]. The Georgia Institute of Technology has developed a visual target designation and tracking system algorithm for an autonomous helicopter platform [10]. The *WITAS* Laboratory at the Linkoping University aims to develop algorithms that would enable an *UAV* to be autonomously deployed to monitor traffic networks. This would enable the *UAV* to navigate in such environment and identify vehicles and their behaviors and react accordingly [11].

Furthermore, some *UAV*s are small enough to be flown through windows to be used for indoor surveillance of human subjects or objects of interest. These *UAV*s do not have access readily to *GPS* data for flying indoors. In this situation, alternative means of providing reference points is necessary. The use of lasers on an *UAV* is one way of solving this problem [12]. Another way of solving this problem would be to use vision algorithms.

Vision is an essential sensory component for *UAV*s to carry out surveillance missions. A Teleos Corp. Advanced Vision Platform (*AVP*) in combination with *GPS* was used to control a *UAV* in [13]. They used the color segmentation ability of the *AVP* to identify and subsequently track objects. Saripalli et al. used vision to identify a helipad and control their *UAV* to land on the helipad [14]. Zufferey described an algorithm capable of efficient course stabilization and collision avoidance using optic flow and inertial information [15]. Erdinc et al. used a pose estimation algorithm in their work, i.e. blobs on the base of the Quadrotor to determine an estimate of x,y, and z positions in the image plane. The image of blobs on the *UAV* was captured by setting up a camera pointing upwards at the *UAV* base. They also used the blobs to determine the estimate of yaw, roll and pitch angles. Their technique was sensitive to noise as a result of using the number of pixels per unit area of the camera [16]. A visual servo tracking controller which makes *UAV* stay at a fixed relative position and orientation was developed in [17]. By comparing the feature points on the fixed target and a corresponding feature points in the template, projective geometric relationships are exploited to construct a Euclidean homograph. In [3], a mean shift algorithm for object tracking was used. It also presents a way of re-acquiring an object by using camera rotations and helicopter movement.

Our research group has experience on the building of autonomous control systems for quadruped robots to play football and for robotic fishes to navigate in water. The successful implementation of these autonomous control systems indicates that we can build an autonomous control system for *UAV*s to carry out surveillance missions as well. The methodology used is to build a behavior based architecture. A two-layer architecture is sufficient for complicated missions. The low level layer is responsible for the atomic actions used for controlling *UAV*s in a 3D space. The atomic actions include angle control of yaw, roll, and pitch, altitude control, and 2D position control. With the implementation of these atomic actions, *UAV*s will have the capability to motion in a 3D space. The high level layer is termed as behavior layer which contributes a group of orthogonal behaviors used for surveillance missions. These behaviors are constructed based on the atomic actions and are targeted to meaningful tasks, including takeoff, object tracking, hovering, landing, trajectory following, and obstacle avoidance. A tracking behavior using either vision or *GPS*-like co-ordinates could be used to stalk and investigate a moving object of interest. If the object becomes stationary, a hovering behavior is used to maintain a fixed position for continuous observation. Obstacle avoidance behavior is used to avoid obstacles in a flight path of the *UAV* while the trajectory tracking behavior is used to maintain a flight path or trajectory precisely. Landing behavior can be used for landing in a static spot or in a movable platform.

Most *UAV*s rely on its *IMU* unit for providing sensory inputs to the angle control of atomic actions, and rely on *GPS* or other sensors for providing sensory inputs to the altitude control and 2D position control. In this investigation, all the atomic actions rely on an *GPS*-like indoor positioning system, the *VICON* positioning system, which can provide precise $3D$ location information to the QuadRotor. As the behavior layer is built on the atomic actions, all the behaviors also rely on the *VICON* positioning system. In addition, some behaviors also require the use of vision information. For example, object tracking behavior, landing behavior, and hovering behavior use vision information to identify a target, estimate its position, and eventually track it based on the estimated result.

The SURF algorithm is widely used in many applications, including object recognition, robotic mapping and navigation, image stitching, 3D modelling, gesture recognition, video tracking, and match moving due to faster computation and comparison capability. Its advantage also includes the robustness to viewpoint angle change, scale change, and illumination change. Furthermore, it has a high repeatability even when the image is blurred [21, 22]. A visual SURF algorithm is used to identify an object of interest in our system. Based on the result of the *SURF* algorithm, a Kalman filter is developed to estimate a moving target and the result estimated from the Kalman filter is used for corresponding behaviors.

In the following, we will introduce the DraganFlyer Quadrotor platform used in this research, our system structure, and atomic actions in Sect. 10.2. In Sect. 10.3, we will present our proposed behavior based control architecture and a group of surveillance behaviors. The vision processing algorithms are discussed in Sect. 10.4. All the proposed algorithms have been implemented in our lab. And the experimental result was obtained and they are presented in Sect. 10.5. We conclude our work and discuss future work in Sect. 10.6.

10.2 Platform and Atomic Actions

10.2.1 UAV Platform

The platform used during this investigation was a DraganFlyer Quadrotor as shown in Fig. 10.1. Its power supply and physical size made it possible to conduct experiments in a space constrained indoor facility—the Robotics Arena in University of Essex. The platform is easy to maintain and repair—things that are difficult to do on other types of rotary platforms due to the use of mechanical linkages.

As shown in Fig. 10.1, the Quadrotor we used is made up of four electric motors. By controlling the speed of the motors, it is possible to control the Quadrotor to fly in a $3D$ space. This platform is highly mobile and unstable and is not humanly possible to control without the use of onboard embedded system. This is partly due to its highly dynamic and coupled nature. However, even with the onboard embedded system, controlling this platform remotely is not a trivial task neither as the human operator has to keep making minute adjustments to make the platform hover at a particular location or follow a path directly. This leads to tiredness and sore thumbs [18]. As a result, we were not able to fly this platform manually during our experiments.

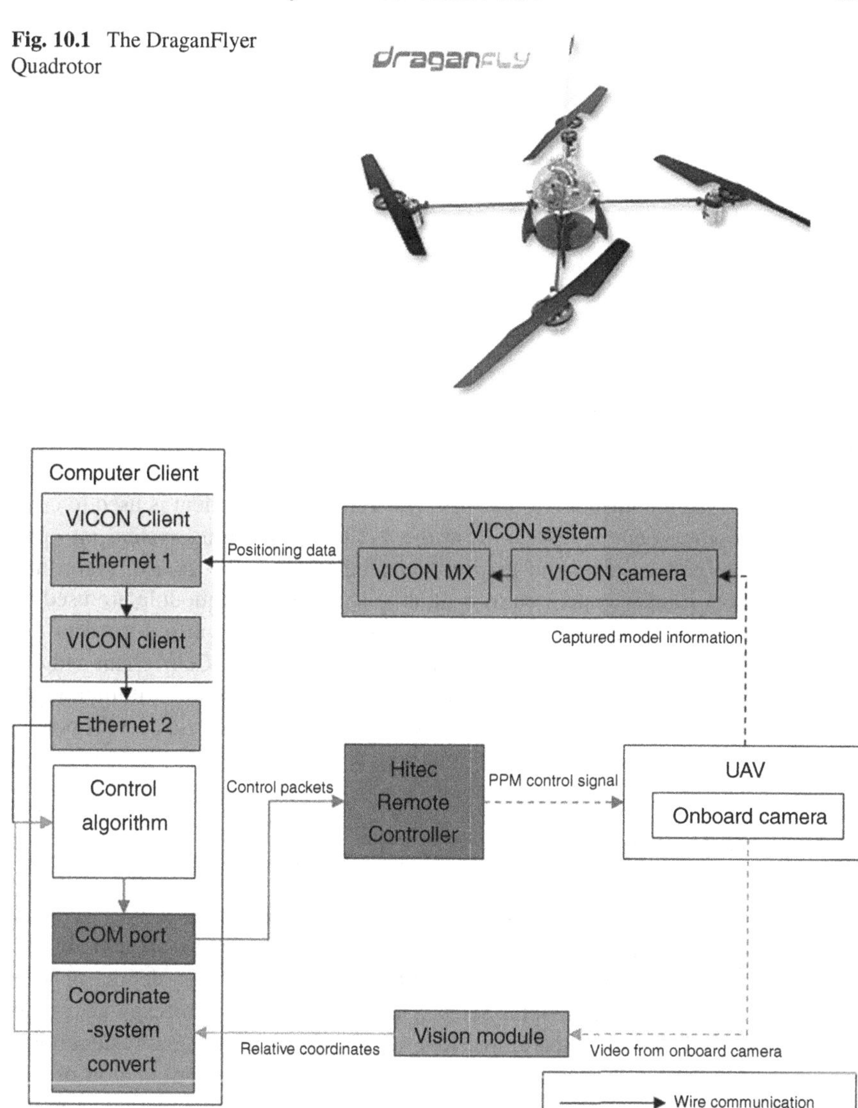

Fig. 10.1 The DraganFlyer Quadrotor

Fig. 10.2 System structure

10.2.2 System Structure

Our system structure is shown in Fig. 10.2. The computer client on the left of the figure sits on a ground station computer. It obtains the positioning data of objects from the *VICON* positioning system. It obtains the vision information from the onboard camera of the DraganFlyer via wireless communication. It sends out the con-

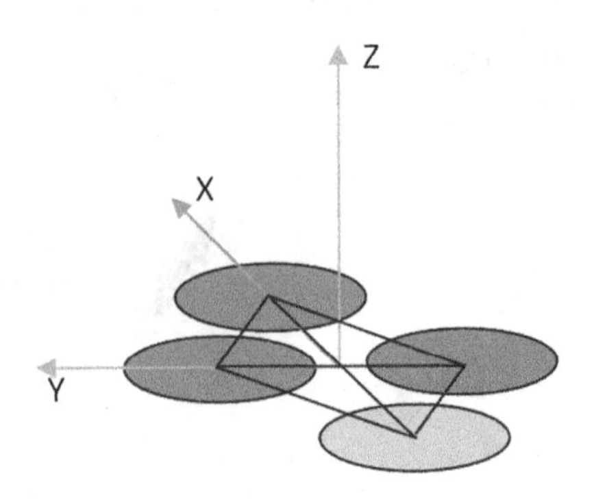

Fig. 10.3 Local coordinate frame xyz

trol packets to a remote controller which communicates with the DraganFlyer via wireless communication.

Inside the computer client, the core part is the control algorithm module which implements all the atomic actions and behaviors. The *VICON* client is used to convert the positioning data of objects from the *VICON* positioning system into position, altitude, and attitude information and feed them to the control algorithm module. The coordinate system convert module and the vision module are used to identify objects and estimate their positions. The estimated result is feeded to the control algorithm module. Based on the information from the *VICON* client and the vision module, atomic actions and behaviors inside of the control algorithm module produce a control packet and it is passed to the remote controller via a COM port.

10.2.3 Atomic Actions

Basic motion manoeuvre in a $3D$ space is fundamental to *UAV*s missions. A local coordinate frame xyz is defined as shown in Fig. 10.3. The angles rotated around x, y, and z axes are called roll, yaw, and pitch, respectively. Zero degree of yaw angle corresponds to the horizontal plane. The height in z axis corresponds to the altitude of the *UAV*. In this research, we identify six atomic actions for basic motion maneuver of an *UAV*s in order to decouple $3D$ motion into a combination of $1D$ motion. They are three angle control actions (yaw, roll, pitch), one altitude control action, and two $2D$ position control actions in x and y directions. Based on these atomic actions, the *UAV* is able to achieve complex behaviors.

PID controllers are used to implement these atomic actions because of its robustness as observed in [19, 20]. Six *PID* controllers are running in parallel to achieve the atomic actions. These *PID* controllers are shown in Figs. 10.4, 10.5, and 10.6.

In all the feedback loops of the *PID* controllers, the position, altitude and attitude information were obtained by using the *VICON* positioning system operating at a frequency of 50 Hz. This positioning system emulates using a *GPS* indoors and allows us carry out experiments similar to that would be carried out outdoors if *GPS* information is available. For the angle controllers, it should be note that the feedback signals may also be obtained from *IMU* for those *UAV*s equipped

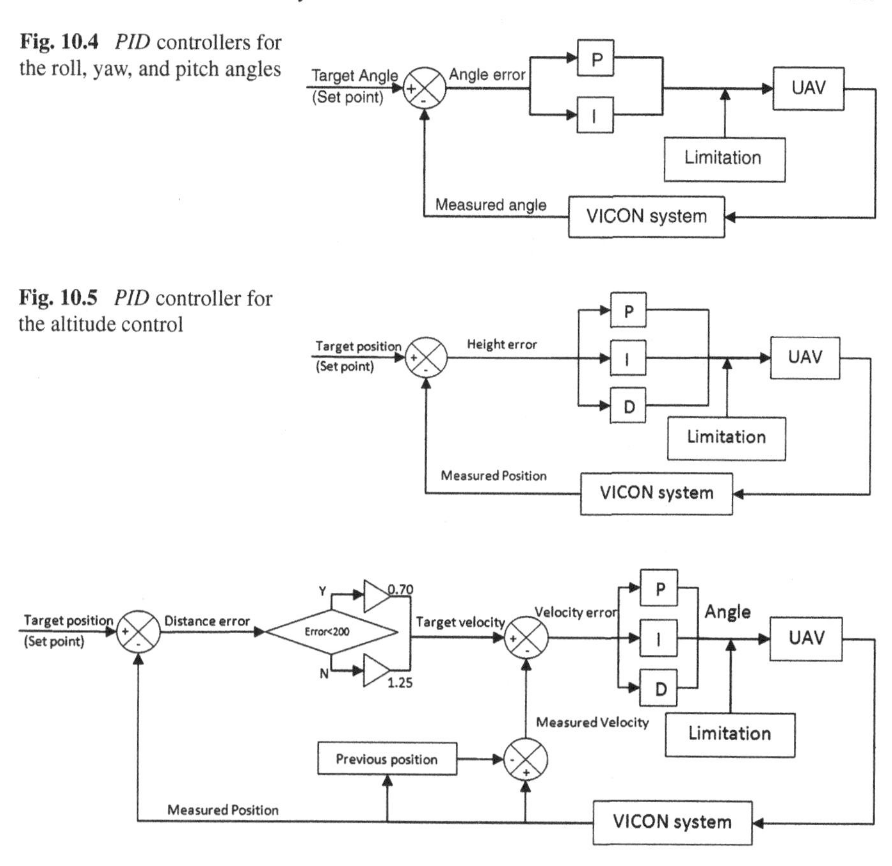

Fig. 10.4 *PID* controllers for the roll, yaw, and pitch angles

Fig. 10.5 *PID* controller for the altitude control

Fig. 10.6 *PID* controllers for $2D$ position control in x and y directions

with *IMU*. Three angle controllers and altitude controller are designed straightforwardly. However, the $2D$ position controllers are implemented by controlling roll and pitch angles. The controllers are adaptive according to the distance to target position.

10.3 Software Architecture and Behavior Development

10.3.1 Software Architecture

The control algorithm module shown in Fig. 10.2 consists of atomic actions, surveillance behaviors, and vision identification and estimation of objects. Its internal structure is shown in Fig. 10.7.

The low level layer is shown at bottom of the figure where four controllers for yaw, roll, pitch, and throttle control are included. They are corresponding to yaw,

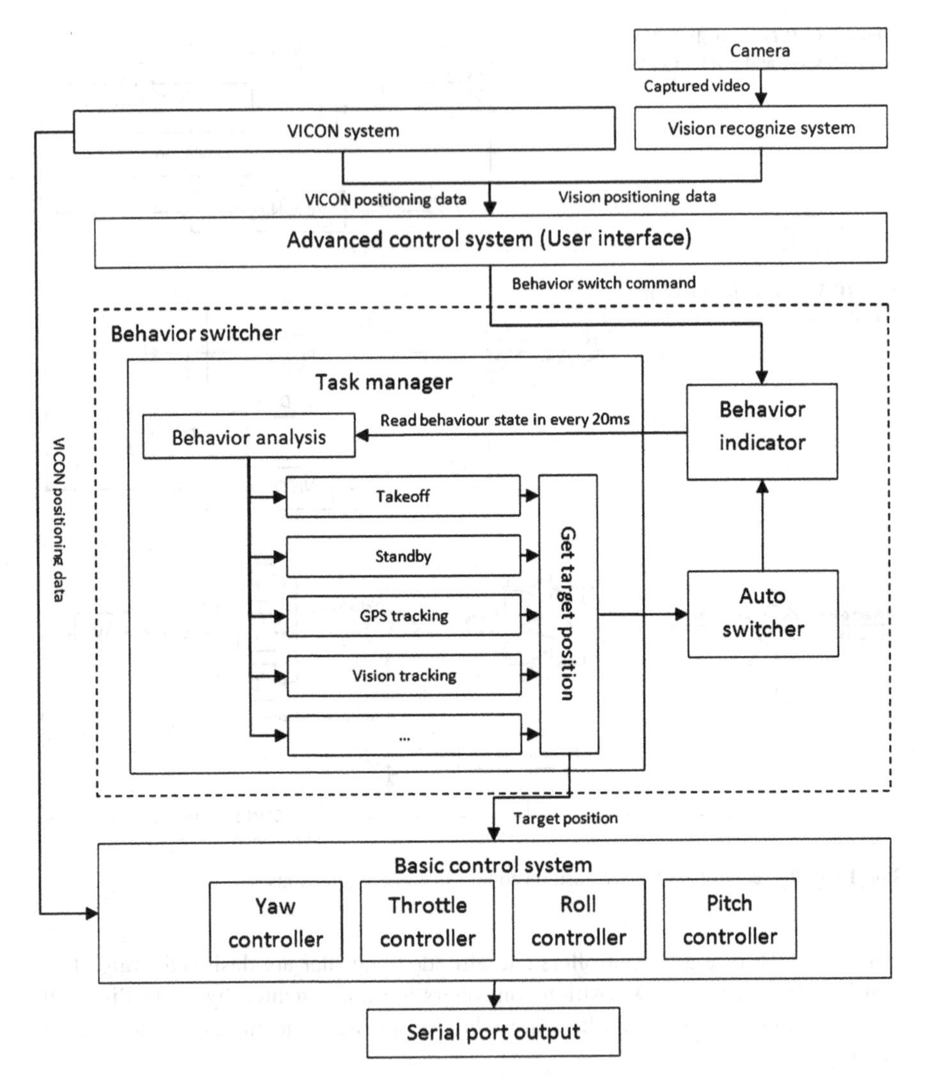

Fig. 10.7 Internal architecture of control algorithm module

roll, pitch, and altitude atomic actions. The position control atomic actions are implemented by a combination of these four controllers. The *VICON* positioning system provides accurate altitude and attitude information of DraganFlyer to this layer.

The high level layer is shown in the middle of the figure where a group of behaviors are stacked together and managed by a task manager. The task manager will read the behavior setting from an indicator every 20 milliseconds and then select one behavior to run. The behavior indicator contains necessary information for a behavior to run and the information can be modified from two different sources. One source is from the end-user, which comes from the advanced control system or user interface, while the other source is from the internal control mechanism - the auto switcher. Every time the task manager reads from the behavior indicator, it searches for a behavior which matches to the requirement, and then generates a new target position for low level atomic actions according to some rules. The new target position information will be sent to the low level layer and the auto switcher. The auto switcher is responsible for checking whether the DraganFlyer moves to the new target position or not. Then the behavior indicator can be modified accordingly.

On the top of the behavior layer, there is an advanced control system, which is also named as user interface. It offers encapsulated function modules, and each function module may contain a serial of behaviors. With this encapsulation design, the end-user can ignore the internal operating procedures, which make the system easy to use and protected from an incorrect manipulation.

The feedback from the vision camera provides the position information of object being tracked. This information is used as the reference for the vision tracking and vision landing behaviors. The vision processing module will be explained further in Sect. 10.4.

10.3.2 Behavior Development

Several behaviors have been developed using the system discussed above. With these behaviors as the base, a customizable behavior based control system can be built up easily with the auto switcher mechanism. The following subsections give details on ten developed behaviors for surveillance missions.

10.3.2.1 Ground Behavior

Ground behavior is the default state of behavior indicator. This behavior cuts off the throttle of DraganFlyer and initializes all the control variables.

10.3.2.2 Takeoff Behavior

Takeoff behavior is a transition behavior, which enables the DraganFlyer takeoff from the ground and reaches to a predefined height. After that, the auto switcher will change the behavior indicator into a setting to trigger the hovering behavior.

10.3.2.3 Hovering Behavior

Hovering behavior is one of the fundamental behaviors of our system. With this behavior, it is possible for the DraganFlyer to maintain a position precisely whilst flying. This behavior is achieved by using the altitude *PID* controller shown in Fig. 10.5 to maintain a height of the DraganFlyer and the $2D$ position *PID* controllers shown in Fig. 10.6 to maintain the $2D$ position of the DraganFlyer.

10.3.2.4 *GPS* Landing Behavior

Landing behavior is needed for autonomous landing after completing a mission. This landing behavior relies on the use of the *VICON* system to find the landing position. The main atomic action used in the landing behavior is the altitude *PID* controller, along with the position controllers. The landing behavior implements an adaptive altitude *PID* control strategy by changing the *PID* parameter values according to the height of the DraganFlyer. The adaptive controller can produce a fast and safe landing behavior. When the DraganFlyer is far from the ground, it should go down fast. When it is close to the ground, it should go down slowly to avoid large overshoot and crashing into the ground.

10.3.2.5 Vision Landing Behavior

Vision landing behavior relies on the use of the vision module to find the landing position and uses the similar strategy as in the *GPS* landing behavior for landing. However, the landing position is prone to be lost due to blurred images during the landing process. Here we take a step by step strategy: each time when the vision module finds the landing target with certain confidence, the behavior controls the DraganFlyer to get lower in a small step, otherwise the behavior controls the DraganFlyer to get higher and tries to find the target again.

10.3.2.6 Emergency Landing Behavior

Emergency landing behavior uses information from the *VICON* system to land. This behavior aims to minimize the time taken to land and thus is faster for landing than the normal landing behavior. It is designed for emergency situations, such as low battery, huge disturbance in the environment. As a tradeoff, the accuracy of this behavior for landing at a particular position is worse than the normal landing behavior.

10.3.2.7 *GPS* Tracking Behavior

GPS tracking behavior can track a marked moving target with the assistant of *VICON* system. It is built based on the hovering behavior by providing a moving target position to the hovering behavior. A remote control car is used as the moving object. The *VICON* system identifies the car and provides its position information to this behavior.

10.3.2.8 Vision Tracking Behavior

Vision tracking behavior is similar to the *GPS* tracking behavior, but the target position is provided from the vision module.

10.3.2.9 Obstacle Avoidance Behavior

Obstacle avoidance behavior is developed by using a potential filed approach. An extra input about the distance to the obstacle is necessary to this behavior. In current experimental setting up, this information is provided by an simulated obstacle inside of the ground station computer. This behavior is programmed to move away from the obstacle with a distance of 1000 mm. Aiming at this distance away from the obstacle enables the low level *PID* controllers to recover from any overshoots so that the DraganFlyer would be able to fly without hitting the obstacle.

10.3.2.10 Trajectory Tracking Behavior

Trajectory tracking behavior is to follow a predefined trajectory precisely. The predefined trajectory is represented by using minute way-points. This behavior simply moves through the way-points sequently by repeating the atomic actions—2*D* position controllers argumented with way-points. However some obstacles may block the predefined trajectory. In order to avoid obstacles, the final trajectory tracking behavior combines both the trajectory tracking behavior without obstacles and the obstacle avoidance behavior.

The combination of two behaviors is implemented by a fuzzy logic approach where the final output of trajectory tracking behavior with collision avoidance capability is a weighted sum of outputs from the trajectory tracking behavior without obstacle and the obstacle avoidance behavior. The weights are generated from triangle fuzzy logic membership functions of the distance to way-points and the distance to obstacle. Figure 10.8 shows the mechanism of this combination.

10.4 Vision Module Development

The vision module captures images from the onboard camera and yields an estimated position of object being tracked. It consists of three stages: object identification, coordination transformation, and position estimation. In the first stage, *SURF* algorithm is used to detect the object and generates the position of object in image plane. In the second stage, the position of object in image plane is transformed into the position in the local coordinate frame xyz and then in global frame. Finally a Kalman filter is used to filter the position of object in the global frame.

10.4.1 SURF Algorithm

SURF algorithm works by first matching each key feature in the present image independently to a database of key features extracted from a desired stored target image. The best candidate match for each key feature is found by identifying its nearest

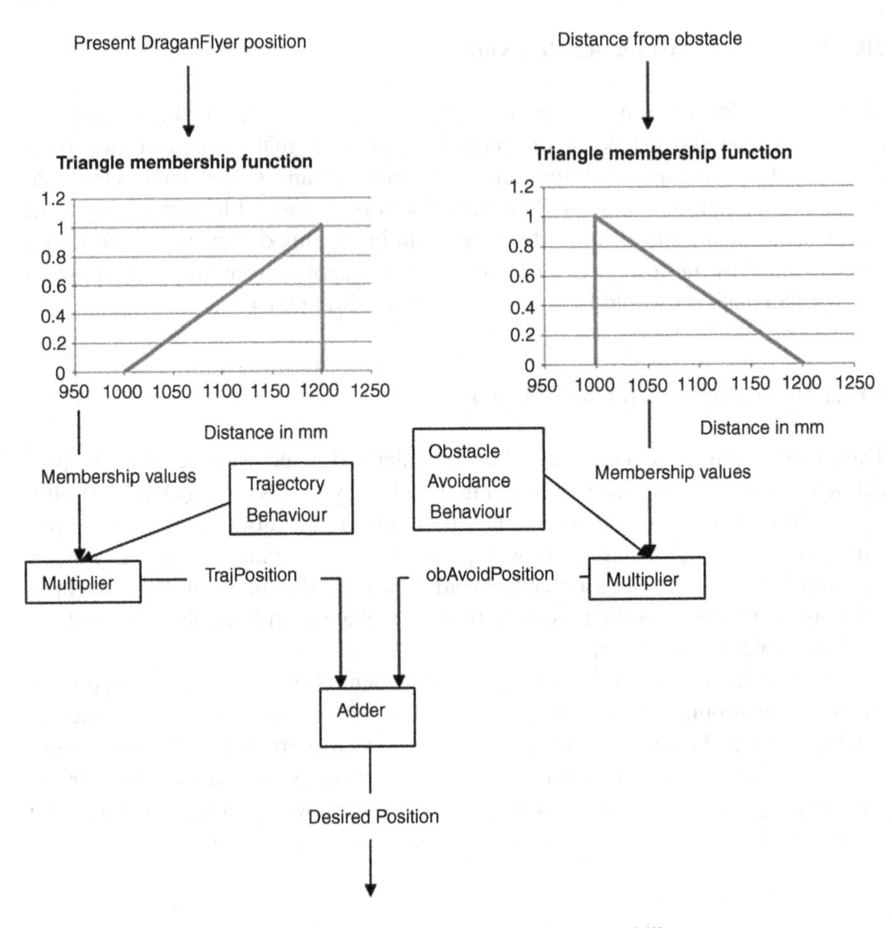

Fig. 10.8 Trajectory tracking behavior with collision avoidance capability

neighbor in the database. The nearest neighbor is defined as the key feature with the minimum Euclidean distance for the invariant descriptor vector [23]. For more information on the details of the SURF algorithm, the reader is advised to refer to the reference [23].

Figure 10.9 shows how the SURF algorithm matches key features of an input image to the key features of a stored image in the database. The point of intersection of the two catercorners denotes the target center. Figure 10.10 shows the accuracy of the point of intersection when the corners are not well defined.

10.4.2 Coordination Transformation

The coordinate transformation projects the coordinate information from the image plane into the local coordination frame of the DraganFlyer. As a result, the global

Fig. 10.9 Feature matching

Fig. 10.10 Feature matching for tilted image

position of object can be obtained by summing the local coordinates and the DraganFlyer's present coordinates together. The coordinate transformation is shown in Fig. 10.11 where H is the height of the DraganFlyer, x^L is the local coordinate of object, h is the focus of camera, x is the object coordinate in image plane, and α is a angle of the DraganFlyer. The coordinate transformation equation is as follows:

$$x^L = H \cdot \tan\alpha + \frac{x \cdot \frac{H}{h}}{\cos\alpha(1 - x \cdot \frac{\sin\alpha}{h})}. \tag{10.1}$$

10.4.3 Kalman Filter

Although the *SURF* algorithm can produce a result of object position, it is noise and unstable due to the motion of the on-board camera. Kalman filter can minimize the estimation error and generate an unbiased estimation. It is used in the vision module

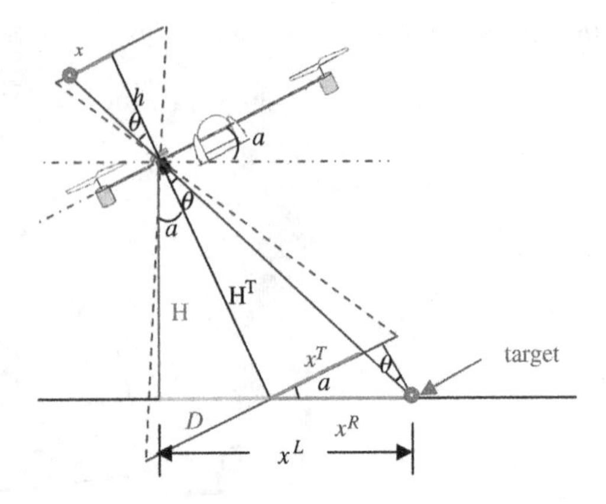

Fig. 10.11 Coordination transformation

of our system. In the Kalman filter, the object is modelled as a discrete time linear time-invariant system defined by the system equation:

$$q_{t+1} = Aq_t + Bw_t$$

where q_t is the object position. w_t is the state noise and its mean and covariance matrix are zero and $Q \geq 0$, respectively. A and B are defined as follows.

$$A = \begin{bmatrix} 1 & 0 & T & 0 \\ 0 & 1 & 0 & T \\ 0 & 0 & 1 & 0 \\ 0 & 0 & 0 & 1 \end{bmatrix}, \qquad B = \begin{bmatrix} T^2/2 & 0 \\ 0 & T^2/2 \\ T & 0 \\ 0 & T \end{bmatrix},$$

where T is the sample interval.

The *SURF* algorithm generates an observation of object and it is denoted as p_t. The measurement equation is defined as

$$p_t = Cq_t + v_t$$

where $C = \left[\begin{smallmatrix} 1 & 0 & 0 & 0 \\ 0 & 1 & 0 & 0 \end{smallmatrix}\right]$. v_t is the Gaussian noise and its mean and covariance matrix are zero and $R > 0$, respectively.

The Kalman filter based on the above equations is a standard iterative process. It goes through a prediction step and updating step. The final result is the estimated position of object.

10.5 Experiment Results

10.5.1 Hovering Behavior

The DraganFlyer was commanded to hover at $(x, y, z) = (-500, 0, 1020)$ mm for approximately 100 seconds. Its height was recorded and shown in Fig. 10.12. Its x

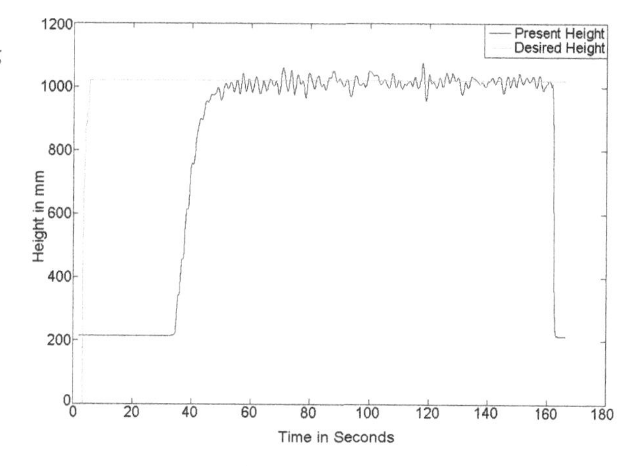

Fig. 10.12 Height of the DraganFlyer during hovering behavior

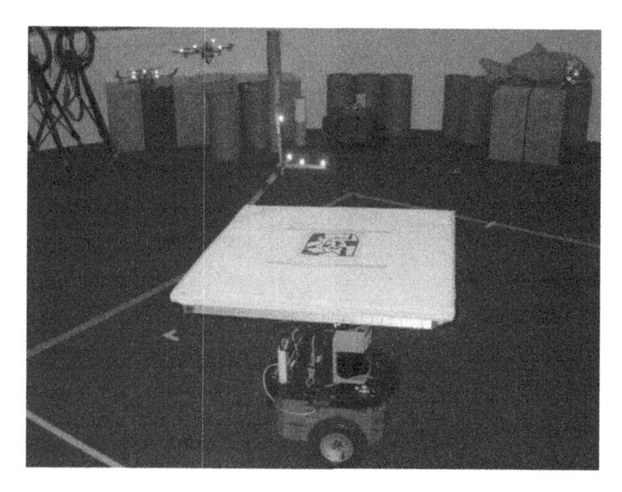

Fig. 10.13 Vision tracking experiment

and *y* position were maintained within 700 mm and 300 mm. These results show the Draganflyer was able to hover on the spot steadily.

10.5.2 Vision Tracking Behavior

A moving unmanned ground vehicle (*UGV*) with a pattern on it was tracked by the Draganflyer using camera as shown in Fig. 10.13. The *UGV* was commanded to move in a circle during data collection. We tested the tracking performance by using *UGV* speed of 150 mm/s and the result is shown in Fig. 10.14. Three trajectories are the *UGV* trajectory obtained from the *VICON* system, estimated trajectory obtained from the vision module, and the Draganflyer's trajectory obtained from the *VICON* system. From these trajectories, it can be seen that the tracking behavior was successfully completed without losing the tracking target.

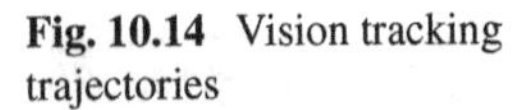

Fig. 10.14 Vision tracking trajectories

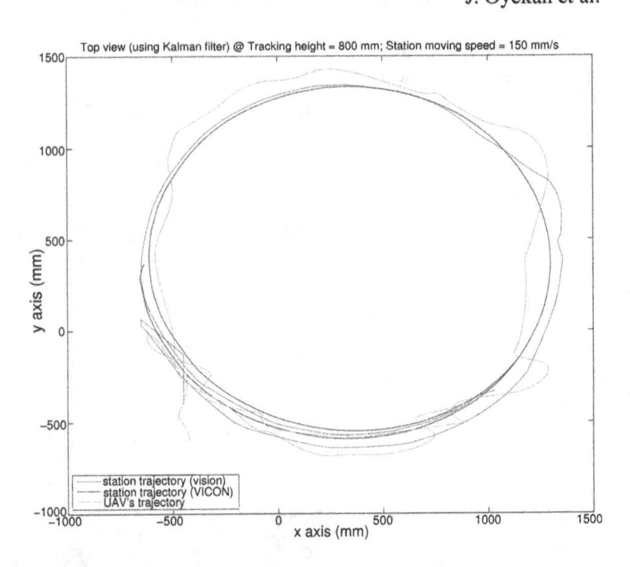

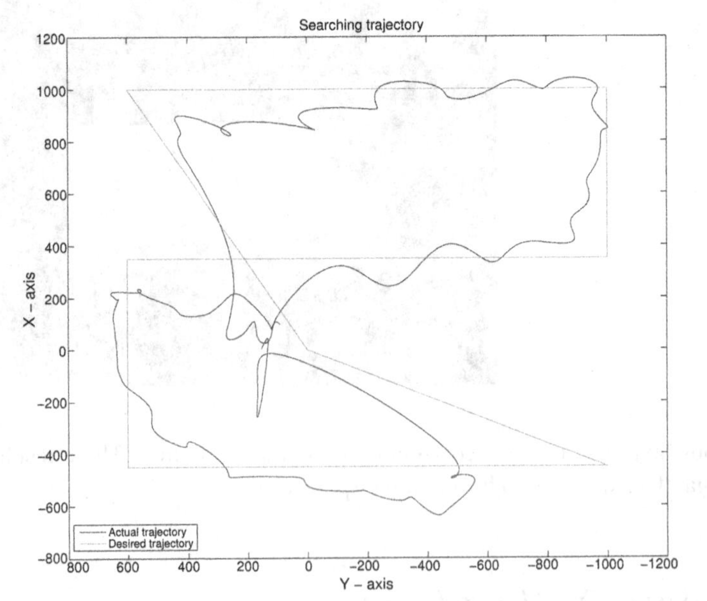

Fig. 10.15 Trajectory tracking behavior

10.5.3 Trajectory Tracking Behavior

Trajectory behavior was investigated by simulating a snake like trajectory in the Robot arena. A few way-points were defined at corners of the snake like trajectory as shown in Fig. 10.15. The result shows both the actual trajectory of the DraganFlyer and the snake like trajectory. As can been seen from this figure, the DraganFlyer was able to follow a complex trajectory path in the Arena.

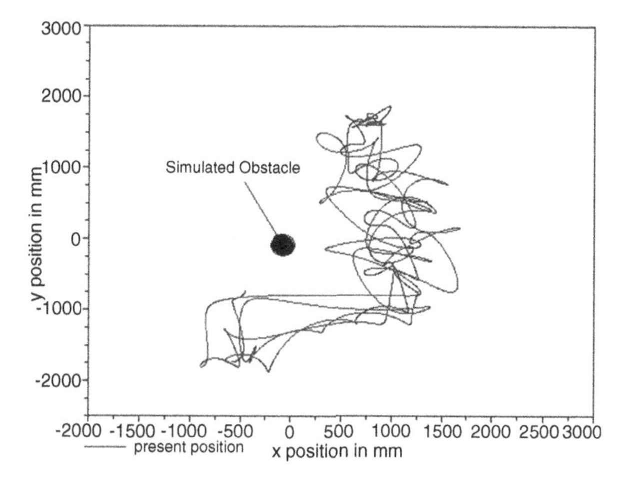

Fig. 10.16 Trajectory tracking and obstacle avoidance

10.5.4 Trajectory Tracking Behavior with Obstacle Avoidance capability

An obstacle was simulated by using coordinates $(x, y) = (0, 0)$ mm. The DraganFlyer was commanded to go from the home way-point $(x, y, z) = (500, 1600, 0)$ mm to a destination way-point of $(x, y, z) = (-500, -1600, 1020)$ mm and then back home. This was repeated three times to prove that the behavior was effective in avoiding the obstacle. The trajectory from the *VICON* system in Fig. 10.16 shows the DraganFlyer was far enough from the obstacle and there was no collusion occurring.

10.5.5 GPS Landing Behavior

Figure 10.17 shows the height of the DraganFlyer during its landing process. The DraganFlyer took off first and then reached at a height of 1000 mm. After 150 s, it started to land. When the height of the DraganFlyer was within 600 mm or 320 mm, the behavior outputted a reduced throttle value so that the thrust generated by the rotors were slightly below the weight of the DraganFlyer. In this way, the DraganFlyer can land to the ground slowly and accurately.

10.5.6 Vision Landing Behavior

Vision landing behavior was tested when it was used for landing on a moving target with speed of 50 mm/s. The result is shown in Fig. 10.18. It takes approximately 40 seconds to land.

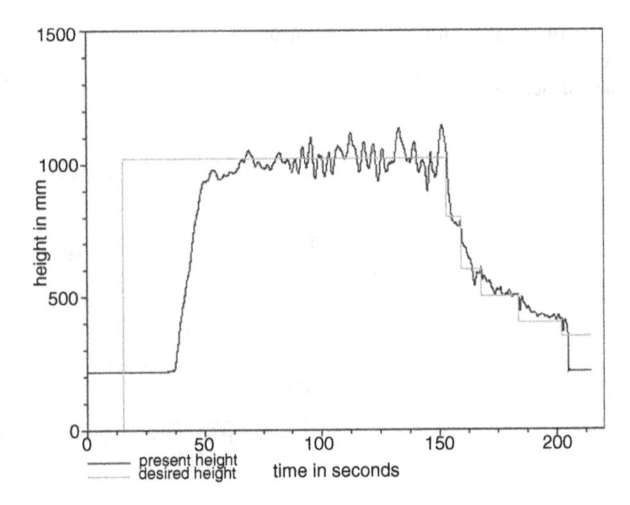

Fig. 10.17 The height changes in the landing behavior

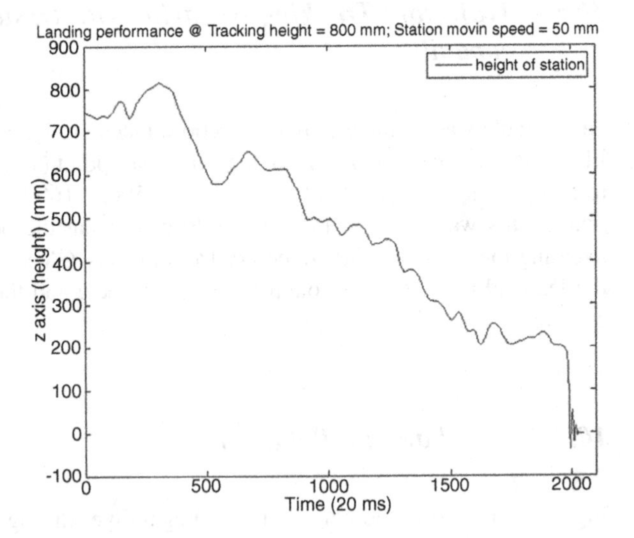

Fig. 10.18 The height changes during landing at a moving *UGV*

10.6 Conclusion and Future Work

In this chapter, we have developed a behavior based control architecture for *UAV* surveillance mission. This architecture contains two layers: atomic action layer and behavior layer. We have also developed six atomic actions and ten behaviors for these layers. Various techniques have been used in the development, including adaptive *PID* controller, fuzzy logic controller, *SURF* algorithm, and Kalman filter. All the behaviors have been tested by using a real DraganFlyer in our lab.

Our future work would extend the work to control a swarm of both *UAV*s and *UGV*s for autonomous multiple target tracking, continuous surveillance and coverage of an area.

References

1. Oyekan, J.O., Hu, H.: Towards autonomous patrol behaviours for UAVs. In: Proceedings of UK EPSRC Workshop on Human Adaptive Mechatronics, Staffordshire University, Stafford, UK, 15–16 January 2009
2. Nelson, D.R., Barder, D.B., McLain, T.W., Beard, R.W.: Vector field path following for small unmanned air vehicle. In: Proceedings of the 2006 American Control Conference, Minneapolis, MN, pp. 5788–5794 (2006)
3. Helble, H., Cameron, S.: OATS: Oxford aerial tracking system. In: Towards Autonomous Robotic Systems (TAROS), Surrey University, Guildford, UK, September 2006, pp. 72–77 (2006)
4. STARMAC, Stanford Test Bed of Autonomous Rotorcraft for Mult-Agent Control. http://hybrid.stanford.edu/starmac/overview (06 March 2008)
5. Hoffmann, G., Rajnarayan, D.G., Waslander, S.L., Dostal, D., Jang, J.S., Tomlin, C.J.: The Stanford testbed of autonomous rotorcraft for MultiAgent control (STARMAC). In: The 23rd Digital Avionics Systems Conference, DASC 04, vol. 2, pp. 12.E.4–121-10(2004)
6. RCHelicopter.com, DraganFlyer RC helicopters in VECPAV autonomous control system at Vanderbilt university, NV, USA. http://www.rchelicopter.com/2007/11/22/draganyer-uav-vanderbilt-university-drone/ (06 March 2008)
7. How, J.P., Bethke, B., Frank, A., Dale, D., Vian, J.: Real-time indoor autonomous vehicle test environment. IEEE Control Syst. Mag. **28**(2), 51–64 (2008)
8. Shim, D.H., Chung, H., Sastry, S., Kim, H.J.: Autonomous exploration in unknown urban environments for unmanned aerial vehicles. In: American Institute of Aeronautics and Astronautics (AIAA) Guidance, Navigation, and Control Conference and Exhibit, San Francisco, CA, 15–18 August 2005
9. Amidi, O., Kanade, T., Miller, R.: Vision-base autonomous helicopter research at Carnegie Mellon Robotics Institute 1991–1997. In: American Helicopter Society International Conference, Heli, Japan (1998)
10. Cardoze, D.E., Arkin, R.C.: Development of visual tracking algorithms for an autonomous helicopter. Proc. SPIE **2591**, 145–156 (1995)
11. Doherty, P., Granlund, G., Kuchcinske, K., Sandewall, E., Nordberg, K., Skarman, E., Wiklund, J.: The WITAS unmanned aerial vehicle project. In: ECAI 2000, Proceedings of the 14th European Conference on Artificial Intelligence. IOS Press, Amsterdam (2000)
12. Achtelik, M., Bachrach, A., He, R., Prentice, S., Roy, N.: Autonomous navigation and exploration of a quadrotor helicopter in GPS-denied indoor environments. In: First Symposium of Indoor Flight Issues. Association for Unmanned Vehicle Systems International (2009)
13. Rock, S.M., Frew, E.W., Jones, H., LeMaster, E.A., Woodley, B.R.: Combined CDGPS and vision-based control of a small autonomous helicopter. In:Proceedings of the American Control Conference, vol. 2, pp. 694–698 (1998)
14. Saripalli, S., Montgomery, J.F., Sukhatme, G.S.: Vision-based autonomous landing of an unmanned aerial vehicle. In: Proceedings of IEEE International Conference on Robotics and Automation, vol. 3, pp. 2799–2804 (2002)
15. Zuffery, J.C., Floreano, D.: Fly-inspired visual steering of an ultralight indoor aircraft. IEEE Tran. Robot. **22**(1) (2006)
16. Altug, E., Ostrowki, J.P., Mahony, R.: Control of a quadrotor helicopter using visual feedback. In: Proceedings of the 2002 IEEE International Conference on Robotics and Automation, Washington, DC, May 2002
17. Jian, C., Dawson, D.M.: UAV tracking with a monocular camera. In: IEEE Conference on Decision and Control, San Diego, CA, pp. 3873–3873 (2006)
18. McKerrow, P.: Modelling the DraganFlyer four-rotor helicopter. In: Proceedings of the IEEE International Conference on Robotics and Automation, vol. 4, pp. 3596–3601 (2004)
19. Bouabdallah, S.: Design and control of quadrotors with application to autonomous flying. PhD thesis, Ecole Polytechnique Federale, Lausanne (2007)

20. Gurdan, D., Stumpf, J., Achtelik, M., Doth, K., Hirzinger, G., Rus, D.: Energy efficient autonomous four rotor flying robot controlled at 1 kHz. In: IEEE International Conference on Robotics and Automation, Roma, Italy (2007)
21. Bay, H., Tuytelaars, T., Van Gool, L.: Surf: speeded up robust features. In: European Conference on Computer Vision, pp. 404–417 (2006)
22. Lowe, D.G.: Distinctive image features from scale-invariant keypoints. Int. J. Comput. Vis. **60**(2), 91–110 (2004)
23. Brown, M., Lowe, D.G.: Invariant features from interest point groups. In: British Machine Vision Conference, Cardiff, Wales, pp. 656–665 (2002)

Chapter 11
Hierarchical Composite Anti-Disturbance Control for Robotic Systems Using Robust Disturbance Observer

Lei Guo, Xin-Yu Wen, and Xin Xin

Abstract A new anti-disturbance control strategy is presented for a class of nonlinear robotic systems with multiple disturbances. This strategy is named hierarchical composite anti-disturbance control (HCADC). Two types of disturbances are studied. One is generated by an exogenous system with uncertainty and the other is described by an uncertain vector with the bounded H_2 norm. The hierarchical control strategy is established which includes a disturbance observer based controller (DOBC) and an H_∞ controller, where DOBC is used to reject the first type of disturbance and H_∞ controller is used to attenuate the second. Stability analysis for both the error estimation systems and the composite closed-loop system is provided. Simulations for a two-link manipulator system show that the desired disturbance attenuation and rejection performances can be guaranteed.

11.1 Introduction

Robots can be modeled by nonlinear systems with disturbances from various sources, un-modeled dynamics and parametric uncertainties (see [1] and references therein). And, actually the un-modeled dynamics and parametric uncertainties can also be described by an 'equivalent' disturbance. Thus, the highly precise control for robotic systems can be regarded as a typical anti-disturbance control problem for

L. Guo (✉)
School of Instrument Science and Opto-Electronics Engineering, Beihang University, Beijing 100191, China
e-mail: lguo@buaa.edu.cn

L. Guo and X.-Y. Wen
Research Institute of Automation, Southeast University, Nanjing 210096, China

X. Xin
Faculty of Computer Science and Systems, Okayama Prefectural University, 111 Kuboki, Japan

H. Liu et al. (eds.), *Robot Intelligence*,
Advanced Information and Knowledge Processing,
DOI 10.1007/978-1-84996-329-9_11,

systems with disturbances. Theoretically, analysis and synthesis for nonlinear control systems with disturbances have been one of the most active research fields in the past decades. Up to date, many feasible anti-disturbance control methodologies have been presented for nonlinear systems with unknown disturbances, such as the output regulation theory, H_∞ control, adaptive control theory, nonlinear constructive theory, and active disturbance rejection (see e.g. [2–4] and references therein). Among them, mainly there are two different types of anti-disturbance control methodologies. One is the disturbance attenuation method such as H_∞ control where the influence of the disturbance can be decreased for the reference output [2, 3]. It is generally a conservative anti-disturbance control method, but feasible for a class of bounded uncertain disturbances. On the other hand, the disturbance rejection methods can realize the compensation of the disturbance with the internal mode controllers or the disturbance observers (see e.g. [4–7]). However, these methods may require some strict confinements on the systems model or involve the solution of partial differential equations (PDEs). Alternatively, the disturbance-observer-based control (DOBC) strategies, which was established in the late of 1980s for linear frequency-domain systems, have a simple structure and is easily implemented in engineering (see surveys [8, 9] and references therein). In [10], the DOBC of linear robotic systems was studied and compared with the passivity-based approaches. In [11], friction compensation was studied by using of a reduced-order observer for a linear robotic model. Recently, DOBC has been extended from linear systems to nonlinear systems (see also [12–16]). In [12], a novel DOBC approach in state space has been firstly presented for a class of multiple-input-multiple-output (MIMO) nonlinear systems, where the disturbances were generalized by a linear exogenous system as in the nonlinear regulation theory. In [13], DOBC for single-input-single-output (SISO) nonlinear robotic systems with well-defined disturbance relative degree were investigated in the presence of constant disturbances.

However, most of the above previous works only focused on plants with a single "equivalent" disturbance, for which either disturbance attenuation or rejection approaches have been investigated. In practical, together with the rapid development on sensor and data processing technologies, the disturbances or noises from different sources (e.g. sensor and actuator noise, friction, vibration, etc.) can be characterized by different models. This is why multiple types of disturbances should be considered for a controlled plant. For such a kind of complex systems, either the pure disturbance attenuation or the pure disturbance rejection strategy is difficult to realize the desired anti-disturbance control performance. As such, the composite anti-disturbance control with both attenuation and rejection performance have been presented respectively for flight control systems with different multiple types of disturbances in ([17, 18]). It is noted that the proposed hierarchical anti-disturbance controller has attractive of flexible structure and is easily combined with other controller to deal with various types of disturbances.

As a class of strong-coupled nonlinear system, the robotic system models have measurement error, friction, varying load, and un-modeled dynamics, which can be characterized as different types of disturbances. Some classic robust control strategy shows lager conservativeness and results in high gain when high-frequency and low-frequency disturbance appears synchronically. For the above purposes, a hierarchical composite anti-disturbance control (HCADC) strategy is firstly designed for a class of robotic systems in this chapter. Following the recent developments in DOBC fields, the disturbance considered in this chapter includes two parts. One part is the bounded vector in H_2-norm context. The other part is supposed to be generated by an exogenous system. Especially, the exogenous system is also supposed to have the modeling perturbation. On the other hand, we also consider the nonlinear uncertainties in the mass matrices, which actually corresponds to a kind of uncertain neutral systems that have been less studied in robust control before. This approach can avoid the heavy computation of Hamilton-Jacobian partial differential equations or the inverse of mass matrix. Simulations on a two-link robotic manipulator demonstrate the advantages of the proposed control scheme. Comparisons have been given with the classic H_∞ control and DOBC strategy.

11.2 Formulation of the Problem

The robotic models considered in this chapter is described as

$$\begin{aligned}\dot{x}(t) + F_{00} f_{00}(\dot{x}(t), t) &= A_0 x(t) + F_{01} f_{01}(x(t), t)) + B_0(u(t) + d_0(t)) + B_1 d_1(t),\\ z(t) &= Cx(t)\end{aligned} \tag{11.1}$$

where $x \in R^n$, $u \in R^m$ and $z \in R^q$ are the state vector, the control input, and the reference output respectively. A_0, B_0, B_1, F_0, and F_{01} are the compatible coefficient matrices. $d_1(t) \in R^p$ is the external disturbance in the H_2-norm caused by parametric uncertainties and the un-modeled dynamics. $d_0(t) \in R^m$ is supposed to be described by an unknown exogenous system as shown in Assumption 11.1, which can represent the constant or the harmonic noises.

Assumption 11.1 The disturbance $d_0(t)$ in the control input path can be formulated by the following exogenous system

$$\begin{cases} \dot{w}(t) = Ww(t) + B_2\delta, \\ d_0(t) = Vw(t) \end{cases} \tag{11.2}$$

where $W \in R^{r\times r}$, $B_2 \in R^{r\times l}$ and $V \in R^{m\times r}$ are proper known matrices. $\delta(t) \in R^l$ is the additional disturbance resulting from the perturbations and uncertainties in the exogenous system. It is also supposed that $\delta(t)$ is bounded in the H_2-norm.

$f_0(\dot{x}(t), t)$, $f_{01}(x(t), t))$ are nonlinear functions, which are supposed to satisfy bounded conditions described as Assumption 11.2.

Assumption 11.2 For any $\dot{x}(t) \in R_n, x(t) \in R_n$, the nonlinear functions $f_{00}(\dot{x}(t), t)$, $f_{01}(x(t), t)$ satisfy $f_{00}(0, t) = 0$, $f_{01}(0, t) = 0$, $\forall t \in R$, and

$$\|f_0(\dot{x}(t), t)\| \le \|U_0\dot{x}(t)\|, \quad \forall \dot{x}(t) \in R,\ \forall t \in R,$$

$$\|f_{01}(x(t),t)\| \leq \|U_{01}x(t)\|, \quad \forall x(t) \in R, \ \forall t \in R$$

where U_{01} and U_0 are given constant weighting matrices.

The following assumption is a necessary condition for the DOBC formulation.

Assumption 11.3 (A_0, B_0) is controllable and (W, B_0V) is observable.

If $d_0(t)$ is considered as a part of the augmented state, then a reduced-order observer is needed provided that the system state is available. In this chapter, we construct the reduced-order observer for $d_0(t)$ and then design a hierarchical composite controller with the disturbance observer and a H_∞ controller so that the multiple disturbances can be rejected and attenuated, simultaneously. And, the stability of the resulting composite system can also be guaranteed.

11.3 Hierarchical Composite Anti-Disturbance Control (HCADC)

In this section, we propose the HCADC scheme with combining DOBC and H_∞ Control Strategy.

Suppose that $f_{01}(x(t),t)$ is given and all states of the system are available. We firstly concern with the following disturbance observer

$$\begin{cases} \hat{d}(t) = V\hat{w}(t), \\ \hat{w}(t) = v(t) - Lx(t) \end{cases} \tag{11.3}$$

where $v(t)$ is the auxiliary vector as the state of the observer satisfying

$$\dot{v}(t) = (W + LB_0V)(v(t) - Lx(t)) + L(A_0x(t) + B_0u(t) + F_{01}f_{01}(x(t),t)).$$

The estimation error is denoted as $e_w(t) = w(t) - \hat{w}(t)$. Based on (11.1), (11.2) and (11.3), it is shown that the error dynamics satisfies

$$\dot{e}_w(t) = (W + LB_0V)e_w(t) + LB_1d_1(t) + B_2\delta(t) - LF_0f_{00}(\dot{x}(t),t)).$$

The objective of disturbance rejection can be achieved by designing the observer gain L such that $e_w(t)$ satisfies the desired stability and robustness performance. The hierarchical structure of the controller is formulated as $u(t) = -\hat{d}_0(t) + Kx(t)$, then closed-loop system is described as

$$\begin{cases} \dot{\bar{x}}(t) + Ff_0(\dot{\bar{x}}(t),t) = A\bar{x}(t) + F_1f_1(\bar{x}(t),t) + Bd(t), \\ z(t) = C\bar{x}(t) \end{cases} \tag{11.4}$$

where

$$\bar{x}(t) := \begin{bmatrix} x(t) \\ e_w(t) \end{bmatrix}, \qquad d(t) := \begin{bmatrix} d_1(t) \\ \delta(t) \end{bmatrix},$$

$$f_0(\dot{\bar{x}}(t),t) = f_{01}(\dot{x}(t),t), \qquad f_1(\bar{x}(t),t) = f_{01}(x(t),t),$$

and

$$A := \begin{bmatrix} A_0 + B_0 K & B_0 V \\ 0 & W + L B_0 V \end{bmatrix}, \qquad B := \begin{bmatrix} B_1 & 0 \\ L B_1 & B_2 \end{bmatrix}, \qquad C := \begin{bmatrix} C_1 & C_2 \end{bmatrix},$$

$$F := \begin{bmatrix} F_0 \\ L F_0 \end{bmatrix}, \qquad F_1 := \begin{bmatrix} F_{01} \\ 0 \end{bmatrix}, \qquad F_0 := \begin{bmatrix} F_{00} & 0 \end{bmatrix}, \qquad F_{01} := \begin{bmatrix} F_{001} & 0 \end{bmatrix}.$$

The following two results can be seen derived based on the results in [3] and their proofs are listed in Appendixes.

Lemma 11.1 *Consider system*

$$\begin{aligned} \dot{x}(t) + F f(\dot{x}(t), t) &= A x(t) + F_1 f_1(x(t), t) + B d(t), \\ z(t) &= C x(t) \end{aligned} \tag{11.5}$$

where $\|f(\dot{x}, t)\|^2 \le \|U x\|^2$, $\|f_1(x, t)\|^2 \le \|U_1 x\|^2$. *If for given* $\lambda > 0$, $\lambda_1 > 0$, *there exits* $P > 0$ *satisfying*

$$\begin{bmatrix} A^T P + P A & P B_H & C_H^T \\ B_H^T P & -I & D_H^T \\ C_H & D_H & -I \end{bmatrix} < 0 \tag{11.6}$$

where

$$B_H = \begin{bmatrix} \lambda F & \lambda_1 F_1 \end{bmatrix}, \qquad C_H = \begin{bmatrix} \frac{1}{\lambda} U A & \frac{1}{\lambda_1} U_1 \end{bmatrix}, \qquad D_H = \begin{bmatrix} U F & \frac{\lambda_1}{\lambda} U F_1 \\ 0 & 0 \end{bmatrix} \tag{11.7}$$

then system (11.5) *is asymptotic stable.*

Proof See Appendix A. □

Lemma 11.2 *For given* $\lambda > 0$, $\lambda_1 > 0$, $\gamma > 0$, *if there exits* $P > 0$ *satisfying*

$$\begin{bmatrix} A^T P + P A & P B_H & C_H^T & P B & C^T \\ B_H^T P & -I & D_H^T & 0 & 0 \\ C_H & D_H & -I & E_H & 0 \\ B^T P & 0 & E_H^T & -\gamma^2 I & 0 \\ C & 0 & 0 & 0 & -I \end{bmatrix} < 0 \tag{11.8}$$

where B_H, C_H, D_H *is denoted by* (11.7), *and* $E_H = \begin{bmatrix} \frac{1}{\lambda} U B \\ 0 \end{bmatrix}$, *then system* (11.5) *is asymptotically stable in the absence of the disturbance* $d(t)$, *and satisfies* $\|z\|_2 < \gamma \|d(t)\|_2$.

Proof See Appendix B. □

With the above two results, we can transform the HCADC problem to a robust H∞ control problem for an uncertain neutral system.

Theorem 11.1 *For given* $\lambda > 0$, $\lambda_1 > 0$, $\gamma > 0$, *if there exit* $Q_1 > 0$, $P_2 > 0$, R_1, R_2 *satisfying* (11.9)

$$\begin{bmatrix} \prod_1 & \lambda F_0 & \lambda_1 F_{01} & \prod_3 & \frac{1}{\lambda}Q_1U_{01}^T & B_1 & 0 & Q_1C_1^T & B_0V \\ * & -I & 0 & F_0^TU_0^T & 0 & 0 & 0 & 0 & \lambda F_0^TR_2^T \\ * & * & -I & \frac{\lambda_1}{\lambda}(U_0F_{01})^T & 0 & 0 & 0 & 0 & 0 \\ * & * & * & -I & 0 & \frac{1}{\lambda}U_0B_1 & 0 & 0 & \frac{1}{\lambda}U_0B_0V \\ * & * & * & * & -I & 0 & 0 & 0 & 0 \\ & * & * & * & * & -\gamma^2 I & 0 & 0 & B_1^TR_2^T \\ * & * & * & * & * & * & -\gamma^2 I & 0 & (P_2B_2)^T \\ & * & * & * & * & * & * & -I & C_2 \\ & * & * & * & * & * & * & * & \prod_2 \end{bmatrix} < 0 \tag{11.9}$$

where $\prod_1 = \mathrm{sym}(A_0Q_1 + B_0R_1)$, $\prod_2 = \mathrm{sym}(P_2W + R_2B_0V)$, $\prod_3 = \frac{1}{\lambda}[Q_1A_0^TU_0^T + R_1^TB_0^TU_0^T]$. *When* $K = R_1Q_1^{-1}$, $L = P_2^{-1}R_2$, *the composite system* (11.4) *is asymptotically stable in the absence of the disturbance* $d(t)$, *and satisfies* $\|z(t)\|_2 < \gamma\|d(t)\|_2$.

Proof For composite system (11.4), denote

$$P = \begin{bmatrix} P_1 & 0 \\ 0 & P_2 \end{bmatrix} = \begin{bmatrix} Q_1^{-1} & 0 \\ 0 & P_2 \end{bmatrix} > 0.$$

Based on Lemma 11.2 and applying

$$A = \begin{bmatrix} A_0 + B_0K & B_0V \\ 0 & W + LB_0V \end{bmatrix}, \qquad B = \begin{bmatrix} B_1 & 0 \\ LB_1 & B_2 \end{bmatrix},$$

$$C = \begin{bmatrix} C_1 & C_2 \end{bmatrix}, \qquad F = \begin{bmatrix} F_0 \\ 0 \end{bmatrix}, \qquad F_1 = \begin{bmatrix} F_{01} \\ 0 \end{bmatrix}$$

and $U = [U_0\ 0]$, $U_1 = [U_{01}\ 0]$ into (11.8), then it can be seen that (11.13) holds. Exchanging of rows and columns, pre-multiplied and post-multiplied simultaneously by $\mathrm{diag}\{Q_1, I, I, I, I, I, I, I, I\}$, thus (11.9) is obtained. □

11.4 Applications to a Two-Link Robotic System

To show the efficiency of the proposed scheme, a two-link robotic manipulator is considered in this section. The model of a two-link robotic manipulator can be represented by (see e.g. [19])

$$M(q)(\ddot{q}) + C(\dot{q}, q)q + G(q) = \tau + f \tag{11.10}$$

where q is the $n \times 1$ vector of joint positions, τ is a $n \times 1$ vector of torques applied to the joints, $M(q)$ is the $n \times n$ symmetric bounded positive definite mass (or inertia) matrix, q is the $n \times 1$ vector of joint angle, $C(q, \dot{q})\dot{q}$ is the $n \times 1$ vector of centrifugal

Table 11.1 The parameters in two-link manipulate system

Parameter	Symbol	Value
Mass of the i_{th} arm	$m_i\ (i=1,2)$	8 kg
The i_{th} arm's moment inertia term	$I_i\ (i=1,2)$	0.4 kg m^2
Length of the i_{th} arm	$l_i\ (i=1,2)$	0.5 m

and Coriolis terms, and $G(q)$ is the $n \times 1$ gravity vector, f is a disturbance torque or force vector. For the sake of simplicity, the mass matrices is give as follows, and is readily extended to the more general case.

$$M(q) = \begin{bmatrix} p_1 + p_2 + 2p_2\cos(q_2) & p_3 + p_2\cos(q_2) \\ p_3 + p_2\cos(q_2) & p_3 \end{bmatrix}, \quad q = \begin{bmatrix} q_1 \\ q_2 \end{bmatrix}.$$

The vector of centrifugal and Coriolis terms and gravity may be expressed as

$$h(q,\dot{q}) = C(q,\dot{q})\dot{q} + G(q), \qquad h(q,\dot{q}) = \begin{bmatrix} h_1 \\ h_2 \end{bmatrix} = \begin{bmatrix} -p_3(2\dot{q}_1\dot{q}_2 + \dot{q}_2^2)\sin q_2 \\ p_3\dot{q}_1^2\sin q_2 \end{bmatrix}$$

and $p_1 = (m_1/4 + m_2)l_1^2 + I_1$, $p_2 = m_2 l_1 l_2/2$, $p_3 = m_2 l_2^2/4 + I_2$.

The significance of the parameters in (11.10) are given in Table 11.1 as in [19]. The exogenous system f is exogenous disturbance caused by actuator, including nonlinear function $\Phi\sin(4t+\varphi)$ and measurement noise δ. As Φ and φ are unknown, it can be seen as (11.3), with unknown $w(0)$. To construct nonlinear observer for f, the nonlinear vector $M(q)\ddot{q}$ and $h(\dot{q},q)$ is decomposed as

$$M(q)\ddot{q} = M_0\ddot{q} + M_1(q)\ddot{q} + \Delta M \tag{11.11}$$

where

$$M_0 = \begin{bmatrix} p_1 + p_2 & p_3 \\ p_3 & p_3 \end{bmatrix}, \qquad M_1(q) = \begin{bmatrix} 2p_2\cos(q_2) & p_2\cos(q_2) \\ p_2\cos(q_2) & 0 \end{bmatrix},$$
$$h(\dot{q},q) = h_1(\dot{q},q) + \Delta h.$$

ΔM and Δq represents system uncertainty or measurement noise, and without loss of generalize, they are supposed to be bounded. Then (11.10) can be rewritten as

$$M_0\ddot{q} + M_1(q)\ddot{q} + \Delta M + h_1(\dot{q},q) + \Delta h = \tau + f.$$

For a given reference trajectory $q_d = \left[\begin{smallmatrix} q_{d1} \\ q_{d2} \end{smallmatrix}\right]$, set

$$\tau = M_0(q)\ddot{q}_d + M_1(q)\ddot{q}_d + M_0 u + h_1(\dot{q},q),$$

$$\ddot{q} - \ddot{q}_d + M_1(q)(\ddot{q} - \ddot{q}_d)/M_0 = u + f/M_0 - \Delta M/M_0 - \Delta h/M_0.$$

Then the dynamic model above may be transferred to former of (11.4) and the corresponding coefficient can be got easily, where

$$x = \begin{bmatrix} q_1 - q_{d1} \\ q_2 - q_{d2} \\ \dot{q}_1 - \dot{q}_{d1} \\ \dot{q}_2 - \dot{q}_{d2} \end{bmatrix}.$$

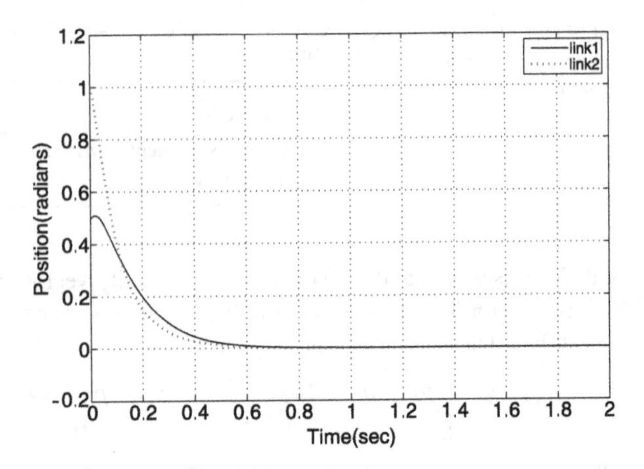

Fig. 11.1 Error of position with no model uncertainty

Furthermore, according to characteristic of $M_1(q)$, Assumption 11.2 can be satisfied.

The acceleration signal is not available in many robotic manipulators, and it is also difficult to construct the acceleration signal from the velocity signal by differentiation due to measurement noise. Although tracking differentiator can be used to provide a high-quality differential signal for effective and robust performance in the presence of measurement noise, it is difficult to select the tracking parameter theoretically. The proposed disturbance observer (DO) is implemented on the two-link robotic manipulator to avoid the calculation of the acceleration signal which is necessary for construction of DO in general. Simulation results for a computed torque controller with and without the DO are compared.

The HCADC structure combines a computed torque H_∞ controller with the DO, such that the effect of the disturbance is compensated by the outputs of the DO. As mentioned above, the disturbance observer is constructed as (11.3). Based on Theorem 11.1, it can be obtained that

$$K = \begin{bmatrix} -280.7153 & 2.0490 & -42.5783 & -0.3533 \\ 2.0595 & -284.8095 & -0.3511 & -41.8709 \end{bmatrix},$$

$$L = \begin{bmatrix} 0 & 0 & -16640 & -3734 \\ 0 & 0 & -3783 & -4244 \end{bmatrix}.$$

Firstly, we verified the proposed HCADC algorithm for the case of no model uncertainty for the disturbance. Figures 11.1 and 11.2 demonstrate that the steady tracking error disappears.

Nextly, we focus on the case that there is model uncertainty for the disturbance. Suppose that there is δ as a random signal varying in $(-2, 2)$, and parameter of p_i $(i = 1, 2, 3)$ varies $+10\%$. It has been shown in [12] that when the frequency of the harmonic noise has perturbations, the classical DOBC method is unfeasible practically. The disturbance estimation error by using the proposed method is plotted in Fig. 11.3, where it is shown that the satisfied tracking performance can be obtained comparing with the DOBC method.

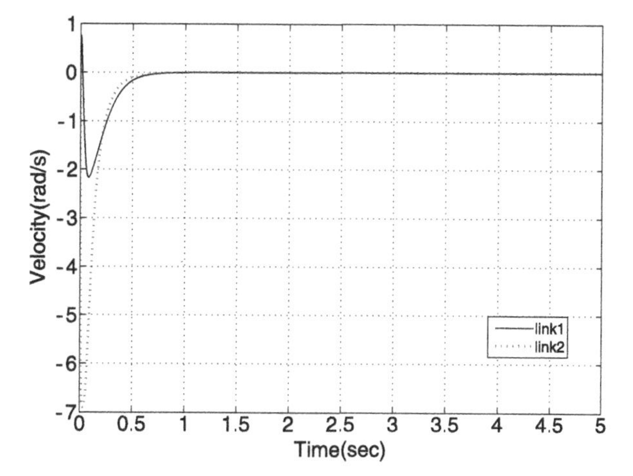

Fig. 11.2 Error of velocity with no model uncertainty

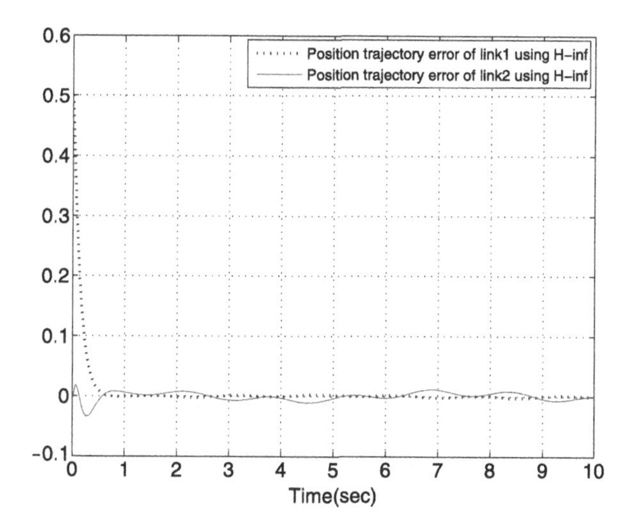

Fig. 11.3 Error of position trajectory of link1 and link2 using H_∞

The performance by using the H_∞ control can be seen in Figs. 11.3 and 11.4. The tracking performance by using the proposed HCADC method can be seen in Figs. 11.5, 11.6, 11.7 and 11.8, which illustrates that the proposed method is more effective than the previous DOBC and H_∞ control methods.

11.5 Conclusions

In this chapter, a new anti-disturbance controller is presented for a class of robotic systems with multiple disturbances. A hierarchical composite anti-disturbance control (HCADC) strategy is designed with enhanced disturbance attenuation and rejection performance. There are the following features of the proposed results combining with the previous works. The disturbance considered in this chapter has multiple

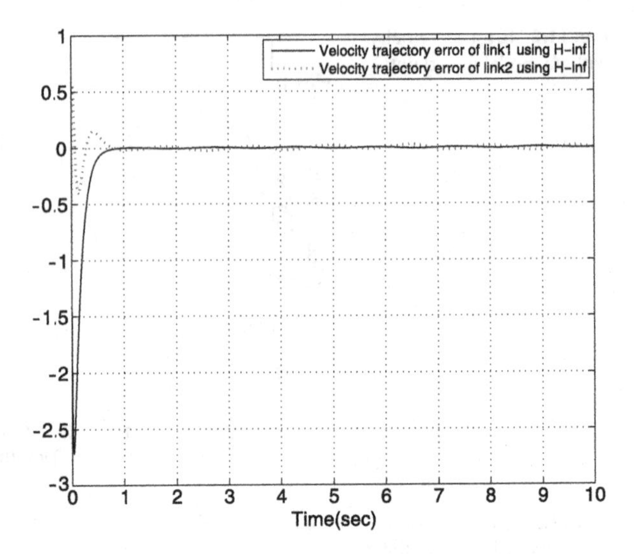

Fig. 11.4 Error of velocity trajectory of link1 and link2 using H_∞

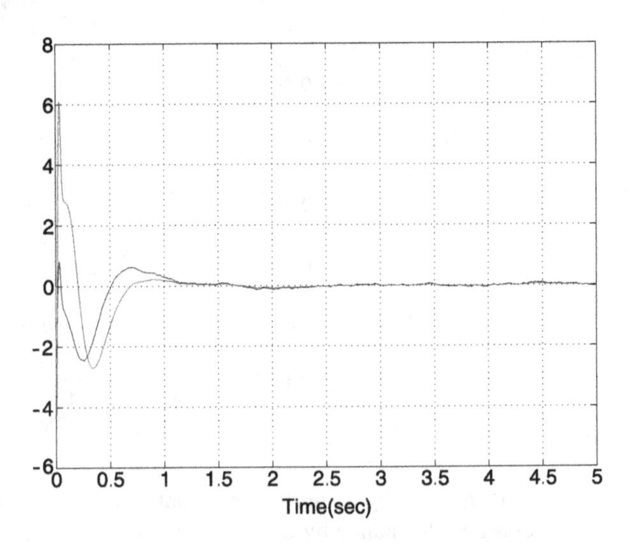

Fig. 11.5 The estimation error of d_1 and d_2

types, which can be divided into two parts. One part is the bounded vector in H_2-norm context. The other part is supposed to be generated by an exogenous system. Especially, the exogenous system is also supposed to have a perturbation. Also, we considered the nonlinear uncertainties in the mass matrices and transformed to a kind of uncertain neutral systems. The hierarchical control strategy consisting of disturbance observer and a H_∞ controller, where the first type of disturbances can be estimated and rejected, and another can be attenuated. Simulations on a two-link robotic manipulator demonstrate the advantages of the proposed control scheme. It is expected that the HCADC approach can be used to many other robotic systems with multiple disturbances.

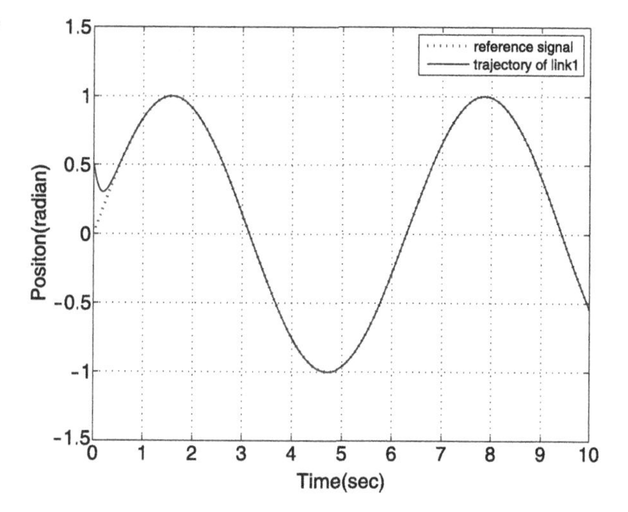

Fig. 11.6 Position trajectory of link1 using HCADC

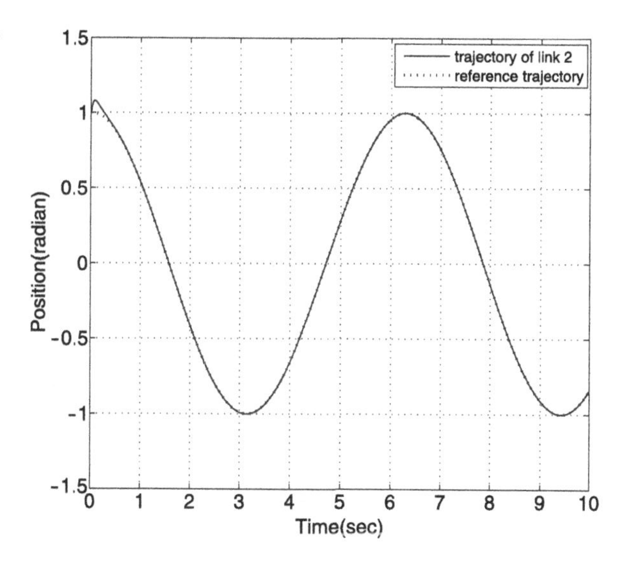

Fig. 11.7 Position trajectory of link2 using HCADC

Acknowledgement This work was supported by 863 programme, 973 programme and National Science Foundation of China and the JSPS fellowship.

Appendix A: Proof of the Lemma 11.1

Denoting a Lyapunov candidate

$$\Pi(x,t) = x^T Px + \frac{1}{\lambda^2}\int_0^t \left[\|U\dot{x}\|^2 - \|f(\dot{x},\tau)\|^2\right] d\tau$$

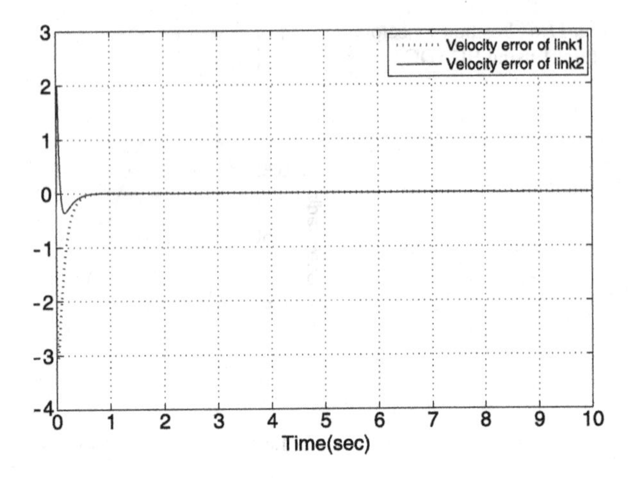

Fig. 11.8 Error of velocity trajectory of link1 and link2 using HCADC

$$+\frac{1}{\lambda_1^2}\int_0^t \left[\|U_1 x\|^2 - \|f_1(x,\tau)\|^2\right] d\tau \tag{11.12}$$

then we have

$$\begin{aligned}
\dot{\Pi}(x,t) &= \dot{x}^T P x + x^T P \dot{x} + \frac{1}{\lambda^2}\left[\|U\dot{x}\|^2 - \|f(\dot{x},\tau)\|^2\right] \\
&\quad + \frac{1}{\lambda_1^2}\left[\|U\dot{x}\|^2 - \|f(\dot{x},\tau)\|^2\right] \\
&= x^T\left[PA + A^T P + \tfrac{1}{\lambda^2}A^T U^T U A + \tfrac{1}{\lambda_1^2}U_1^T U_1\right] x \\
&\quad + x^T\left[\lambda PF + \tfrac{1}{\lambda}A^T U^T U F \quad \lambda_1 P F_1 + \tfrac{\lambda_1}{\lambda^2}A^T U^T U F_1\right] q \\
&\quad + q^T\left[\lambda PF + \tfrac{1}{\lambda}A^T U^T U F \quad \lambda_1 P F_1 + \tfrac{\lambda_1}{\lambda^2}A^T U^T U F_1\right]^T x \\
&\quad + q^T\begin{bmatrix} F^T U^T U F - I & \tfrac{\lambda_1}{\lambda}F^T U^T U F_1 \\ \tfrac{\lambda_1}{\lambda}F_1^T U^T U F & \tfrac{\lambda_1^2}{\lambda^2}F_1^T U^T U F_1 - I \end{bmatrix} q \\
&= \begin{bmatrix} x \\ q \end{bmatrix}^T \begin{bmatrix} PA + A^T P + C_H^T C_H & PB_H + C_H^T D_H \\ B_H^T P + D_H^T C_H & D_H^T D_H - I \end{bmatrix}\begin{bmatrix} x \\ q \end{bmatrix}
\end{aligned} \tag{11.13}$$

where B_H, C_H, D_H is denoted by (11.7), and $q^T = [-(\frac{1}{\lambda}f)^T \ (\frac{1}{\lambda_1}f_1)^T]$. Denote

$$M_1 = \begin{bmatrix} PA + A^T P + C_H^T C_H & PB_H + C_H^T D_H \\ B_H^T P + D_H^T C_H & D_H^T D_H - I \end{bmatrix}.$$

Based on Schur complement, it can be seen that $M_1 < 0 \Leftrightarrow M_2 < 0$ where

$$M_2 = \begin{bmatrix} PA + A^T P & PB_H & C_H^T \\ B_H^T P & -I & D_H^T \\ C_H & D_H & -I \end{bmatrix}.$$

Thus $\dot{\Pi}(x,t) < 0 \Leftrightarrow M_2 < 0$, i.e. if $M_2 < 0$ holds, system (11.1) is asymptotic stable.

Appendix B: Proof of the Lemma 11.2

For system (11.5), select a Lyapunov candidate as (11.12), and denote the following auxiliary function (known as the storage function)

$$J := \int_0^t \left[z^T z - \gamma^2 w^T w + \dot{\Pi}(x,t) \right] dt.$$

It can be verified that

$$\begin{aligned}
& z^T z - \gamma^2 w^T w + \dot{\Pi}(x,t) \\
&\quad = x^T C^T C x - \gamma^2 w^T w + \dot{x}^T P x + x^T P \dot{x} + \frac{1}{\lambda^2}\left[\|U\dot{x}\|^2 - \|f(\dot{x},\tau)\|^2 \right] \\
&\qquad + \frac{1}{\lambda_1^2}\left[\|U\dot{x}\|^2 - \|f(\dot{x},\tau)\|^2 \right] \\
&\quad = x^T \left[PA + A^T P + \frac{1}{\lambda^2} A^T U^T U A + \frac{1}{\lambda_1^2} U_1^T U_1 + C^T C \right] x \\
&\qquad + w^T \left[+ \frac{1}{\lambda^2} B^T U^T U B - \gamma^2 I \right] w \\
&\qquad + q^T \begin{bmatrix} F^T U^T U F - I & \frac{\lambda_1}{\lambda} F^T U^T U F_1 \\ \frac{\lambda_1}{\lambda} F_1^T U^T U F & \frac{\lambda_1^2}{\lambda^2} F_1^T U^T U F_1 - I \end{bmatrix} q \\
&\qquad + q^T \begin{bmatrix} \frac{1}{\lambda} F^T U^T U F_1 \\ \frac{\lambda}{\lambda} F_1^T U^T U B \end{bmatrix} w + w^T \begin{bmatrix} \frac{1}{\lambda} F^T U^T U B \\ \frac{\lambda_1}{\lambda^2} F_1^T U^T U B \end{bmatrix}^T q \\
&\qquad + x^T \left[\lambda P F + \frac{1}{\lambda} A^T U^T U F \quad \lambda_1 P F_1 + \frac{\lambda_1}{\lambda^2} A^T U^T U F_1 \right] q \\
&\qquad + q^T \left[\lambda P F + \frac{1}{\lambda} A^T U^T U F \quad \lambda_1 P F_1 + \frac{\lambda_1}{\lambda^2} A^T U^T U F_1 \right]^T x \\
&\qquad + x^T \left[PB + \frac{1}{\lambda^2} A^T U^T U B \right] w + w^T \left[PB + \frac{1}{\lambda^2} A^T U^T U B \right]^T x \\
&\quad = \begin{bmatrix} x \\ q \\ w \end{bmatrix}^T \begin{bmatrix} PA + A^T P + C_H^T C_H + C^T C & PB_H + C_H^T D_H & PB + C_H^T E_H \\ B_H^T P + D_H^T C_H & D_H^T D_H - I & D_H^T E_H \\ B^T P + E_H^T C_H & E_H^T D_H & E_H^T E_H - \gamma^2 I \end{bmatrix} \\
&\qquad \times \begin{bmatrix} x \\ q \\ w \end{bmatrix}.
\end{aligned} \tag{11.14}$$

Denote M_3 as

$$M_3 = \begin{bmatrix} PA + A^T P + C_H^T C_H + C^T C & PB_H + C_H^T D_H & PB + C_H^T E_H \\ B_H^T P + D_H^T C_H & D_H^T D_H - I & D_H^T E_H \\ B^T P + E_H^T C_H & E_H^T D_H & E_H^T E_H - \gamma^2 I \end{bmatrix} \tag{11.15}$$

where B_H, C_H, D_H is also denoted by (11.7), and

$$E_H = \begin{bmatrix} -\frac{1}{\lambda} UB \\ 0 \end{bmatrix}, \qquad q^T = \begin{bmatrix} -(\frac{1}{\lambda} f)^T & (\frac{1}{\lambda_1} f_1)^T \end{bmatrix}.$$

Based on Schur complement, it can be seen that $M_3 < 0 \Leftrightarrow M_4 < 0$ where

$$M_4 = \begin{bmatrix} PA + A^T P & PB_H & PB & C_H^T & C^T \\ B_H^T P & -I & 0 & D_H^T & 0 \\ B^T P & 0 & -\gamma^2 I & E_H^T & 0 \\ C_H & D_H & E_H & -I & 0 \\ C & 0 & 0 & 0 & -I \end{bmatrix}.$$

Exchanging of rows and columns yields that $M_4 < 0 \Leftrightarrow M_5 < 0$, where

$$M_5 = \begin{bmatrix} PA + A^T P & PB_H & C_H^T & PB & C^T \\ B_H^T P & -I & D_H^T & 0 & 0 \\ C_H & D_H & -I & E_H & 0 \\ B^T P & 0 & E_H^T & -\gamma^2 I & 0 \\ C & 0 & 0 & 0 & -I \end{bmatrix}.$$

Since $M_5 < 0$ implies that $M_2 < 0$, system (11.5) is asymptotically stable in the absence of the disturbance $d(t)$ with zero initial condition. Furthermore, since $J = \int_0^\infty [z^T z - \gamma^2 w^T w] dt + \Pi(x,t)$, then $J < 0$ holds when $M_5 < 0$ holds, which implies that $\|z\|_2 < \gamma \|d\|_2$.

References

1. Spong, M.W., Vidyasagar, M.: Robot Dynamics and Control. Wiley, New York (1989)
2. Basar, T., Bernhard, P.: H_∞-Optimal Control and Related Minimax Design Problems: a Dynamic Game Approach. Springer, Berlin (1995)
3. Guo, L.: H_∞ output feedback control for delay systems withnonlinear and parametric uncertainties. IEE Proc., Control Theory Appl. **149**, 226–236 (2002)
4. Byrnes, C.I., Delli Priscoli, F., Isidori, A.: Output Regulation of Uncertain Nonlinear Systems. Birkhauser, Basel (1997)
5. Ding, Z.T.: Asymptotic rejection of asymmetric periodic disturbances in output-feedback nonlinear systems. Automatica **43**, 555–561 (2007)
6. Nikiforov, V.O.: Nonlinear servocompensation of unknown external disturbances. Automatica **37**, 1647–1653 (2001)

7. Serrani, A.: Rejection of harmonic disturbances at the controller input via hybrid adaptive external models. Automatica **42**, 1977–1985 (2006)
8. Guo, L., Feng, C., Chen, W.: A survey of disturbance-observer-based control for dynamic nonlinear system. Dyn. Contin. Discrete Impuls. Syst., Ser. B, Appl. Algorithms **13E**, 79–84 (2006)
9. Radke, A., Gao, Z.: A survey of state and disturbance observers for practitioners. In: Proceedings of American Control Conference, Minneapolis, 14–16 June 2006
10. Bickel, R., Tomizuka, R.: Passivity-based versus disturbance observer based robot control:equivalence and stability. ASME J. Dyn. Syst. Control Meas. **121**, 41–47 (1999)
11. Mallon, N., van de Wouw, N., Putra, D., Nijmeijer, H.: Friction compensation in a controlled one-link robot using a reduced-order observer. IEEE Trans. Control Syst. Technol. **14**, 374–383 (2006)
12. Guo, L., Chen, W.-H.: Disturbance attenuation and rejection for systems with nonlinearity via DOBC approach. Int. J. Robust Nonlinear Control **15**, 109–125 (2005)
13. Chen, W.: Disturbance observer based control for nonlinear systems. IEEE/ASME Trans. Mechatron. **9**(4), 706–710 (2004)
14. She, J., Ohyama, Y., Nakano, M.: A new approach to the estimation and rejection of disturbances in servo systems. IEEE Trans. Control Syst. Technol. **13**, 378–385 (2005)
15. Yang, Z., Tsubakihara, H.: A novel robust nonlinear motion controller with disturbance observer. IEEE Trans. Control Syst. Technol. **16**(1), 137–147 (2008)
16. Back, J., Shimb, H.: Adding robustness to nominal output-feedback controllers for uncertain nonlinear systems: a nonlinear version of disturbance observer. Automatica **44**, 2528–2537 (2008)
17. Guo, L., Wen, X.-Y.: Hierarchical anti-distance adaptive control for nonlinear systems with composite disturbances. Trans. Inst. Meas. Control (2009, to appear)
18. Wei, X., Guo, L.: Composite disturbance-observer-based control and H_∞ control for complex continuous models. International Journal of Robust and nonlinear Control (2009, to appear). doi:10.1002/rnc.1425
19. Xu, J.M., Zhou, Q.J., Leung, T.P.: Implicit adaptive inverse control of robot manipulators. In: Proceedings of the IEEE Conference on Robotics and Automation, Atlanta, pp. 334–339 (1993)

Chapter 12
Autonomous Navigation for Mobile Robots with Human-Robot Interaction

James Ballantyne, Edward Johns, Salman Valibeik, Charence Wong, and Guang-Zhong Yang

Abstract Dynamic and complex indoor environments present a challenge for mobile robot navigation. The robot must be able to simultaneously map the environment, which often has repetitive features, whilst keep track of its pose and location. This chapter introduces some of the key considerations for human guided navigation. Rather than letting the robot explore the environment fully autonomously, we consider the use of human guidance for progressively building up the environment map and establishing scene association, learning, as well as navigation and planning. After the guide has taken the robot through the environment and indicated the points of interest via hand gestures, the robot is then able to use the geometric map and scene descriptors captured during the tour to create a high-level plan for subsequent autonomous navigation within the environment. Issues related to gesture recognition, multi-cue integration, tracking, target pursuing, scene association and navigation planning are discussed.

12.1 Introduction

As demands for mobile robots continue to increase, so does the pursuit for intelligent, autonomous navigation. Autonomous navigation requires the robot to understand the environment, whether static or dynamic, and to interact with people seamlessly. In practice, there are several key components that enable a robot to behave intelligently. They include localization and mapping, scene association, human-robot interaction, target pursuing and navigation. Localization and mapping is a well studied topic in robotics and autonomous vehicles for dealing with both known and unknown environment whilst keeping track of the current location. For

J. Ballantyne (✉), E. Johns, S. Valibeik, C. Wong, and G.-Z. Yang
Institute of Biomedical Engineering, Imperial College of London, London, UK
e-mail: james.ballantyne@imperial.ac.uk; edward.johns09@imperial.ac.uk; salman.valibeik05@imperial.ac.uk; charence.wong05@imperial.ac.uk; gzy@doc.ic.ac.uk

H. Liu et al. (eds.), *Robot Intelligence*,
Advanced Information and Knowledge Processing,
DOI 10.1007/978-1-84996-329-9_12,

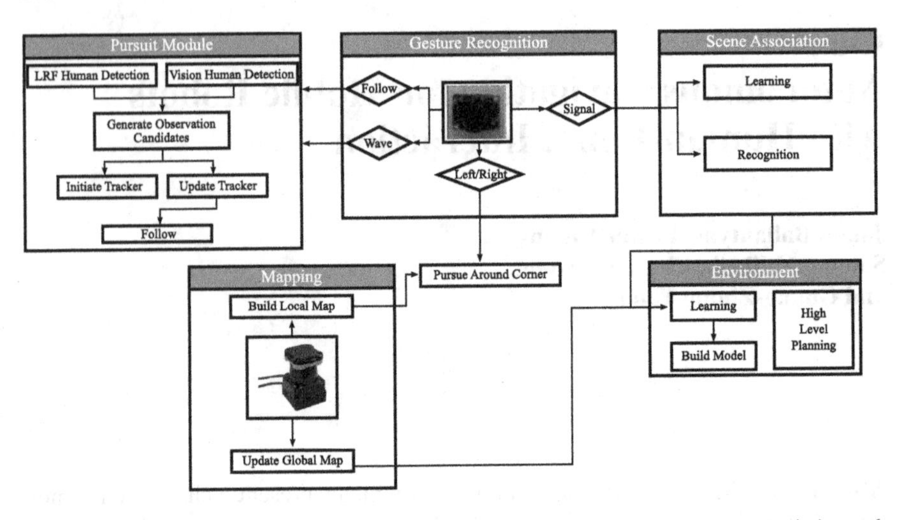

Fig. 12.1 A schematic illustration of the use of gesture recognition, pursuit, scene association and environment mapping for human guided navigation. Under this scheme, the gesture recognition system detects commands issued by a guide, which then activates a specific component based on the detected gesture. The pursuit component is activated on specific gestures and when an "*attention*" gesture is detected, a scene descriptor is built, which is then integrated with the environment model

purposeful navigation, it also requires learning and scene association to build progressively the surrounding environment. For complex scenes, such as those encountered in a crowded indoor setting, gesture recognition is necessary to ensure seamless human-robot interaction so that they can follow specific commands or pursue relevant tasks. Figure 12.1 outlines an example configuration when these components are required to work together for autonomous navigation within an indoor environment.

In terms of human robot interaction, vision based approaches represent a key technique for establishing natural and seamless interaction. For understanding human gesture or intention, static or dynamic hand gestures and facial expression can be used [9, 14, 23, 28, 42]. Static gesture normally relies on identifying different postures whereas dynamic gestures include interpreting cascade of events through different time space. In other words, static gestures are extracted by analyzing the contextual information at each time instance, whereas dynamic gestures are recognized by analyzing the temporal information across consecutive time periods. Effective use of human-robot interaction enables a person to initiate various tasks for the robot to carry out. In this chapter, we will use human guided exploration for a robot in a novel environment as an example. The technical details for gesture recognition are described in Sect. 12.2. Key to any successful gesture recognition system is the incorporation of natural, socially acceptable gestures similar to those used in human-human interaction. The technical details for gesture recognition are described in Sect. 12.2.

Following a guide also requires the robot to maintain and keep track of the location of the person continuously. To this end, a tracking system as described in

Sect. 12.3 is proposed. The method is based on the use of multiple cues from two main sensor modalities based on vision and laser scanning systems. The visual cues from each sensor are fused to create a robust map of people within the environment. Once the location of the guide is obtained, the robot is able to follow the guide through the environment whilst avoiding visible obstacles autonomously visible obstacles. The basic approaches used are also described in Sect. 12.3. Even with human guidance, However, situations may arise when the person goes outside the field-of-view of the robot. In this situation, the robot needs to predict where the guide may end up and autonomously navigate to the position and re-establish visual tracking.

In order to build a global map of the new environment through a guided tour, qualitative localization is necessary. We have proposed in Sect. 12.4 a concept called scene association, which enables the robot to identify salient features of different locations as it navigates around. This information is then incorporated with the internal map generated at relevant locations. The proposed scene association framework uses visual data to learn key features of a scene, which are distinctive but can be consistently identified from different viewpoints.

After the guide has taken the robot through the environment and indicated the points of interest via hand gestures, the robot is then able to use the geometric map and scene descriptors captured during the tour to create a high-level plan for subsequent autonomous navigation within the environment. In Sect. 12.5, an A* graph search algorithm is used to plan the route of the robot for goal directed navigation and localization. We will also discuss how learning techniques can be used to improve the robot's ability for autonomous navigation. Throughout this chapter, the examples used are for indoor environments with people moving around. So the proposed framework is ideally suited for museum, office, home-care and hospital wards. The theoretical concepts of using directed navigation to reinforce vision based autonomous localization and mapping can also be extended to other environment. To this end, human gesture recognition can be replaced by other signalling methods, but the basic concept of scene association and high-level planning can remain the same.

12.2 Human-Robot Interaction

In this section, we will describe a robust gesture recognition framework suitable for human guided navigation in normal indoor environment including crowded scenes. For this purpose, the method proposed in [35] is to be used. In this approach, vision based dynamic hand gestures are derived for robotic guidance. The gestures used include "*hello*" (wave gesture for initialization), "*turn left*", "*turn right*", "*follow forward*" and "*attention*" (for building new scene descriptors). The overall system structure is depicted in Fig. 12.2.

Hitherto, three distinctive factors are commonly employed for extracting hands for gesture recognition. These include skin color, hand motion and shape. We have

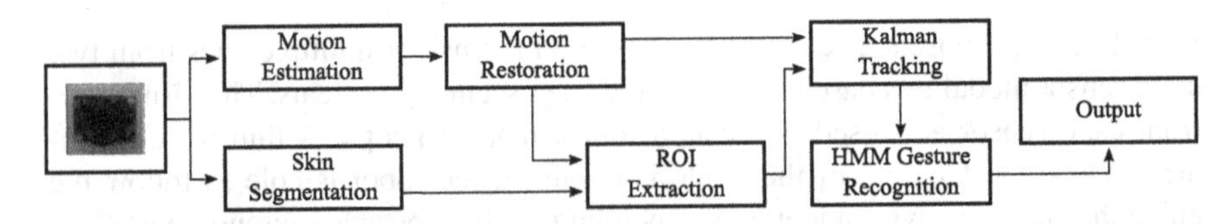

Fig. 12.2 Key processing components for the proposed gesture recognition sub-system. Raw images taken from the vision sensor are used to extract low-level cues such as motion information and skin colored objects. Regions of Interest (ROI) consisting of dynamic skin regions are then extracted and tracked with Kalman Filters. Finally, in the high-level reasoning phase, a Hidden Markov Model (HMM) is used to extract specific gestures

integrated the first two components, since hand shape is not robust enough for systems using wide angle cameras. In the proposed method, skin segmentation is first performed to identify skin-colored objects which consist of faces, hands or any other skin colored regions. Subsequently, motion segmentation is used to prune out background objects, which are mostly static. The remaining skin-colored objects now mostly consist of hands and faces.

In order to extract temporal information suitable for dynamic gesture recognition, a robust tracking algorithm is required. To this end, Least Median Square Error (LMeds) motion restoration is performed to remove outliers due to rapid illumination changes, partial motion occlusions and depth discontinuities. In practice, object deformation is also important to consider, which is more evident for tracking non-rigid objects. In this work, hand tracking is mainly used to extract dynamic hand gestures. With the proposed framework, pose variability and occlusions are also taken into account by incorporating multiple cues to associate the extracted regions of interest at each time instance to previous measurements. Kalman-filter is used to provide robust tracking across time [35].

Once the hand motion trajectories are accurately tracked, the next step is to perform detailed motion analysis to evaluate if the extracted trajectory is similar to pre-defined gestures. For this purpose, Hidden Markov Models (HMMs) are used. HMM is particularly suitable for modeling time series. The main advantage is that it is based on a probabilistic framework and is beneficial when multiple gestures are evaluated for the same sequence.

To demonstrate the practical value of the proposed gesture recognition framework, Fig. 12.3 demonstrates some example results when different subjects are asked to perform the aforementioned gestures for human guided navigation. Initially, the wave gesture is used to attract the attention of the robot. Subsequently, the robot can be guided by using "*move forward*", "*turn left*" or "*turn right*" commands.

In Fig. 12.3, all the tracked objects are color coded and the recognized gestures are illustrated. After successful gesture recognition, the next challenge is to identify and maintain the commanding person within the field-of-view in a crowded environment. This requires the robot to keep track of the person's location at all times once engaged. Furthermore, the robot must be able to follow the guide through the environment and build a detailed map of the environment for future navigation purposes. In the next section, we will introduce the tracking and pursuit system and explain how the identified gestures are used to control the robot during navigation.

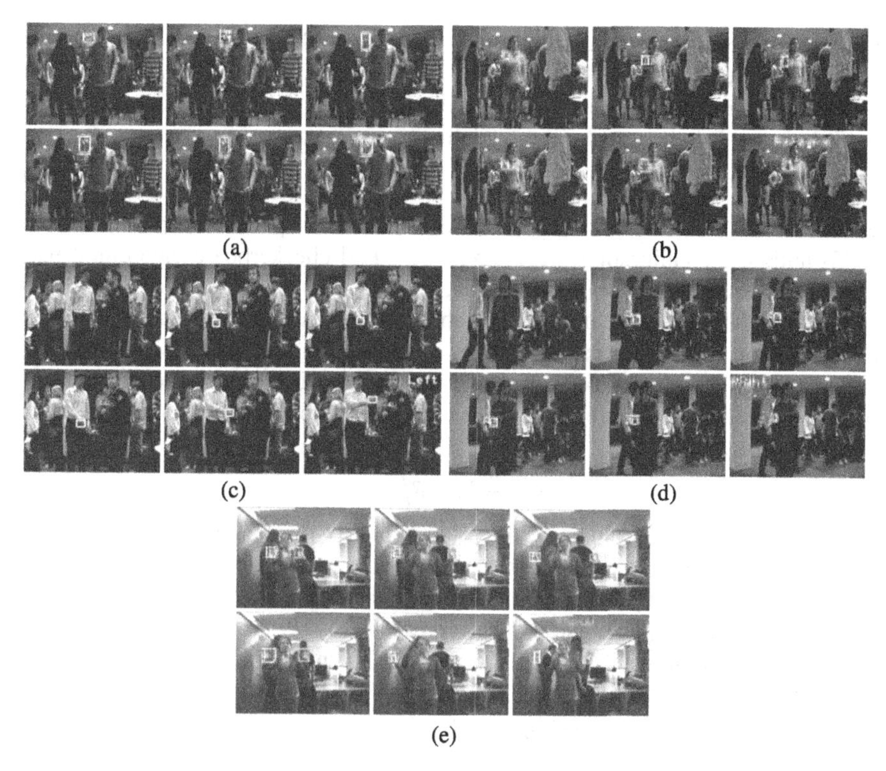

Fig. 12.3 Illustration of different gestures used in the proposed system. Each set of images shows the sequence of motions involved for each gesture. Tracked hand locations are color coded and each detected gesture is indicated in the last image of the sequence; (**a**), (**b**), (**c**), (**d**) and (**e**) show "*wave*", "*follow forward*", "*turn left*", "*turn right*", and "*attention*", respectively

12.3 Subject Following with Target Pursuing

In order to follow the commands given by a person in the scene, the proposed system relies on the robot's ability to detect and track humans based on the sensor data. In this section, we will use information from both vision sensors and Laser Range Finders (LRF) to accurately track and pursue the movement of the person. For human guided navigation, negotiating corners can be problematic as the guiding person can easily move out of the field-of-view. To overcome this problem, the guide can issue a "*left*" or "*right*" signal to activate an autonomous corner manoeuvre during which path planning and obstacle avoidance is performed autonomously.

12.3.1 Correspondence

The proposed framework relies on multiple cues from different sensors to accurately track the guide. However, the cues in the current setup reside in two different camera reference systems. Therefore, calibration is required to fuse both sets of cues into a common reference frame. This establishes a transformation to allow the projection of a laser range point into the vision space. The system utilizes the method defined in [39] to establish the transformation defined as

$$i \approx K(\Phi \cdot P + \nabla) \tag{12.1}$$

where $i = [u, v]^T$, $P = [x, y, z]^T$, and K are the intrinsic parameters of the camera.

To initiate the calibration procedure, a standard checkerboard calibration pattern as proposed by Zhang [39, 40] is used to calibrate the vision sensor. The aim of the procedure is to take multiple instances of the checkerboard within the view of both vision and LRF sensors. The vision calibration provides both the intrinsic parameters, K, and the extrinsic parameters, (R_i, t_i) with respect to each checkerboard location. Furthermore, the extrinsic parameters provide the normal for each checkerboard grid as

$$N = -R_{i,3}(R_{i,3}^T \cdot t_i) \tag{12.2}$$

where $R_{i,3}$ is the third column of the rotation matrix for the ith checkerboard obtained from the extrinsic parameters.

After calibration, the laser points on each checkerboard are collected. These points fall on the xz-plane and can be represented by $P^f = [x, z, 1]^T$. Therefore, a point falling on the calibration plane with surface normal must satisfy the plane equation $N \cdot P = -D$. From (12.1), we have

$$N \cdot \Phi^{-1}(P^f - \nabla) = -D. \tag{12.3}$$

This can be rewritten as

$$\begin{aligned} &N \cdot HP^f = -D, \\ &H = \Phi^{-1} \begin{pmatrix} 1 & 0 & \\ 0 & 0 & -\Delta \\ 0 & 1 & \end{pmatrix}. \end{aligned} \tag{12.4}$$

For each pose of the camera, there exist several linear equations for the unknown parameter H, which can be solved with standard linear least squares algorithms. Once H is determined, the relative orientation and position between the two sensors can be calculated as

$$\begin{aligned} \Phi_R &= [H_1, -H_1 \times H_2, H_2]^T, \\ \Delta &= -[H_1, -H_1 \times H_2, H_2]^T H_3 \end{aligned} \tag{12.5}$$

where Φ_R is the Rodrigues representation of the rotation matrix. To further enhance the accuracy of the rotation and translation parameters, a non-linear optimization technique can be used [39]. It aims to minimize the Euclidean distance from the laser points to the checkerboard grids by using the following equation:

$$\sum_i \sum_j (N_i \cdot (\Phi^{-1}(P_{ij}^f - \Delta)) + D)^2. \tag{12.6}$$

The final result provides the optimized transformation that allows for accurate projection of the laser data points into the image space.

12.3.2 Multi-Cue Integration

One of the most common methods for identifying people in vision is to locate the head in each image [10, 11, 13, 16, 20, 38, 41]. These techniques suffer from three major issues; (1) the assumption of the availability of a priori background information; (2) the requirement of large size silhouettes; (3) the need for a controlled environment in terms of illumination changes. Due to the relative positioning of the LRF, laser scanning systems usually identify humans using leg detection schemes. One common approach is to search for local minima [3, 8] in the scan data. This has shown promising results for relatively simple environments. However, as soon as the environment becomes cluttered, the detection results become unreliable and error prone [33]. A second common approach is to use motion detection to identify humans [7, 17] as people are often the only moving objects in most environments. These methods usually compare the current and previous scans to determine the dynamics objects within the environment. The areas from the current scan, which are not found in the previous scan, are considered as the moving objects. The very nature of the algorithm means that the system is not able to detect stationary persons in the environment.

To overcome these drawbacks, we propose to utilize cues from each sensor for person identification. A person is identified if it is evident from both the vision and laser systems. The vision system uses the head detection approach employed by Viola and Jones [36]. In addition, a cascade of adaptive boosting classifiers is used to quickly prune the background and place more emphasis on potential targets. In the examples shown in this chapter, 32 cascades of classifiers are used to provide accurate localization with minimal number of false positives. Furthermore, about 1,399 heads with different orientation, poses and illumination conditions along with 800 background images have been gathered for training. By using adaptive boosting of Haar-like features, a multi-pose head detection classifier has been created. To increase sensitivity, a Kalman filter based tracking system is employed. Head position is updated using Shi-Tomasi features [29]. The method proposed by Valibeik and Yang [35] is used to measure the correlation between newly detected regions with the tracked ones.

The cues from the laser scanning system are formed using a new approach for human detection [4]. The system aims to identify people by searching for three patterns associated with the presence of a person, which are typically found in laser scans. These patterns include split leg (LSA), forward straddle (FS), and two legs together (SL) as illustrated in Fig. 12.4.

The patterns are detected by finding the correct left and right edge sequences where an edge is defined as a segment between two points $\{x_i, x_{i+1}\}$ such that the distance between them is greater than a predefined threshold. An edge is defined as a left edge if $x_i > x_{i+1}$ and a right edge if $x_i < x_{i+1}$. The edges are generated and stored in a list $\sum = \{e_1, e_2, \dots, e_n\}$. The aim of the algorithm is to find a subset of the edges that follow one of the three patterns with constraints on the size between each segment. The patterns are defined as:

1. LA pattern with quadruplet $\{L, R, L, R\}$.

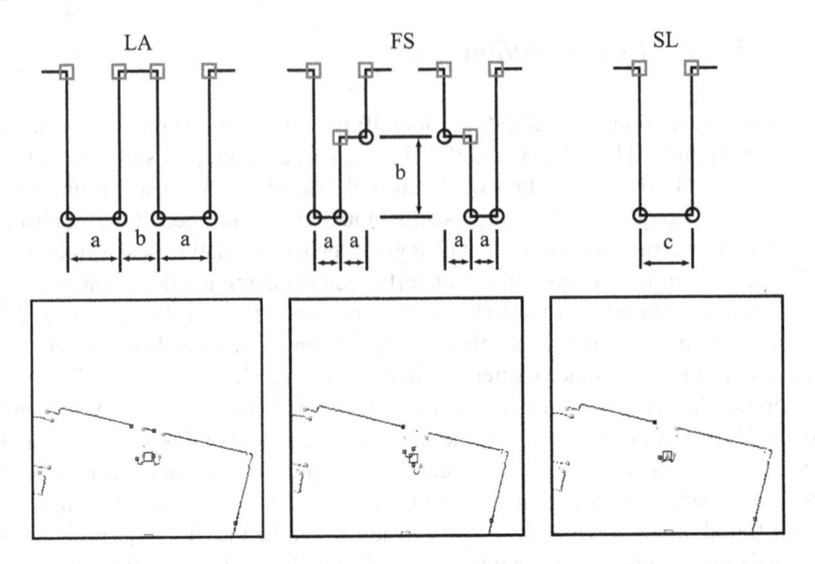

Fig. 12.4 Illustration of the three patterns used to detect human legs in the laser scan data. The top row illustrates the patterns with the three constraints used. In the proposed system, constraint (a) is between 10 cm and 40 cm, constraint (b) is under 40 cm, and constraint (c) is also under 40 cm. The bottom row shows a real example from the laser range finder for each type of pattern, where the blue squares represent the right edges, while green squares represent left edges

2. FS pattern with triplet $\{L, L, R\}$ or $\{L, R, R\}$
3. SL pattern with doublet $\{L, R\}$

A single edge from the list can only belong to a single pattern and is thus removed from further consideration. Furthermore, each pattern is searched sequentially, *i.e.*, the edge list is initially search for LA patterns, then for FS patterns, and finally for SL patterns. This is to help reduce the number of false positives.

In order to reduce the number of false positives from each sensor individually, the system fuses both sets of candidates into a single list. To this end, all people detected in the laser scan data are projected into image space using the transformation found in (12.6). Only those candidates that fall into the horizontal view of the camera are considered. The remaining candidates are then matched with the head detected in the vision system using a nearest-neighbor approach. Therefore, the final list consists of matched pairs from each sensor.

12.3.3 Robust Tracking

The previous section provides a way of identifying humans in the environment based on cues from both vision and LRF sensors. To maintain a continuous estimate of the location of the commanding person, a temporal tracking system is used. Traditional systems have used cues from multiple sensors to identify targets [2, 12, 18]. The proposed system attempts to handle the tracking problem in a similar fashion by using cues from the two sensor modalities, *i.e.*, vision and laser. As mentioned in Sect. 12.2, the system is activated when a "*hello*" command is received from a person in the environment. Upon receiving the gesture, the robot identifies the most likely candidate from the observation data set by choosing the head most likely to be part of the arm giving the gesture. This observation is used to initialize the tracking system. For tracking, an Interacting Multiple Model (IMM) [27] filter equipped with three motion models is used to deal with unpredictable movement of people. The IMM filter has been shown to provide more accurate tracking results than using a Kalman filter on its own [27]. The three motion models used assume constant acceleration, constant velocity, or a stationary motion model. The system tracks the location of the guide on the xz-plane with a weighted model to provide the most likely estimate of the location of the guide. The key component for ensuring accurate tracking is data association. Potential observations come from the fused information obtained from the two sensors as described in Sect. 12.3.2. To help limit the number of potential observations, the minimum gate of the three motion models [6] is used which is defined using a distance metric:

$$d = \sqrt{(y_m - z_i)^T \sum_m^{-1} (y_m - z_i)} \tag{12.7}$$

where $(y_m - z_i)$ is the measurement residual vector and $\sum_m^{-1}$ is the measurement residual covariance matrix. Finally, only observations that fall in the χ^2 distribution with a probability of 99% of the gate are considered. The tracking system ensures that there is always an estimate of the location of the guide, enabling the robot to follow the guide through the environment when requested.

12.3.4 Pursuing

The tracking system provides the robot with the necessary information to follow the commanding person through the environment. In the example presented in this chapter, the robot uses the given position to ensure:

1. Robot is required to face the guide at all times;
2. Robot is required to maintain a distance of roughly 1.5 m to the guide;
3. Movements can only be performed if all objects are avoided.

To adhere to these three goals, the robot follows the pursuit movement as defined in [24]. The steering behavior for each frame determines the necessary velocity vector (rotation and translation) that the robot should follow to adhere to Rules (1) and (2). The velocity vector is determined by the current predicated location of the guide and the robot's current velocities.

$$\begin{aligned} v_{desired} &= norm(pos_r - pos_t) \cdot v_{max}, \\ v_{actual} &= v_{desired} - v_{current}. \end{aligned} \tag{12.8}$$

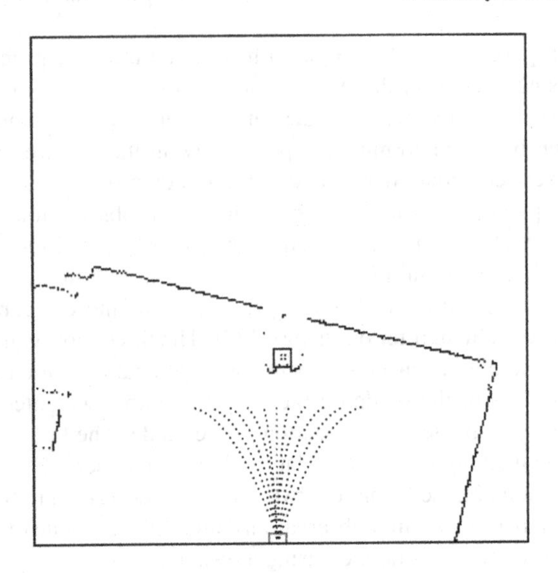

Fig. 12.5 Illustration of the pursuit strategy employed by the robot for following the guide through an indoor scene. The system chooses the best set of velocities that will allow the robot to approach the guide without colliding with any of the objects in the environment. In the above example, the central path is chosen, which is highlighted in *dotted lines*. This path allows the robot to arrive at the desired location at about 1.5 m from the guide

Upon arriving at the actual velocity to use, the robot must ensure that Rule (3) is preserved. To this end, the robot selects a series of velocities within a window of v_{actual} and generates the curves that the robot would follow at the selected velocities. The robot then chooses the velocity that allows the robot to arrive at the desired location, 1.5 m from the guide while avoiding all obstacles. Figure 12.5 illustrates the potential paths that the robot could take to approach the guide marked with a blue square. In this example, the red, dotted path is chosen because it brings the robot closest to the desired position of 1.5 m in front of the guide.

The secondary pursuit goal is to handle situations when the guide leaves the field-of-view of the robot. This situation arises when the guide either goes around a corner or enters into a room through a doorway. To circumvent these problems, the guide is able to direct the robot either with a "*left*" or "*right*" signal depending on the traveling direction. When the robot receives one of the two gestures, it predicts the future location of the guide around the corner in the desired direction. To do this, the robot projects the current location of the guide to the left or right of the field-of-view.

When the projected location has been found, the robot plans a path to the location that avoids all obstacles. An example situation is shown in Fig. 12.6 when the user is leaving a room and moving to the right and down the hallway. The robot chooses a position about 1 m to the right of the latest predicted location of the guide, creating a path through the doorway to the goal location shown in red.

12.3.5 Mapping

During a guided tour of an environment, the robot has the ability to use the laser data to build a geometric estimate of the environment. To this end, an occupancy

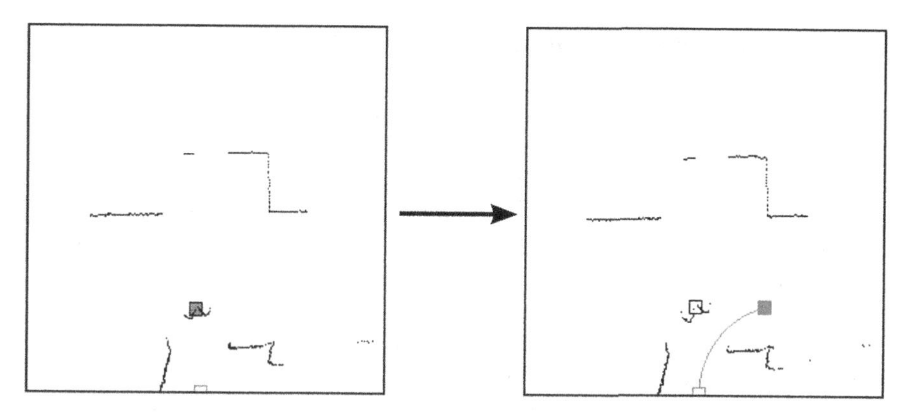

Fig. 12.6 Illustration of the method used by the robot to turn around a corner. The *left image* presents a guide standing in a hallway, while the robot is still in the room. After the guide has issued a "*turn right*" gesture, the robot assumes the guide will leave the field-of-view and begins to perform a corner manoeuvre autonomously. The *right image* illustrates the path in a *solid line* connecting two *box* labelled positions, generated to allow the robot to arrive at the projected future location of the guide

Fig. 12.7 A sample occupancy grid generated during a tour of our lab at Imperial College London. The *square* illustrates the current location of the robot in the environment

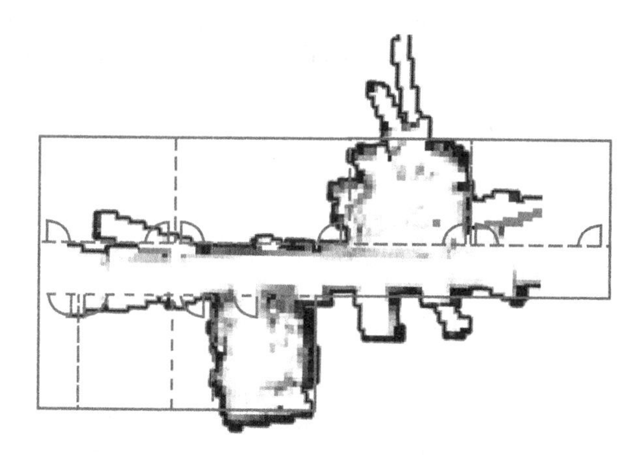

grid is constructed, while the location of the robot in the environment is tracked using a scan-matching system. In the proposed framework, an implementation based on the "*vasco*" scan-matching system, which is part of the Carnegie Mellon Robot Navigation Toolkit [22], is used. A sample map generated for our lab at Imperial College London is shown in Fig. 12.7.

The map generated during the tour of the environment provides only a geometric perspective of the environment. This introduces ambiguities as locally, many of the architecture features are very similar across a building. The robot can easily become confused to its actual location. To address this limitation, a vision based scene descriptor is used to build a global perspective based on appearance. These descriptors are generated after receiving a "*signal*" gesture from the guide during the tour.

12.4 Qualitative Localization

The laser mapping system presented so far is capable of calculating the quantitative location of the robot in a defined coordinate system. In practice, however, its robustness is far from perfect, and it is important to provide complementary location information. To this end, our proposed system relies on visual information to qualitatively identify the scene that the robot is currently in view of, working independently of the mapping system, and ultimately with both systems reinforcing each other.

For indoor scenes, rooms are often geometrically similar, and this can cause problems with the proposed mapping system when only building a rough geometrical view of the environment. Visually, however, these rooms are often very distinctive. Features of a room, such as pictures on the wall or lights on the ceiling, present information that a laser system is unresponsive to. Visual information can help reveal the room and significantly reduce the search space for localizing the robot.

The ability of the robot to understand in which room it is located also adds to the pervasive nature of the system. For example, should the robot be required to relay its location for repair, simply stating the name of the room to the engineer is more meaningful than providing a series of numbers representing its location. In addition, visual representations of a scene lend contextual information that laser systems cannot provide. This is of great benefit when the robot is required to interact with its environment.

It is also worth noting that no navigation system should rely entirely upon one type of sensor. Combining visual and laser sensing provides both depth and content information, which presents a sound framework upon which to build a robust navigation system. This overcomes the malfunctioning of a sensor and/or an environment poorly suited to a single sensor.

The compliment to the laser mapping system in the proposed framework is based on scene recognition. Scene recognition for robotic applications is a field that can often be considered as a special case of image matching. Finding the image in a database most similar to a candidate image has been widely addressed in literature. Many approaches represent images by a distribution of features such as SIFT [19], with matches between features based upon similarity in feature descriptors, as well as the spatial relationships of features [25, 26, 30]. The transfer of these techniques to robot localization must deal with the problems associated with indoor scenes. Such scenes often have a lower presence of discriminating features, and instead contain a large number of uniform regions representing commonly-occurring bodies such as walls, floors and ceilings. This results in images not only having fewer features to match, but those features found are often present in other similar rooms. A further issue is that viewpoint changes in indoor environments are often large relative to images of outdoor scenes, which is generally the focus of the above techniques. As such, most approaches for indoor scenes use more advanced methods such as supervised learning [34], probabilistic matching [15], feature grouping [1], or a combination of both global and local features [37].

12.4.1 Scene Association

The ability to recognize specific features of a scene is important for a robot to interact with and navigate within a scene. For this purpose, our method of scene association allows a scene to be recognized from a number of viewpoints, whilst still identifying specific features. We propose that we call features that are viewpoint-invariant and are consistently detected across different viewpoints as *association features.*

In order to extract the association features from a scene, several images of the scene are captured from varying viewpoints, and features which occur across all images are retained. In this work, SIFT features are used. During the training phase, a match is tested between each feature in an image, and all features in the other images of the scene. Those features which are found in all images are retained as association features. In our equations, f_a represents an association feature and f_c represents a candidate feature which we are attempting to match to an association feature. To determine whether a match is made, three steps are taken, and steps with the least computational expense and most likely to eliminate the greatest number of false matches, are handled first.

In the examples shown in this chapter, it is assumed that the robot maintains an upright position, such that the features will only vary by small amounts due to affine viewpoint changes and not absolute camera rotations. Thus a candidate feature is firstly discarded if its orientation differs to that of an association feature by more than a threshold, t_θ:

$$abs(f_a(\theta) - f_c(\theta)) > t_\theta. \tag{12.9}$$

Then, for any candidate feature that is not eliminated by (9), the difference in descriptors between f_a and f_c is calculated, by summing the dimension-by-dimension differences between the SIFT descriptors, $d_1 \cdots d_{128}$. The feature is discarded if this difference is more than t_{sift}:

$$\sum_{i=1}^{128} abs(f_a(d_i) - f_c(d_i)) > t_{sift}. \tag{12.10}$$

For those candidate features not eliminated by (10), elementary graph theory is then used to eliminate matches that are not verified by the local neighborhood. The neighborhood of a feature is defined as the 10 spatially-closest features captured in the same image, as proposed in [25]. Then, a feature $f_c^{(n)}$ in the neighbor-hood of f_c is considered a neighborhood match, if there exists a feature $f_a^{(m)}$ in the neighborhood of f_a, which has a similar orientation and descriptor to $f_c(n)$, as defined in (9) and (10). Additionally, the angle between f_c and $f_c^{(n)}$ must differ to the angle between f_a and $f_a^{(m)}$ by no more than t_φ. Then the candidate feature f_c is discarded if the number of neighborhood matches to is f_a less than N:

$$\sum_{n=1}^{10} NumMatches(f_c^{(n)}, f_a) < N \tag{12.11}$$

where

$$NumMatches(f_c^{(n)}, f_a) = \begin{cases} 1 & \text{if } \sum_{m=1}^{10} IsMatch(f_c^{(n)}, f_a^{(m)}) \geq 1, \\ 0 & \text{otherwise} \end{cases} \tag{12.12}$$

where

$$IsMatch(f_c^{(n)}, f_a^{(m)}) = \begin{cases} 1 & \text{if } abs(f_c^{(n)}(\theta) - f_a^{(m)}(\theta)) < t_\theta \\ & \text{and } \sum_{i=1}^{128} abs(f_c^{(n)}(d_i) - f_a^{(m)}(d_i)) < t_{sift} \\ & \text{and } abs(\varphi(f_c^{(n)}, f_c) - \varphi(f_a^{(m)}, f_a)) < t_\varphi, \\ 0 & \text{otherwise.} \end{cases} \tag{12.13}$$

In the above equation, $\varphi(f_1, f_2)$ represents the orientation of the line connecting features f_1 and f_2. If a candidate feature satisfies all these criteria, then it is considered a match between the two images. It is then passed on to the next image of the scene to determine whether the same feature is found again. Once an association feature is found across all images, its descriptor is calculated by computing the dimension-by-dimension average of the descriptors of all the features contributing to this association feature.

In the example shown below in Fig. 12.8, three images of each scene are used, and an association feature is recorded if it is present in all three images. Using more images can significantly reduce the number of detected association features, thus affecting its ability to perform scene association on a captured image in later stages. The top row shows all the originally detected features, and the bottom row showing only the association features, which were found in all three of the top row images.

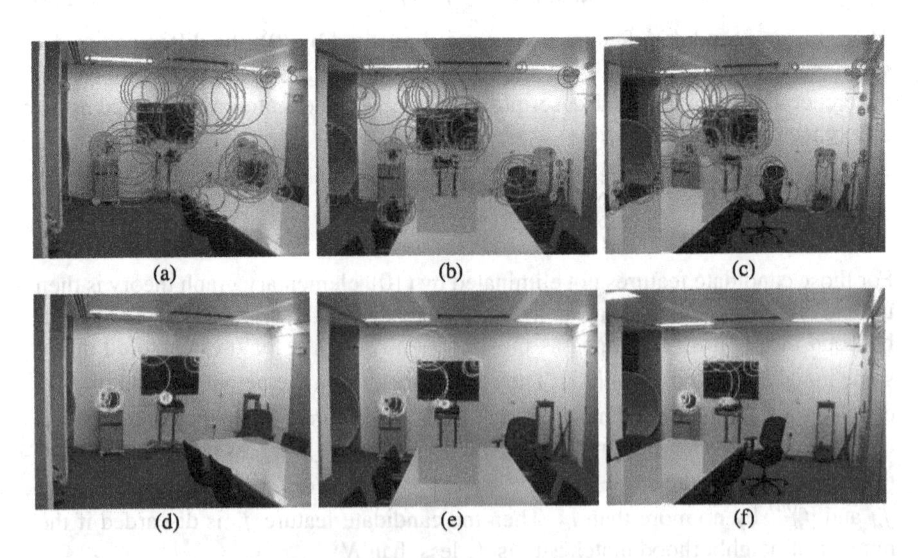

Fig. 12.8 Example of the training phase during which association features are detected for a scene. Images **(a)**–**(c)** are taken from different viewpoints of the same scene. SIFT features are then detected in the images and highlighted. Those features that are found in all three images **(a)**–**(c)** are memorized as association features and highlighted in images **(d)**–**(f)**

In Fig. 12.8, it is worth noting that the association features all form part of the background of the images, and all features on foreground objects are eliminated. There are two reasons for this. First, a background feature across all three images has a similar incident viewpoint than foreground features, and hence the feature descriptor varies less between the viewpoints. Second, background features which lie against a wall have no background clutter to confuse the feature descriptor, whereas the descriptor for foreground features varies as different elements of the background come into view behind the feature.

12.4.2 Scene Recognition

With association features learnt for each room, the next task is to match features from a captured image as the robot navigates through the environment, to those association features stored in the robot's memory. This is done in a similar manner as during the training phase. First, candidate SIFT features, f_c, are extracted from the latest captured image. Then, for every association feature, f_a, in each scene in memory, a match is attempted to every candidate feature, f_c. A match is classified as positive if it is similar in orientation to f_c, has a similar descriptor to f_c, and is verified by the local neighborhood of f_c. This is identical to the process of learning association features in Sect. 12.4.1, except that we now use a smaller value for t_{sift}. This adjustment is necessary because in the training phase, features are only compared to those from a small number of images of the same scene. However, during the recognition stage, features are compared to features from all scenes in the database, and so descriptors are required to be closer to have sufficient confidence of a match.

Choosing the actual values of t_{sift} in the two phases is a compromise between feature discrimination, and viewpoint invariance. In our example results, we found that for the training phase, $t_{sift} = 25$ was an optimum value, generating a large number of positive matches and leaving only 10% false positive matches, which were then all eliminated during neighborhood verification. For the recognition phase, t_{sift} can be tweaked in accordance with the number of rooms in the environment and for the examples shown in this chapter, $t_{sift} = 45$. With a smaller value, the same feature detected across large viewpoints was often eliminated, and with a larger value, too many false positive matches were found that could not be eliminated by neighborhood verification.

If a match between an association feature and a candidate feature is positive, the algorithm attempts to find a match to the next association feature. For each scene, the percentage of association features which have been matched then enters the scene into a ranking system, where the scene with the highest percentage of association feature matches is output as the scene within which the robot is located.

Figure 12.9 demonstrates a typical arrangement within the boundaries of a room where the proposed scene association is used. At each location, the robot captures a series of 8 images at 45° intervals to form a panoramic sequence, and computes

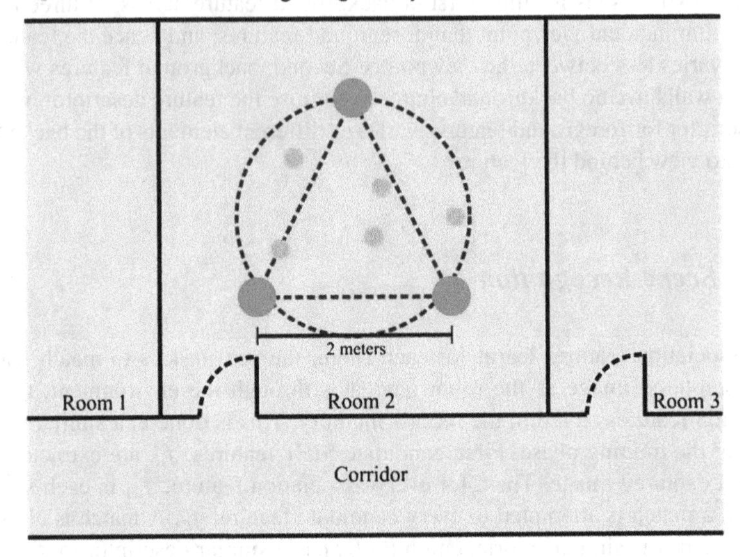

Fig. 12.9 Arrangement of robot locations within a room during training and recognition phases. During the training phase, scenes are captured at three points of the triangle, whereas during the recognition phase, scenes are captured randomly within the circle tangential to the triangle. At each location, the robot rotates to capture multiple images to form a panoramic sequence

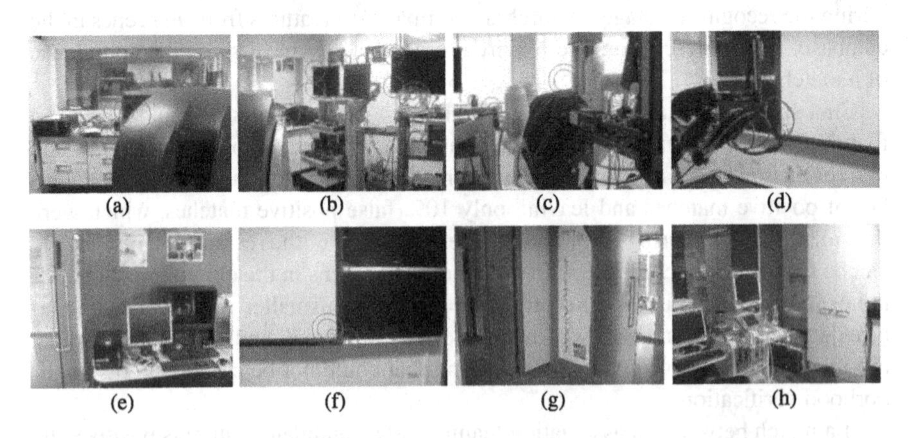

Fig. 12.10 Panoramic images with SIFT features highlighted for scene association. Images (**a**)–(**h**) are captured at 45° intervals as the robot rotates within a room. This is performed in both the training and recognition phases

SIFT features for each image. Figure 12.10 shows the panoramic images with all detected features highlighted. An image matched which match to an association feature in memory increases the likelihood of it being associated that scene.

Table 12.1 Recognition accuracy by using the scene association method proposed for a laboratory scene consists of 7 rooms. Bold numbers are percentage of true positive feature matches, non-bold numbers are the percentage of false positive feature matches. Parameters used in (9)–(11): $t_\theta = 20$, $t_\varphi = 45$, $N = 1$, $t_{sift} = 45$ for training phase, 25 for recognition phase

	% Feature matches in each room						
Room number	1	2	3	4	5	6	7
1	**60**	4	17	1	9	17	29
2	2	**48**	2	1	0	0	4
3	13	11	**82**	16	27	16	1
4	2	45	0	**55**	42	17	3
5	8	7	6	37	**72**	25	1
6	7	18	3	14	31	**36**	0
7	0	10	7	4	7	3	**63**

During the training phase, the robot is initially instructed by hand gestures to capture panoramic images in 7 rooms of the building. In each room, the robot learns the association features by capturing images at each of the three locations in Fig. 12.9.

During the recognition stage, the robot captures one set of panoramic views and calculates the percentage of matches to association features for each scene. In this experiment, 93% of the test scenes were identified with the correct room, by considering the highest percentage features matches across all scenes in the database. Table 12.1 shows the recognition performance across the seven rooms visited, showing the average results across multiple recognition attempts for each room. The numbers in bold represent the percentage of association features recognized in the correct room (true positives), whilst the non-bold numbers represent the percentage of association features recognized in all the other incorrect rooms (false positives).

It is evident that some rooms have generated a higher confidence in their correct identification. For example, Rooms 1, 2, 3, 5 and 7 have large differences between the most likely and second most likely rooms, whereas with Rooms 4 and 6, the system is less confidence that the most likely room was indeed identified correctly. This is largely due to the presence of similar objects in different rooms, such as television screens, whose features are similar across different scenes, and who also have similar features in the local neighborhood, drawn from the same object.

Nonetheless, with a 93% positive scene identification, this vision system is well equipped to work in tandem with the laser mapping system, and integrates appropriately with the gesture-recognition task. The final challenge is then to incorporate both the qualitative and quantitative localization data, into a system that is able to autonomously navigate between rooms, as instructed by the user.

12.5 Planning and Navigation

As the robot is guided around the environment, laser data is collected in order to build a geometric map of its surroundings. As mentioned earlier, the guide indicates points of interest within the environment by performing an "*attention*" gesture. The tour is to enable the robot to map the environment using quantitative and qualitative

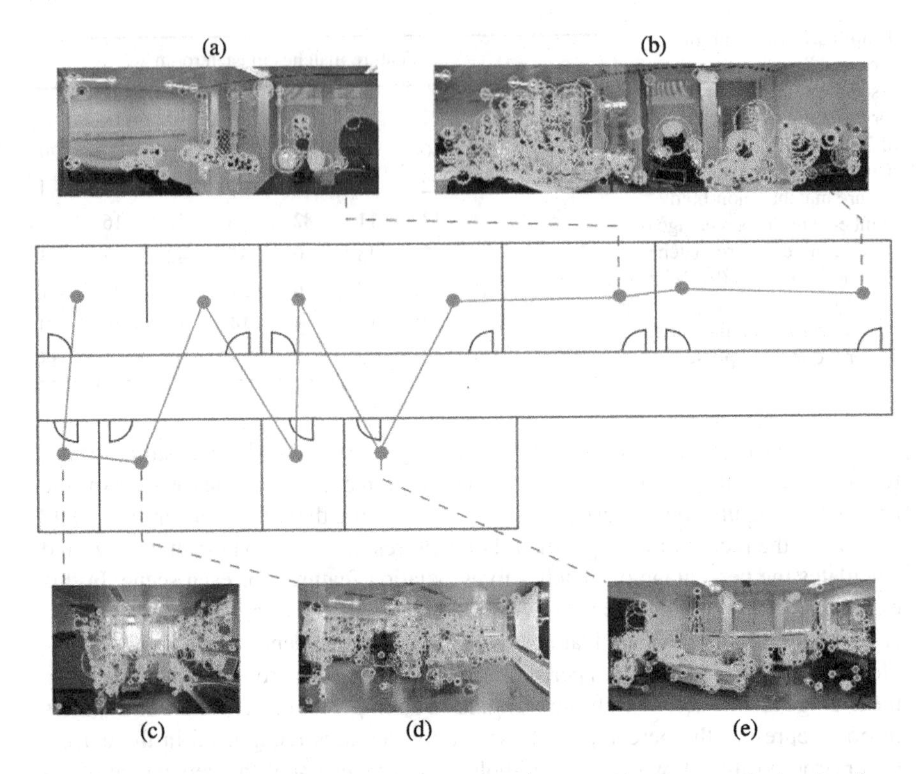

Fig. 12.11 A topological map of the environment storing scene descriptors at the key locations indicated by the user during the tour

localization techniques; incorporating scene association improves localization and also the high-level planning used for navigation.

After mapping and localization, in order to autonomously navigate towards a goal, there needs to be a plan. A plan can be described as a sequence of moves or re-actions which lead towards the goal [21]. Formulating a plan when the environment map is discrete is simpler since classical graph-searching algorithms such as A* and Dijkstra can be used [21]. The two main approaches for discetizing the environment is to either store it as a grid, grid-based (metric) paradigm, or as a graph, topological paradigm [32]. By using the laser mapping system and scene association descriptors, we can integrate both grid-based and topological paradigms to allow for fast path planning on the easy to construct occupancy map, utilizing the advantages of each representation, as mentioned by Thrun and Bücken [32].

During a guided tour, the robot constructs the occupancy map of its environment and, when gestured by the user, records a scene descriptor for its current location, which is mapped onto the occupancy map as shown in Fig. 12.11. In addition to the scene descriptors created, a key location is also indicated. It would also be useful, for navigation purposes, if descriptors are captured automatically through-out the tour as waypoints, since this will allow the topology of the environment to be captured more accurately.

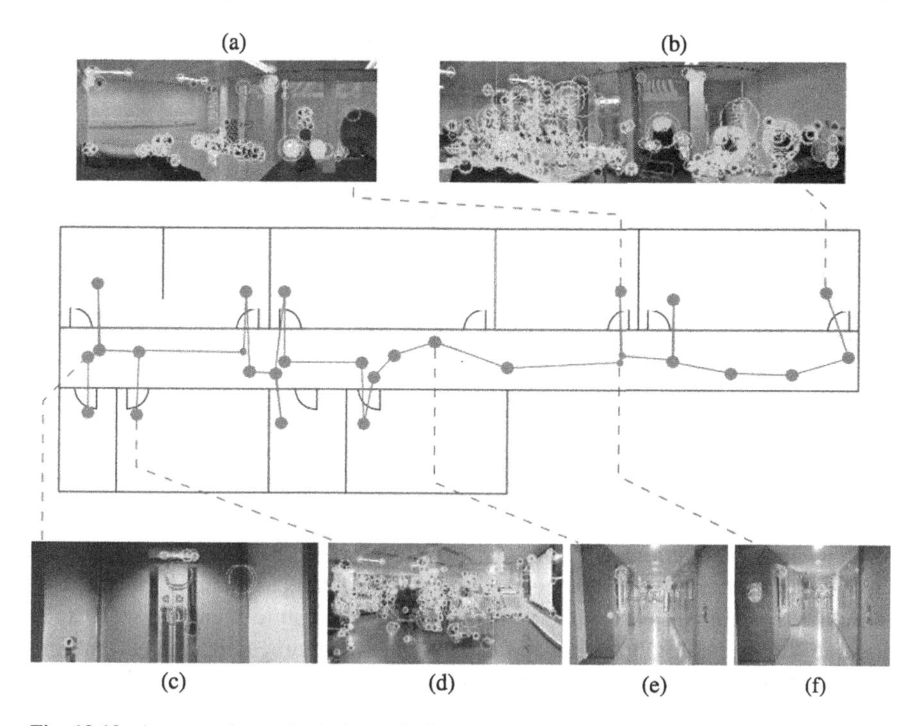

Fig. 12.12 An example topological map built during a guided tour. Capturing scene descriptors periodically during the guided tour allows the robot to build a more detailed topological map of the environment, better recording the path taken by the user

Scene descriptors are periodically captured during the tour, allowing the graph-based map to also contain information about the path taken by the guide, and not just the points of interest; we captured these waypoints when turns greater than 5 m were made to ensure the robot would be able to later retrace the path taken during automated runs as shown in Fig. 12.12.

Once the robot localizes itself on the occupancy map, we can plan a route to the target locations using the topological map, starting from the current nearest node. This high level planning procedure is done by using the A* graph-search algorithm. The system uses cues from the LRF and camera to recognize when it reaches waypoints or the goal location.

In autonomous systems, learning can potentially provide the flexibility the system needs to adapt to dynamic environments [5]. Consider, for example, that a new optimal path is discovered between two locations, it would be desirable for the robot to update its internal model to reflect this discovery. Based on Thrun's idea of sensor interpretation [31], a learning method which interprets readings from different sensors, such as the laser range finder and color camera, could be utilized for coping with varying environments. For example, in repetitive scenes, such as the corridor shown in Fig. 12.13, the ability for accurate localization using scene recognition would decline dramatically. In such scenarios, it would perhaps be more beneficial if the robot could learn to rely more on other sensory information.

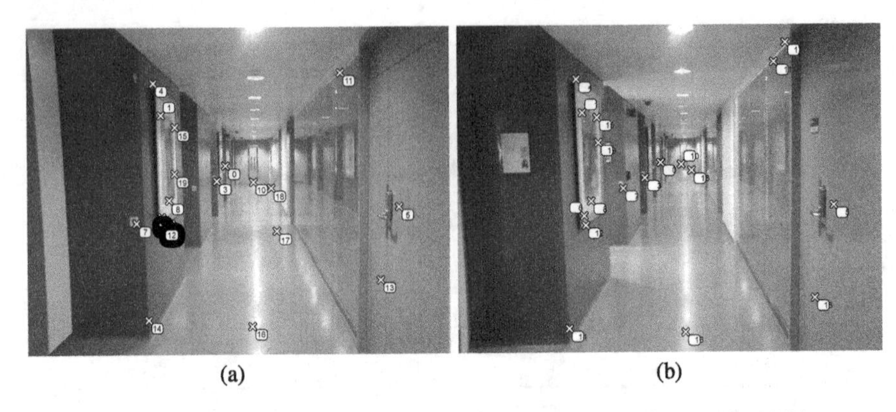

Fig. 12.13 Localization within a scene with repetitive visual features. Scenes (a) and (b) are captured from different locations in the environment, however, many features in both images indicate a match; scene association is not useful in all situations as most of the matches shown are incorrect

Other factors, besides a changing environment, would benefit from updates to the robot's internal model. Graph searching can be a computationally demanding, especially in complex environments. Our focus is to capture scenes to store as a node on the graph-based map automatically when a significant rotational motion is executed or if a large distance has been covered since the last recorded node. Although the topological map allows for faster planning, when compared to the grid-based occupancy map, the robot should seek to further simplify its representation of the environment as shown Fig. 12.14.

The simplified representation of the environment allows the robot to carry out future tasks in an autonomous fashion. Furthermore, the simplified map provides a user-friendly interface for control of the robot. This type of interface allows the proposed system to work in a variety of environments including museums, offices, home-care and hospital wards. Not only is the robot able to identify different rooms in the environment, whether it be to carry out a task or alert an engineer for repair, but also does the proposed system allow the robot to interact with people in the environment, whilst avoiding all obstacles.

12.6 Conclusion

Mobile robots present many opportunities to carry out mundane tasks in everyday life. Before robots are able to perform such tasks, basic intelligence must be developed. In this chapter, we have addressed several key challenges related to robotic navigation and the value of using HRI for environment mapping and scene association. Effective use of HRI allows the user to naturally interact with a mobile robot via gestures, which can be detected using a vision based system. We have demonstrated the practical use of the proposed gesture recognition system for guided exploration in a novel environment. These gestures help the robot in difficult situations and build scene descriptors. Upon being informed to follow, the proposed system used a

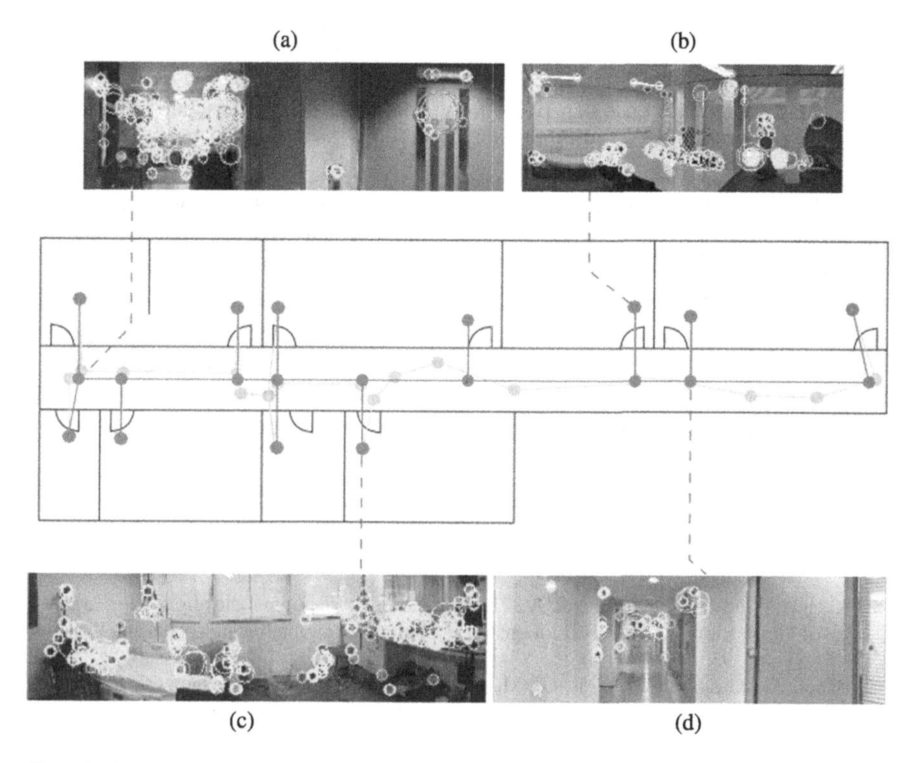

Fig. 12.14 Topological simplification for route planning. A simplified topological map of the environment, in comparison to Fig. 12.12, brings performance benefits for route planning and ease of visualization

multi-cue tracking system to maintain an estimate of the location of the guide at all times.

During a guided tour, the robot uses the laser data to create an occupancy map of the environment. However, there are scenarios where localization using this quantitative approach can be improved by using qualitative data. To this end, a vision based scene association is used to complement the occupancy map by capturing descriptors of particular scenes on the map. These descriptors are built at salient locations of the environment. The visual descriptors consist of distributions of SIFT features, which the robot has learned to memorize as they occur consistently across multiple viewpoints of a scene.

To autonomously navigate within the recorded environment, the robot uses both the geometric occupancy map and topological map of the scene descriptors. Quantitative and qualitative localization techniques are complementary with each other, providing accurate localization in geometrically similar environments. To accurately retrace the path taken by the guide, scene information is captured periodically by the robot during the guided tour. High level path planning is carried out by performing A* search on the topological map from the current scene to the goal destination. In this chapter, we have described our considerations on how autonomous navigation can be improved by incorporating mechanisms that will allow it to cope with a changing environment and uncertainty from sensor readings. We demonstrated how visually similar scenes can potentially cause confusion for scene association and suggest how the robot could adapt its interpretation of sensor data under these conditions.

While the proposed system attempts to handle many of the issues related autonomous navigation, future work will aim to improve the robustness of the system. More sensor modalities could be used to further help the robot understand the environment. For example, 3D time-of-flight cameras could be used to accompany the 2D laser scanner to provide a more detailed view of the environment. This could help the robot identify the exact location of objects in the environment. Furthermore, a more detailed tracking system could help the robot to maintain the motion of all moving objects in the environment for improved planning and obstacle avoidance.

References

1. Ascani, A., Frontoni, E., Mancini, A., Zingaretti, P.: Feature group matching for appearance-based localization. In: IEEE/RSJ International Conference on Intelligent Robots and Systems (IROS), pp. 3933–3938 (2008)
2. Bauer, A., Klasing, K., Lidoris, G., Mühlbauer, Q., Rohrmüller, F., Sosnowski, S., Xu, T., Kühnlenz, K., Wollherr, D., Buss, M.: The autonomous city explorer: towards natural human-robot interaction in urban environments. Int. J. Soc. Robot. **1**(2), 127–140 (2009)
3. Bellotto, N., Hu, H.: Multisensor integration for human-robot interaction. IEEE J. Intell. Cybern. Syst. **1** (2005)
4. Bellotto, N., Hu, H.: Multisensor-based human detection and tracking for mobile service robots. IEEE Trans. Syst. Man Cybern., Part B, Cybern. **39**(1), 167–181 (2009)
5. Buhmann, J., Burgard, W., Cremers, A., Fox, D., Hofmann, T., Schneider, F., Strikos, J., Thrun, S.: The mobile robot Rhino. AI Mag. **16**(2), 31 (1995)
6. Busch, M., Blackman, S.: Evaluation of IMM filtering for an air defense system application. In: Society of Photo-Optical Instrumentation Engineers (SPIE) Conference Series, vol. 2561, pp. 435–447 (1995)
7. Chakravarty, P., Jarvis, R.: Panoramic vision and laser range finder fusion for multiple person tracking. In: Proceedings of IEEE/RSJ International Conference on Intelligent Robots and Systems (IROS), pp. 2949–2954 (2006)
8. Fritsch, J., Kleinehagenbrock, M., Lang, S., Plötz, T., Fink, G., Sagerer, G.: Multi-modal anchoring for human–robot interaction. Robot. Auton. Syst. **43**(2–3), 133–147 (2003)
9. Hasanuzzaman, M., Ampornaramveth, V., Zhang, T., Bhuiyan, M., Shirai, Y., Ueno, H.: Real-time vision-based gesture recognition for human robot interaction. In: IEEE International Conference on Robotics and Biomimetics (ROBIO), pp. 413–418 (2004)
10. Ishii, Y., Hongo, H., Yamamoto, K., Niwa, Y.: Real-time face and head detection using four directional features. In: Proceedings of the Sixth IEEE International Conference on Automatic Face and Gesture Recognition, pp. 403–408 (2004)
11. Jin, Y., Mokhtarian, F.: Towards robust head tracking by particles. In: IEEE International Conference on Image Processing (ICIP), vol. 3, pp. 864–867 (2005)
12. Koenig, N.: Toward real-time human detection and tracking in diverse environments. In: IEEE 6th International Conference on Development and Learning (ICDL), pp. 94–98 (2007)
13. Krotosky, S., Cheng, S., Trivedi, M.: Real-time stereo-based head detection using size, shape and disparity constraints. In: Proceedings of the IEEE Intelligent Vehicles Symposium, pp. 550–556 (2005)

14. Lee, H., Kim, J.: An HMM-based threshold model approach for gesture recognition. IEEE Trans. Pattern Anal. Mach. Intell. **21**(10), 961–973 (1999)
15. Li, F., Kosecka, J.: Probabilistic location recognition using reduced feature set. In: IEEE International Conference on Robotics and Automation, pp. 3405–3410. Citeseer (2006)
16. Li, Y., Ai, H., Huang, C., Lao, S.: Robust head tracking based on a multi-state particle filter. In: 7th International Conference on Automatic Face and Gesture Recognition, pp. 335–340 (2006)
17. Lindstrom, M., Eklundh, J.: Detecting and tracking moving objects from a mobile platform using a laser range scanner. In: Proceedings of the IEEE/RSJ International Conference on Intelligent Robots and Systems, vol. 3, pp. 1364–1369 (2001)
18. Loper, M., Koenig, N., Chernova, S., Jones, C., Jenkins, O.: Mobile human-robot teaming with environmental tolerance. In: Proceedings of the 4th ACM/IEEE International Conference on Human Robot Interaction, pp. 157–164. ACM, New York (2009)
19. Lowe, D.: Distinctive image features from scale-invariant keypoints. Int. J. Comput. Vis. **60**(2), 91–110 (2004)
20. Luo, J., Savakis, A., Singhal, A.: A Bayesian network-based framework for semantic image understanding. Pattern Recognit. **38**(6), 919–934 (2005)
21. Meyer, J., Filliat, D.: Map-based navigation in mobile robots, II: a review of map-learning and path-planning strategies. Cogn. Syst. Res. **4**(4), 283–317 (2003)
22. Montemerlo, M., Roy, N., Thrun, S.: Perspectives on standardization in mobile robot programming: the Carnegie Mellon navigation (CARMEN) toolkit. In: Proceedings of the IEEE/RSJ International Conference on Intelligent Robots and Systems, pp. 2436–2441. Citeseer (2003)
23. Nickel, K., Stiefelhagen, R.: Real-Time Person Tracking and Pointing Gesture Recognition for Human-Robot Interaction. In: Proceedings of the Computer Vision in Human-Computer Interaction: ECCV 2004 Workshop on HCI, Prague, Czech Republic, p. 28. Springer, New York (2004)
24. Reynolds, C.: Steering behaviors for autonomous characters. In: Game Developers Conference. http://www.red3d.com/cwr/steer/gdc99 (1999)
25. Schaffalitzky, F., Zisserman, A.: Automated scene matching in movies. In: Lecture Notes in Computer Science, pp. 186–197 (2002)
26. Schmid, C., Mohr, R.: Local grayvalue invariants for image retrieval. IEEE Trans. Pattern Anal. Mach. Intell. **19**(5), 530–535 (1997)
27. Shalom, Y., Blair, W.: Multitarget-Multisensor Tracking: Applications and Advances, vol. 3. Artech House, Boston (2000)
28. Shan, C., Wei, Y., Tan, T., Ojardias, F.: Real time hand tracking by combining particle filtering and mean shift. In: Sixth IEEE International Conference on Automatic Face and Gesture Recognition, pp. 669–674 (2004)
29. Shi, J., Tomasi, C.: Good features to track. In: IEEE Conference on Computer Vision and Pattern Recognition, pp. 593–600. Citeseer (1994)
30. Sivic, J., Zisserman, A.: Video Google: a text retrieval approach to object matching in videos. In: Proceedings of the ICCV, vol. 2, pp. 1470–1477. Citeseer (2003)
31. Thrun, S.: Exploration and model building in mobile robot domains. In: Proceedings of the IEEE International Conference on Neural Networks vol. 1, pp. 175–180. Citeseer (1993)
32. Thrun, S., Bücken, S.: Learning maps for indoor mobile robot navigation. Carnegie Mellon University, technical report No. CMU-CS-96-121 (1996)
33. Topp, E., Christensen, H.: Tracking for following and passing persons. In: Proceedings of the IEEE/RSJ International Conference on Intelligent Robots and Systems, pp. 70–76. Citeseer (2005)
34. Ullah, M., Pronobis, A., Caputo, B., Luo, J., Jensfelt, P., Christensen, H.: Towards robust place recognition for robot localization. In: Proceedings of the IEEE International Conference on Robotics and Automation, pp. 530–537. Citeseer (2008)
35. Valibeik, S., Yang, G.: Segmentation and tracking for vision based human robot interaction. In: Proceedings of the 2008 IEEE/WIC/ACM International Conference on Web Intelligence and Intelligent Agent Technology, vol. 3, pp. 471–476. IEEE Computer Society, Washington (2008)

36. Viola, P., Jones, M.: Robust real-time face detection. Int. J. Comput. Vis. **57**(2), 137–154 (2004)
37. Wimpey, B., Drucker, E., Martin, M., Potter, W.: A multilayered approach to location recognition. In: Proceedings of the SoutheastCon, pp. 1–7
38. Won, W., Kim, M., Son, J.: Driver's head detection model in color image for driver's status monitoring. In: 11th International IEEE Conference on Intelligent Transportation Systems, pp. 1161–1166 (2008)
39. Zhang, Q., Pless, R.: Extrinsic calibration of a camera and laser range finder (improves camera calibration). In: 2004 IEEE/RSJ International Conference on Intelligent Robots and Systems, Proceedings, vol. 3, pp. 2301–2306 (2004)
40. Zhang, Z.: Flexible camera calibration by viewing a plane from unknown orientations. In: International Conference on Computer Vision vol. 1, pp. 666–673 (1999)
41. Zhang, Z., Gunes, H., Piccardi, M.: An accurate algorithm for head detection based on XYZ and HSV hair and skin color models. In: 15th IEEE International Conference on Image Processing (ICIP), pp. 1644–1647 (2008)
42. Zhu, Y., Ren, H., Xu, G., Lin, X.: Toward real-time human-computer interaction with continuous dynamic hand gestures. In: Proceedings of the Fourth International Conference on Automation Face Gesture Recognition, Grenoble, France, pp. 544–549. IEEE Computer Society, Los Alamitos (2000)

Chapter 13
Prediction-Based Perceptual System of a Partner Robot for Natural Communication

Naoyuki Kubota and Kenichiro Nishida

Abstract This chapter discusses the adaptation of perceptual modules of a partner robot based on classification and prediction through actual interactions with a human. The prediction is very important to extract the perceptual information for natural communication with a human. Therefore we proposed a prediction-based perceptual system composed of four layers: the input layer, clustering layer, prediction layer, and perceptual module selection layer. The proposed system has three main functions; (1) the clustering of perceptual information (i.e., the extraction of spatial patterns), (2) the prediction of transition among the clusters (the extraction of temporal patterns), and (3) selection of perceptual modules (the control of sampling intervals). In this chapter, we apply the proposed method to the actual interaction between a human and a human-like partner robot. Finally, we show experimental results on the interaction with a human to discuss the effectiveness of our proposed method.

13.1 Introduction

Capabilities of social communication are required for human-friendly robots such as pet robots, partner robots, and robot-assisted therapy to realize natural communication with humans [3, 13–16, 19, 20, 22, 24, 27, 30, 34, 35]. These capabilities need to be based on verbal communication and non-verbal communication. Relevance theory is helpful to discuss the social communication between human and

N. Kubota (✉)
Department of System Design, Tokyo Metropolitan University, 6-6 Asahigaoka, Hino, Tokyo 191-0065, Japan
e-mail: kubota@tmu.ac.jp

K. Nishida
Multimedia Soc Development Dept., Toshiba Corporation Semiconductor Company, 580-1, Horikawa-Cho, Saiwai-Ku, Kawasaki 212-8520, Japan
e-mail: Kenichiro.nishida@toshiba.co.jp

H. Liu et al. (eds.), *Robot Intelligence*,
Advanced Information and Knowledge Processing,
DOI 10.1007/978-1-84996-329-9_13,

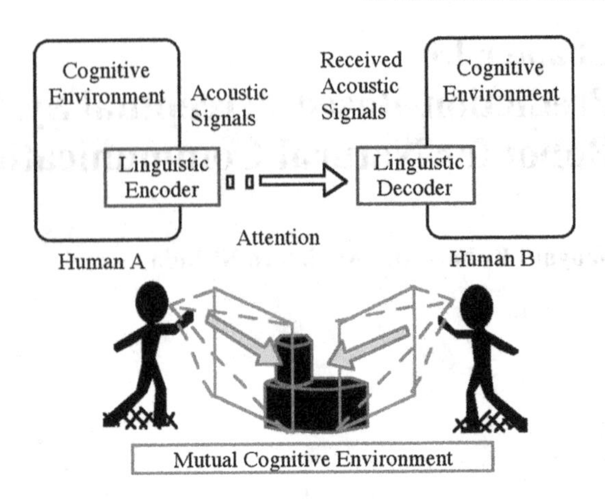

Fig. 13.1 Mutual cognitive environment in communication between humans

robot [29]. In relevance theory, human thought is not transmitted, but is shared between two people. Each person has his or her own cognitive environment (Fig 13.1). Even though two people speak different languages, one person can understand the meaning of an unknown term spoken by the other through communication because the person can consider the unknown term as a symbol corresponding to the percept. Here an important role of utterances or gestures is to make a person pay attention to a specific target object or person. As a result, the cognitive environment of the other can be enlarged by utterances or gestures. The cognitive environment shared between two people is called a mutual cognitive environment. Consequently, a robot also needs to have such a cognitive environment, and keep updating it according to the current perception through interaction with a human in order to realize the natural communication.

In the classical approach of environmental perception in artificial intelligence, a robot tries to extract environmental information and to build a complete environmental model in the robot [26]. When the robot interacts with a human, the robot selects the perceptual information from the built environmental model for natural communication with the human. However, it is very difficult for the robot to extract the perceptual information to be shared with the human beforehand. Furthermore, it is computationally expensive to build a complete environmental model including unnecessary perceptual information for human-robot interaction. Therefore, the robot should have the predictive capability [5, 8] to extract the perceptual information to be shared with the human.

Most of the previous research on prediction has focused on the prediction of given patterns [2, 6]. If patterns to be predicted are given, the robot has only to memorize and recall them. However, the robot does not know exact patterns to be predicted in communication with a human, because each human has his or her own unique behavior patterns. Therefore, in order to predict human behavior patterns, the robot should divide human behaviors into several action segments according to its own criteria, and learn transitions of action segments. Moreover, we can unite a time-series of meaningful action segments into a behavior pattern from the viewpoint of hierarchy and granularity of behaviors. In this study, we have focused on the prediction capabilities based on the spatio-temporal dynamics of behavior patterns [17], because the patterns included in perceptual information required for the communication can vary dependent on the spatio-temporal context or situation.

Next, we discuss the mechanism of prediction for the perception. In general, a human behavior is restricted by the situation, while the meaning of a situation is specified by the series of behaviors. For example, assume that we see other people rushing to the platform. We can easily guess that the next train is coming to the station, and will check the time or a display of the time-table. In such way, we can understand the situation by observing behaviors of other people in a specific environment, and predict what to be perceived. Consequently, the robot should perceive both the human behavior and its related environmental states. Therefore, we proposed a prediction-based perceptual system composed of four layers: the input layer, clustering layer, prediction layer, and perceptual module selection layer [17, 18]. The proposed system has three main functions; (1) the clustering of perceptual information (the extraction of spatial patterns), (2) the prediction of transition among the clusters (the extraction of temporal patterns), and (3) selection of perceptual modules (the control of sampling intervals). To represent a human behavior pattern as the internal state of the robot, the robot should use a set of primitive perceptual information obtained by image processing, voice recognition, and others. The robot extracts the patterns of primitive perceptual information to be paid attention to. According to the transition of perceptual information, the robot selects perceptual modules suitable to the dynamics of human interactions in the next perception. Furthermore, we applied the proposed method to a PC-based partner robot, and discussed the learning capability of the proposed method by using the recorded movies [17]. In this chapter, we discuss the applicability of the proposed method to actual interaction between a human and a human-like partner robot.

This chapter is organized as follows. Sections 13.2 and 13.3 propose a prediction-based perceptual system and image processing methods for visual perception of a robot. Section 13.4 shows experimental results on real-time interaction with a person of a partner robot based on the proposed method and discusses the effectiveness of our proposed method.

13.2 Prediction-Based Perceptual System for A Partner Robot

13.2.1 A Partner Robot; Hubot

We developed a human-like partner robot, Hubot [20] in order to realize the social communication with a human (Fig. 13.2). This robot is composed of a mobile base, a body including two CPUs, two arms with grippers. The robot has various sensors such as a Pan-tilt CCD camera, two line sensors, microphone, ultrasonic sensors, touch sensors in order to perceive its environment. In the previous researches, we proposed (1) a human detection method using a series of images from the CCD

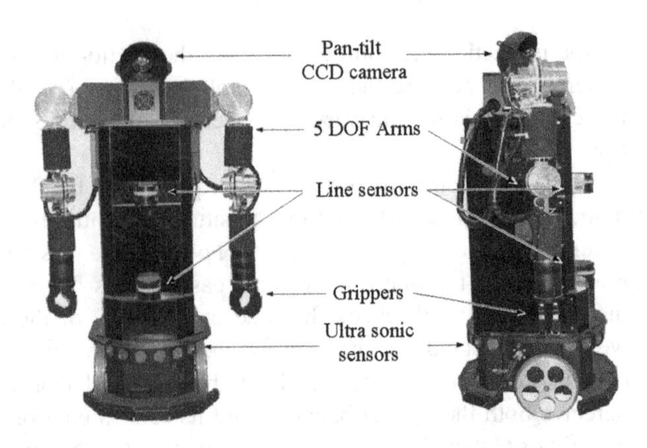

Fig. 13.2 A partner robot; Hubot

camera, (2) an interactive trajectory planning method for a hand-to-hand behavior based on human evaluations [10], (3) utterance and voice recognition for natural communication based on relevance theory [14], (4) reinforcement learning based on multiple value functions, (5) map building through interaction with a human [15], and (6) multi-objective behavior coordination for reproducing acquired behaviors [20]. These methods were proposed for the situation sharing between the robot and a human based on behaviors, but we should discuss the perceptual capabilities of partner robots to realize more natural communication.

In ecological psychology, the smallest unit of analysis must be the perceiving-acting cycle situated in an intentional context [7, 28, 33]. Especially, situated perception enables the prediction suitable to the spatio-temporal context of the environment. Predictions of goal-directed behaviors may arise from knowledge of human cognitive and physical abilities and constraints. The robot should extract the human behavior patterns in finite time, because the prediction of the human behavior patterns is important to interact with the human.

13.2.2 A Prediction-Based Perceptual System

The prediction-based perceptual system is composed of four layers: the input layer (I-layer), clustering layer (C-layer), prediction layer (P-layer), and perceptual module selection layer (S-layer), respectively (Fig. 13.3). The I-layer is composed of spiking neurons used for recognizing a specific state. Here spiking neurons for the I-layer is called SN-I. Each perceptual module outputs the inputs to SN-I from sensory inputs.

Next, the C-layer performs unsupervised classification based on the spike outputs of the SN-I by using reference vectors. As a result of unsupervised classification, each neuron at the C-layer acquires the relationship among the sets of perceptual information. Here, a clustered perceptual state is called a perceptual mode. The dimension of a reference vector in each perceptual mode is the same as the number of

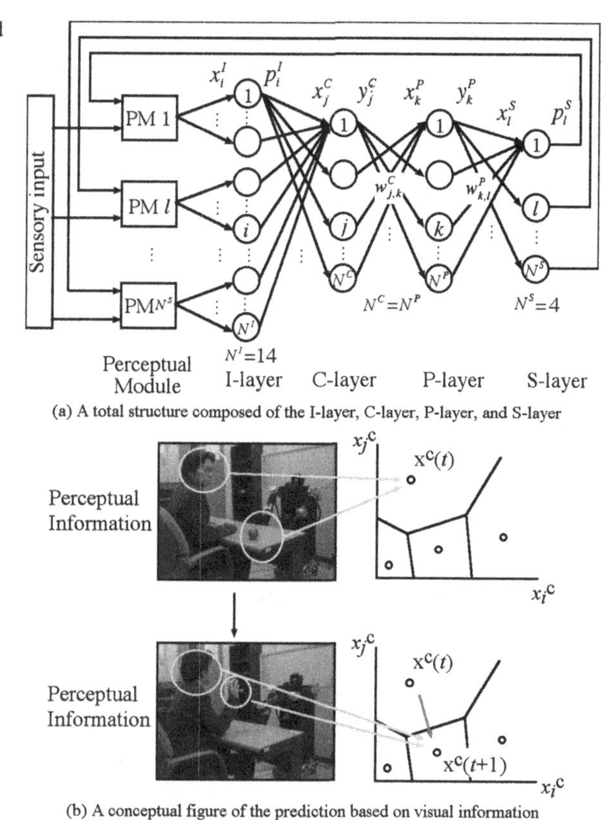

(a) A total structure composed of the I-layer, C-layer, P-layer, and S-layer

(b) A conceptual figure of the prediction based on visual information

Fig. 13.3 A prediction-based perceptual system

the spiking neurons of I-layer. Each perceptual mode relates with a specific combination of perceptual modules to extract its perceptual information.

The transition of perceptual modes can represent the dynamics of human interactions. Therefore, the robot should select perceptual modules suitable to the dynamics of human interactions in the next perception. Next, the robot learns the transition among perceptual modes to select perceptual modules for the next perception. The P-layer calculates the mode transition probability among perceptual modes.

According to the transition probabilities among perceptual modes, the S-layer selects the perceptual modules for extracting perceptual information required in the next perception. The S-layer is composed of spiking neurons used for selecting the perceptual modules and for controlling the sampling interval of each perceptual module. Here spiking neuron for the S-layer is called SN-S. Each perceptual module outputs the inputs to SN-I from sensory inputs according to the spike output of the SN-S corresponding to the perceptual module. Therefore, the time series of spike outputs from the SN-I and SN-S construct the spatio-temporal pattern of the perception. Since the change of the firing patterns indicates the dynamics of perception, the robot can select perceptual modules to be used in the next perception by learning the changing patterns as a result of prediction.

Table 13.1 Perceptual information outputted from the perceptual modules

Perceptual module	Outputs
1st SN-S: Differential extraction	1st SN-I:
	Magnitude of the human motion
2nd SN-S: Human detection	2nd–4th SN-I:
	Whether human is detected or not. Face direction (right and left)
3rd SN-S: Object detection	5th–11th SN-I:
	Whether object is detected or not.
	Color (red, green, and blue)
	Shape (circle, triangle and square)
4th SN-S: Hand motion recognition	12th–14th SN-I:
	Horizontal, vertical and slanting motion

13.2.3 Perceptual Modules

Vision includes much perceptual information for the communication with humans, but image processing takes much time. Therefore, the sampling interval control of perceptual modules is very important to extract perceptual information necessary to the communication with humans. Table 13.1 shows the SN-Ss corresponding to perceptual modules and their output information to SN-Is. An image is taken from the CCD camera attached on the top of the robot. We explain the detail of perceptual modules used for differential extraction, human detection, object detection, and human hand motion recognition in the following subsections.

13.2.4 Differential Extraction

The differential extraction module calculates the difference of the number of pixels between the previous and current images. First, the color difference between pixels at the current discrete time t and the previous time $t-1$ is calculated. If the color difference is larger than the predefined threshold, we assume the pixel is included in the. If the robot does not move, the center of gravity (COG) of the difference represents the location of the moving object. Therefore, the main search area for the human detection can be formed according to the COG for the fast human detection. Furthermore, the magnitude of the difference is used as the perceptual information inputted to the SN-I for measuring the magnitude of human motion.

13.2.5 Human Detection

We use a steady-state genetic algorithm (SSGA) for human detection and object detection as one of search methods, because SSGA can easily obtain feasible solutions through environmental changes with low computational costs. SSGA simulates a continuous model of the generation, which eliminates and generates a few individuals in a generation (iteration) [31]. Here SSGA for human detection is called SSGA-H, while SSGA for object detection is called SSGA-O.

Human skin and hair colors are extracted by SSGA-H based on template matching. Figure 13.4 shows a candidate solution of a template used for detecting a target. A template is composed of numerical parameters of $g_H^{i,1}$, $g_H^{i,2}$, $g_H^{i,3}$, and $g_H^{i,4}$. The number of individuals is G. One iteration is composed of selection, crossover, and mutation. The iteration of SSGA-H is repeated until the termination condition is satisfied. In this chapter, the worst candidate solution is eliminated ("Delete least fitness" selection), and is replaced with the candidate solution generated by the crossover and the mutation. We use elitist crossover and adaptive mutation. The elitist crossover randomly selects one individual and generates an individual by combining genetic information from the randomly selected individual and the best individual. Next, the following adaptive mutation is performed to the generated individual,

$$g_{i,j}^H \leftarrow g_{i,j}^H + \left(\alpha_j^H \cdot \frac{f_{max}^H - f_i^H}{f_{max}^H - f_{min}^H} + \beta_j^H\right) \cdot N(0,1) \tag{13.1}$$

where f_i^H is the fitness value of the ith individual, f_{max}^H and f_{min}^H are the maximum and minimum of fitness values in the population; $N(0,1)$ indicates a normal random variable with a mean of zero and a variance of one; α_j^H and β_j^H are the coefficients ($0 < \alpha_j^H < 1.0$) and offset ($\beta_j^H > 0$), respectively. In the adaptive mutation, the variance of the normal random number is relatively changed according to the fitness values of the population. Fitness value is calculated by the following equation,

$$f_i^H = C_{Skin}^H + C_{Hair}^H + \eta_1^H \cdot C_{Skin}^H \cdot C_{Hair}^H - \eta_2^H \cdot C_{Other}^H \tag{13.2}$$

where C_{Skin}^H, C_{Hair}^H and C_{Other}^H indicate the numbers of pixels of the colors corresponding to human skin, human hair, and other colors, respectively; η_1^H and η_2^H are the coefficients ($\eta_1^H, \eta_2^H > 0$). Therefore, this problem results in the maximization

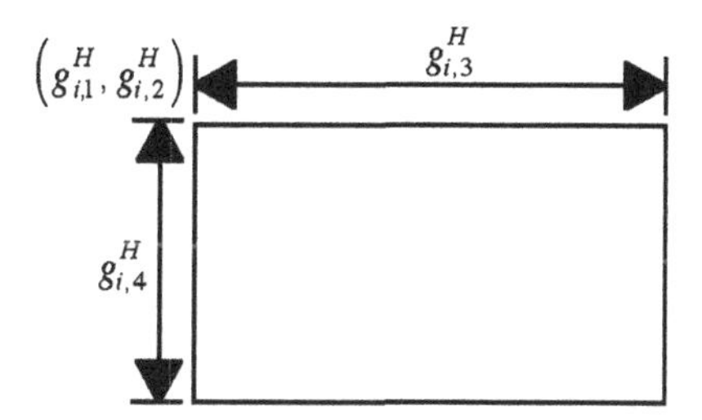

Fig. 13.4 A template used for human detection in SSGA-H

problem. The fitness value of the best individual is used as the perceptual information inputted to SN-I for the perception of the human detection. The facial direction can be approximately extracted by using the relative positions of human hair and human face. We use the relative position of the COG of areas corresponding to human hair and human face. The relative positions of the COG against the central position of the detected face region are used as the input to SN-I for extracting the human facial direction.

13.2.6 Object Detection

We focus on color-based object detection with SSGA-O based on template matching. The shape of a candidate template is generated by the SSGA-O. Figure 13.5 shows a candidate template used for detecting a target where the jth point $g_O^{i,j}$ of the ith template is represented by $(g_O^{i,j} + g_O^{i,j}\cos(g_O^{i,j+1}), g_O^{i,2} + g_O^{i,j}\sin(g_O^{i,j+1}))$, $i = 1, 2, \ldots, n$, $j = 3, \ldots, 2m + 2$; O_i $(= (g_O^{i,1}, g_O^{i,2}))$ is the center of a candidate template on the image; n and m are the number of candidate templates and the searching points used in a template, respectively. Therefore, a candidate template is composed of numerical parameters of $(g_O^{i,1}, g_O^{i,2}, \ldots, g_O^{i,2m+2})$. We used an octagonal template in this chapter ($m = 8$). The fitness value of the ith candidate template is calculated as follows.

$$f_i^H = C_{Skin}^H + C_{Hair}^H + \eta_1^H \cdot C_{Skin}^H \cdot C_{Hair}^H - \eta_2^H \cdot C_{Other}^H \tag{13.3}$$

where η^O is the coefficient for penalty $(\eta^O > 0)$; C_{Target}^O and C_{Other}^O indicate the numbers of pixels of a target and other colors included in the template, respectively. The target color is selected according to the pixel color occupied mostly in the template candidate. The fitness value of the best individual is used as the perceptual information inputted to the SN-I for perception of the object detection. Furthermore, the number of acute angles in a template candidate is used as the perceptual information inputted to SN-I for extracting the shape of the detected color object. We use three SN-Is for extracting the shapes of the circle, triangle, and rectangle.

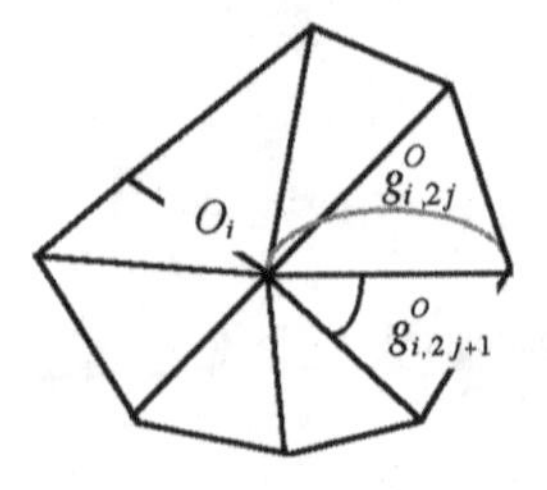

Fig. 13.5 A template used for object recognition in SSGA-O

13.2.7 Hand Motion Recognition

In order to the natural communication with a human, a robot should recognize basic human hand motions to understand what a human performs with objects. Here we assume a human has and moves an object. Consequently, the human hand motions are regarded as the movement of an object. The change of a human hand position, i.e., the velocity extracted on the image, is used as the perceptual information inputted to the SN-I. The SN-I corresponding to a specific moving direction fires according to a human hand motion.

13.3 Architecture of Prediction Based Perceptual System

13.3.1 Input Layer Based on Spiking Neurons

Various types of artificial neural networks have been proposed to realize clustering, classification, nonlinear mapping, and control [1, 4, 9–12, 21, 23, 25, 32]. Basically, artificial neurons are classified into pulse-coded and rate-coded neuron models from the viewpoint of abstraction level [21]. A pulse-coded neuron approximates the dynamics introduced from the ignition phenomenon of a neuron, and simulates the propagation mechanism of the pulses between neurons. A pulse-coded neuron is often called a spiking neuron, and we use the term "spiking neuron" in the following. Hodgkin-Huxley model is one of the classic neuronal spiking models, and uses four differential equations. An integrate-and-fire model with a first-order linear differential equation is known as a neuron model of higher abstraction levels. A spike response model is slightly more general than the integrate-and-fire model, because the spike response model can choose kernels arbitrarily [21]. On the other hand, a rate-coded neuron neglects the pulse structure, and is considered as one of the much higher level of abstraction. A neuron model of McCulloch-Pitts is well known as a famous rate-coded model, and Perceptron was proposed as a rate coded neural network [1].

An important feature of spiking neurons is the capability of temporal coding. In fact, various spiking neural networks have been applied to memorizing spatial and temporal context. Therefore, we apply spiking neurons to representing the time series of perceptual information. And, we use a simple spike response model to reduce computational cost. First of all, the internal state of the ith SN-I, $Ih_i(t)$, is calculated as follows;

$$Ih_i(t) = \gamma_1 \cdot Ih_i(t-1) + Ih_i^{ref}(t) + X_i^I(t) \tag{13.4}$$

where γ_1 is the discount rate ($0 < \gamma_1 < 1.0$). The internal state is calculated by the external input, $X_i^I(t)$, and the refractoriness of the SN-I, $Ih_i^{ref}(t)$. In general, the outputs from other spiking neurons are also used in the calculation of the internal state, but in this model, we don't use the connection among spiking neurons because we focus on the temporal patterns of spike outputs. The number of neurons is N_I ($N_I = 14$). The external input, $X_i^I(t)$ is the perceptual information from perceptual modules. For example, $X_2^I(t)$ is calculated based on fitness value f_i^H in the human detection module according to the spike output from the 2nd SN-S. This value is inputted to the 2nd SN-I corresponding to the perception of the human detection (see Table 13.1 and (2)). When the SN-I is fired, R is subtracted from $Ih_i^{ref}(t)$ in the following,

$$Ih_i^{ref}(t) = \begin{cases} \gamma_2 \cdot Ih_i^{ref}(t-1) - R & \text{if } p_i^I(t-1) = 1, \\ \gamma_2 \cdot Ih_i^{ref} & \text{otherwise} \end{cases} \tag{13.5}$$

where γ_2 is the discount rate ($0 < \gamma_2 < 1.0$); $p_i^I(t)$ is the output of the ith SN-I at the discrete time t. When the internal state of ith SN-I is larger than the predefined threshold for firing, θ_i^I, a spike is outputted as follows;

$$p_i^I(t) = \begin{cases} 1 & \text{if } Ih_i^I(t) \geq \theta_i^I, \\ 0 & \text{otherwise.} \end{cases} \tag{13.6}$$

The presynaptic spike output is transmitted to the next layer of the C-layer according to an excitatory postsynaptic potential (EPSP). The EPSP of the SN-I, $Ih_i^{EPSP}(t)$, is calculated as follows;

$$Ih_i^{EPSP}(t) = \sum_{n=0}^{T} \kappa^n p_i^I(1-n) \tag{13.7}$$

where κ is the discount rate ($0 < \kappa < 1.0$); T is the time sequence to be considered.

13.3.2 Clustering Layer Based on Unsupervised Learning

The spatio-temporal context in the perception can be represented by the simultaneous spikes and sequential spikes of SN-Is. The C-layer performs the clustering based on unsupervised learning of spike outputs. Cluster analysis is used for grouping or segmenting observations into subsets or clusters based on similarity. A self-organizing map (SOM), growing neural gas, K-means algorithm, and Gaussian mixture model have been often applied as clustering algorithms [11, 12]. SOM can be used as incremental learning, while K-means algorithm and Gaussian mixture model need all data in the learning phase (batch learning). SOM is often applied for extracting a relationship among data, since SOM can learn the hidden topological structure from the data. Furthermore, a growing neural gas can update the number of nodes according to the structure of a data set.

The sequence of the spikes from SN-Is is used as the inputs for clustering by the unsupervised learning. The input to a neuron of the C-layer is a set of the EPSP from SN-Is of the I-layer,

$$X^C = (Ih_1^{EPSP}, Ih_2^{EPSP}, \ldots, Ih_{N^I}^{EPSP}) \quad (13.8)$$

where N^I is the number of inputs (the number of SN-Is at the I-layer). Here, a two-dimensional structure is used for representing the neighboring relationship among neurons of the C-layer. Next, the dth neuron minimizing the Euclidean distance between the input vector and the jth reference vector, $r_j = (r_{j,1}, r_{j,2}, \ldots, r_{j,N^C})$, is selected by

$$d = arg \min_j \{\|X^C - r_j\|\}. \quad (13.9)$$

The number of neurons (reference vectors) at the C-layer is N^C. Furthermore, the reference vector of the jth neuron is trained by

$$r_j \leftarrow r_j + \xi^C \cdot \xi_{d,j}^C \cdot (X^C - r_j) \quad (13.10)$$

where ξ^C is the learning rate ($0 < \xi^C < \xi_{max}^C < 1.0$); $\xi_{d,j}^C$ is the neighborhood function ($0 < \xi_{d,j}^C < 1.0$). In the unsupervised learning algorithm, the parameters of ξ^C and $\xi_{d,j}^C$ are gradually reduced toward 0 according to the learning state.

13.3.3 Prediction Layer and Perceptual Module Selection Layer

The robot learns the transition patterns among perceptual modes. The output from the jth neuron of the C-layer is used as the input to the neuron of the P-layer and calculated by,

$$y_j^C(t) = \exp(-\|X^C - r_j\|). \quad (13.11)$$

The number of neurons at the P-layer is N^P. The output of the kth neuron of the P-layer is

$$y_k^P(t) = \sum_{j=1}^{N^C} w_{j,k}^C y_j^C(t), \quad \text{with } x_k^P(t) = y_j^C(t) \quad (13.12)$$

where $w_{j,k}^C$ is the connection weight between the jth neuron of the C-layer and kth neuron of the P-layer. The connection weight $w_{j,k}^C$ indicates the strength of the mode transition.

Next, we explain how to select perceptual modules for the next perception. The internal state of the lth SN-S, $Sh_l(t)$ is calculated as follows,

$$Sh_l(t) = \gamma_3 \cdot Sh_l(t-1) + x_l^S(t), \quad (13.13)$$

Table 13.2 The relationship between inputs and outputs of each layer. $p_i^I(t)$ and $p_l^S(t)$ are spike outputs (0 or 1); the other parameters are continuous values; $X_i^C(t)$ is calculated by EPSP according to spike outputs of $p_i^I(t)$

Layer	Index	Input	Output
I-Layer	i	$X_i^I(t)$	$p_i^I(t)$
C-Layer	j	$X_j^C(t)$	$y_j^C(t)$
P-Layer	k	$X_k^P(t)$	$y_k^P(t)$
S-Layer	l	$X_l^S(t)$	$p_l^S(t)$

$$x_l^S(t) = \xi^{S1} \cdot \sum_{N^P}^{k=1} w_{k,l}^P \cdot y_k^P(t) + q^S \tag{13.14}$$

where γ_3 is the discount rate ($0 < \gamma_3 < 1.0$); ξ^{S1} is the learning rate ($0 < \xi^{S1} < \xi_{max}^{S1} < 1.0$); $w_{k,l}^P$ is the connection weight between the kth neuron of the P-layer and lth neuron of the S-layer; q^S is the regular input to a SN-S ($q^S = 1 - \xi^{S1}$). The 1st term of (13.14) represents the sum of outputs from the P-layer. The number of neurons at the S-layer is N^S ($N^S = 4$). If $Sh_l(t)$ is higher than the threshold θ^S, SN-S outputs $p_l^S(t)$ as the same manner with SN-I (see (6)). As a result of the spike output, its corresponding perceptual module is selected. Table 13.2 shows the relationship between inputs and outputs of each layer.

13.3.4 Update of Learning Rate for Perceptual Module Selection

In the beginning of learning, all of the perceptual modules are easily selected to extract various perceptual information because the learning rate S1 is small and the input to a SN-S q^s ($q^s = 1 - \xi^{S1}$) is nearly equal to 1. The learning rate ξ^{S1} gradually increases according to the learning state. As a result, the internal state of lth SN-S comes to increase or decrease dependent on the outputs from the P-layer. According to this update process, the robot gradually starts to select the perceptual modules needed for extracting necessary perceptual information based on prediction results. Therefore, the robot can control the sampling interval of the perceptual modules and can concentrate on extracting necessary perceptual information.

The learning rate ξ^{S1} is updated according to prediction difference d^P. When the perceptual mode transits to the other mode, the prediction difference d^P is calculated by a Gaussian membership function to evaluate the learning state of SN-S as follows,

$$d^P = 1 - \exp\left(-\frac{\sum_{k=1}^{N^C} w_{j',k}(r_{j'}(t) - r_k(t-1))^2}{2\sigma^2}\right) \tag{13.15}$$

where σ^2 is the variance and $w_{j',k}$ is the connection weight between the j'th neuron of the C-layer and kth neuron of the P-layer. The j' indicates the selected neuron N_O in the C-layer at time $t-1$. The standard deviation σ determines the shape of the curve of the Gaussian membership functions. The value of the standard deviation

σ is determined by preliminary experiments. Since the prediction difference is not an error index, this value does not converge to zero.

The learning rate ξ^{S1} is increased if the prediction difference d^P is lower than the threshold θ^P in the P-layer

$$\xi^{S1} \leftarrow \xi^{S1}/\alpha^{S1}. \tag{13.16}$$

Otherwise, the learning rate is decreased,

$$\xi^{S1} \leftarrow \beta^{S1} \cdot \xi^{S1} \tag{13.17}$$

α^{S1} and β^{S1} are the update rates for the learning rate. The learning phase is stopped when the learning rates converges to a certain value, e.g., ξ^{S1}_{max}. The number of training times till the convergence varies dependent on the complexity and length of human behaviors to be predicted.

13.3.5 Learning for Prediction and Perceptual Module Selection

The learning of connection weights between the C-layer and P-layer is performed by the Hebbian rule. Since the prediction is performed by the changing pattern of clustering state of perceptual modes, the connection weight between the jth neuron of the C-layer and kth neuron of the P-layer is updated according to the temporal input values at t and $t-1$

$$w^C_{j,k} \leftarrow (1-\xi^{S1})w^C_{j,k} + \xi^P \cdot y^C_j(t-1)y^C_k(t) \tag{13.18}$$

where ξ^P is the learning rate ($0 < \xi^P < \xi^P_{max} < 1.0$). The connection weights are normalized as follows,

$$w'^C_{j,k} = \frac{w^C_{j,k}}{\sum_{n=1}^{N^C} w^C_{j,k}}. \tag{13.19}$$

Next, the $w'^C_{j,k}$ is substituted for $w^C_{j,k}$. The learning rate ξ^P is updated according to the learning state of the C-layer because the boundary among clusters is under the learning if the clustering is not efficiently performed. In this way, the learning of the P-layer is performed according to the accuracy of the C-layer.

Next, we explain how to update the connection weights between the P-layer and S-layer. The connection weight, $w^P_{k,l}$ is updated according to the reference vectors as follows,

$$w^P_{k,l} \leftarrow (1-\xi^{S2}) \cdot w^P_{k,l} + \xi^{S2} \cdot \frac{\sum_{n \subseteq PM_l} r_{j',n}}{S_l} \tag{13.20}$$

where PM_l and S_l are the set and the number of features extracted in the lth perceptual module, respectively; ξ^{S2} is the learning rate ($0 < \xi^{S2} < \xi^{S2}_{max} < 1.0$).

13.4 Experimental Results

This section shows experimental results of the proposed system to the interaction between a person and the partner robot; Hubot. The important feature of the proposed method is in the control of perceptual modules. In order to discuss the effectiveness of the proposed method, we conducted several experiments on the learning of the robot based on the human interaction composed of different behavior patterns.

13.4.1 Clustering for Prediction

The clustering is the most important in the system, because the clustering determines the boundaries among perceptual modes used for the prediction in the visual perception. If the clustering result is different, the prediction also becomes different. The size of an image (X,Y) is (160, 120). Table 13.3 shows parameters used in this experiments. First, we conducted a preliminary experiment on the relationship between the number of neurons of the C-layer and the clustering performance. The number of neurons of the C-layer determines the clustering performance which affects the prediction performance directly.

In this experiment, the robot observed human behaviors composed of (a) talking on a green ball in a human hand, (b) writing some sentences concerning with the ball on the whiteboard, and (c) erasing them from the whiteboard. The person repeats the series of behaviors 15 times for 40 minutes (Fig. 13.6), but the person sometimes talks to the robot without having these items (Fig. 13.6(d)). These behaviors are taken into a movie, and the robot learns by observing the movie for the performance comparison at different numbers of neurons. The numbers of neurons in the C-layer are set at 9, 16, 25, 36, and 49 for the performance comparison in the experiment. Figure 13.7 and 13.8 show the change of the learning rates in each case. In cases

Table 13.3 Parameters used in experiments

Parameter	Value
Threshold of the I-layer θ_i^I	1.0
Discount rate of the I-layer γ_1	0.9
Discount rate of the I-layer γ_2	0.1
Discount rate of the I-layer κ	0.96
Maximum of the learning rate in the C-layer ξ_{max}^C	0.3
Maximum of the learning rate in the P-layer ξ_{max}^P	0.3
Threshold of the prediction difference θ^P	0.6
Maximum of the learning rate in the P-layer ξ_{max}^{S1}	0.8
Maximum of the learning rate in the P-layer ξ_{max}^{S2}	0.3
Discount rate of the S-layer γ_3	0.9
Threshold of the S-layer θ^S	0.7
Standard deviation of the prediction difference σ	1.83

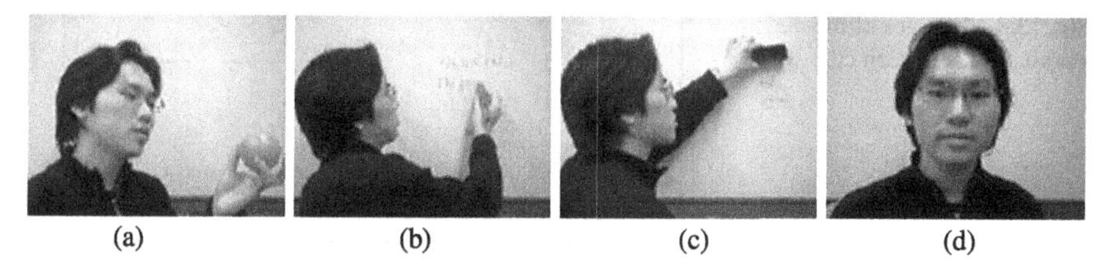

Fig. 13.6 Snapshots of human behaviors

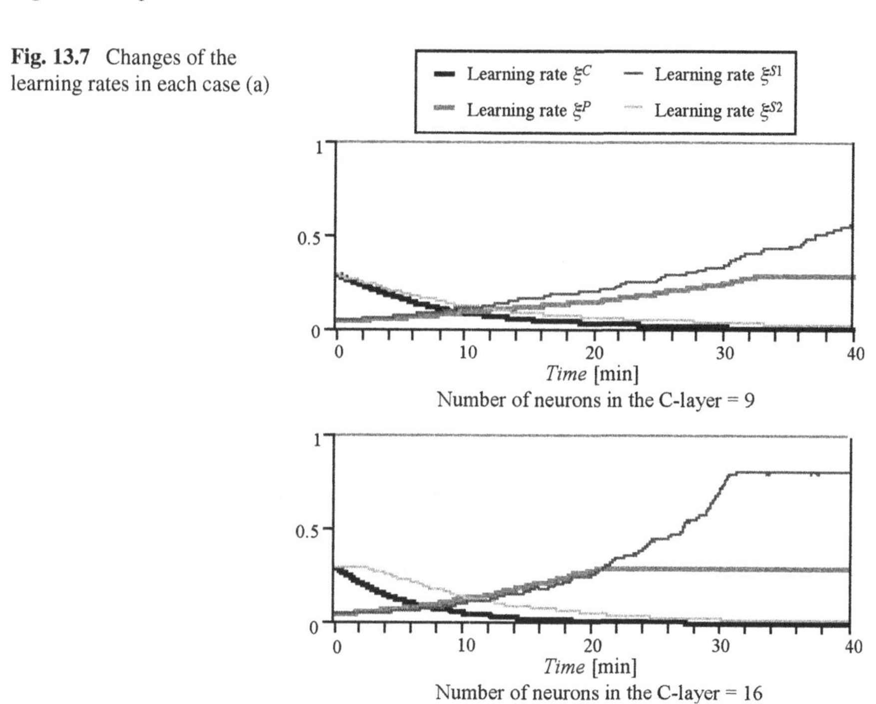

Fig. 13.7 Changes of the learning rates in each case (a)

of 9 and 16 neurons, the time required to reach the convergence is much longer than that of 25, 36, and 49. The learning rate C is increased more as the Euclidean distance in the clustering is shorter. If the number of neurons is less than the minimal number required for the learning, the clustering is not performed well. On the other hand, the learning rates ξ^{S1} and ξ^{S2} used for the perceptual module selection have similar tendency with C, but their convergence are later than that of C, because the learning of the perceptual module selection cannot be performed without the learning of clustering and prediction. Table 13.4 shows the number of neurons not contributed much in the C-layer in the case that the usage rate for the 8 minutes after the learning is less than the value that is 10 divided by the number of neurons. All neurons are used for the prediction of human behaviors in the case of 9 neurons, while the number of not contributed neurons is much larger than other cases. It is obvious that the learning speed is faster if the number of neurons is enough or redundant for the distribution of the data. However, the human action segment used

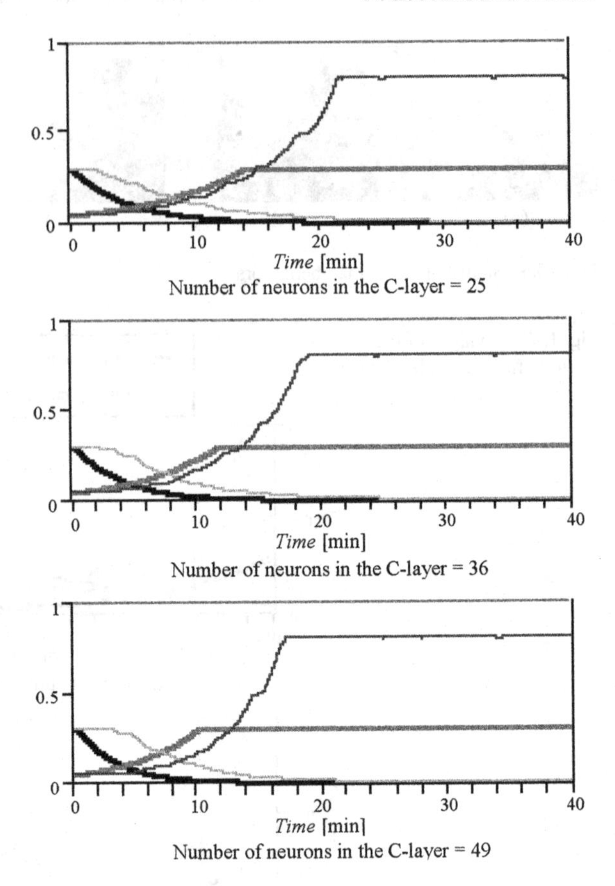

Fig. 13.8 Changes of the learning rates in each case (b)

Table 13.4 Number of neurons not contributed much in the C-layer

Neurons of the C-layer	Neurons not used
9	0
16	2
25	3
36	5
49	11

for prediction might become too small even if the number of neurons is redundant. The suitable number of neurons is 25 in this experiment. The size of human action segments to be predicted depends on the number of neurons in C-layer, but it is very difficult to correspond with the size of action segments in the robot perception level as that in the human perception beforehand.

13.4.2 Real-Time Learning in Interaction

Next, we conducted an interaction experiment by using other human behaviors for 20 minutes. In the experiment, a human showed a ball (Fig. 13.9(a)–(c)) and drank a cup of water (Fig. 13.9(d)–(f)) while speaking English. If we assume this learning as an imitative learning composed of model observation and model reproduction, the robot first performs the model observation (learning phase). As the prediction becomes correct, the robot performs the model reproduction (testing phase). The learning phase ends when the learning rates converge to the predefined value. In the testing phase, the robot has the blue ball in the left hand and the red cup in the right hand. The robot moves its arms based on the prediction if the robot predicts that a human shows a blue ball, or a human lifts a red cup. Otherwise, the robot puts its hands down.

Figure 13.10 shows the changes of the prediction difference and the learning rates. The prediction difference is gradually decreasing, but it does not converge to zero and sometimes becomes large because the human does not repeat the completely same behavior to the robot in the real-time experiment. In fact, the unpredictability is included necessarily in the human interaction. On the other hand, each learning rate converges to the predefined value after 18 minutes learning. In this experiment, the training times of learning phase is about 12,000.

Figure 13.11 shows the reference vectors of neurons in the C-layer where the number of neurons in the C-layer is 9. The 1st neuron memorizes the perceptual mode that a human without any object is detected. The 5th and 8th neurons memorize the perceptual modes that a human face turns to the right and left, respectively.

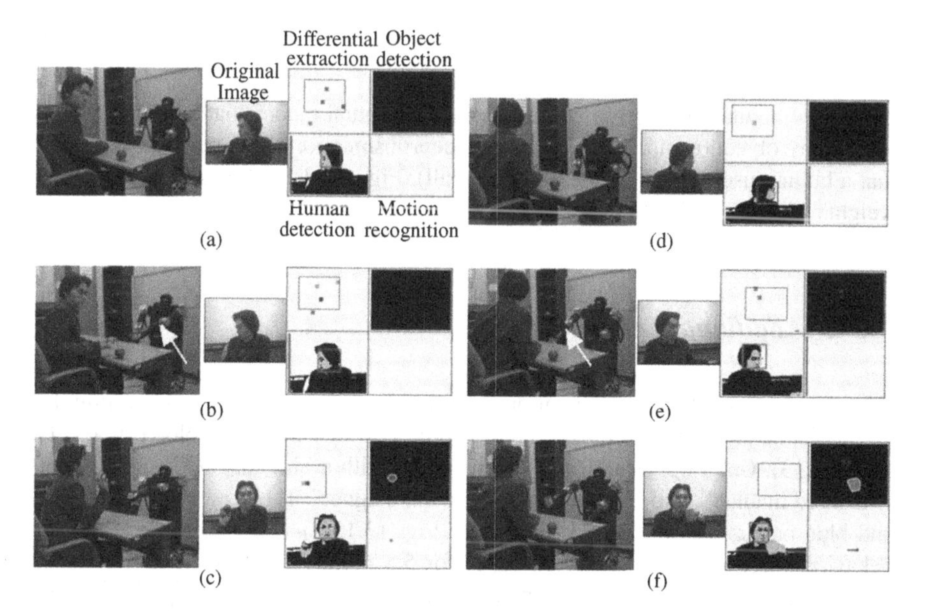

Fig. 13.9 Snapshots of image processing and predictive action by the robot in the learning phase

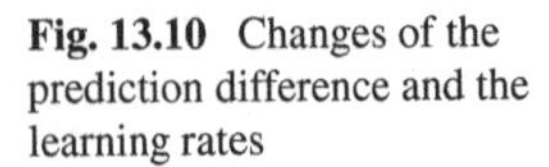

Fig. 13.10 Changes of the prediction difference and the learning rates

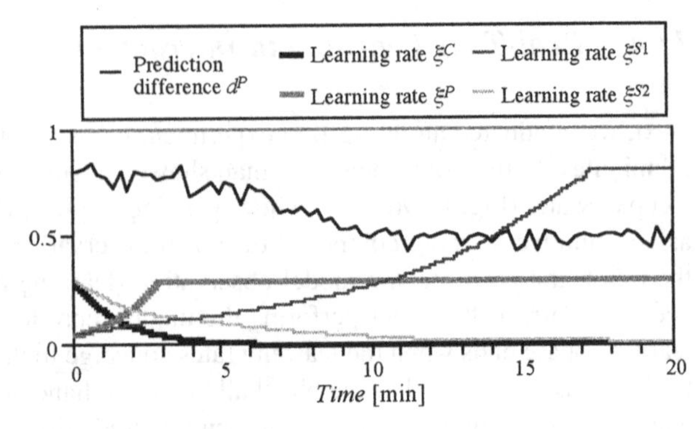

Fig. 13.11 Reference vectors of neurons in the C-layer

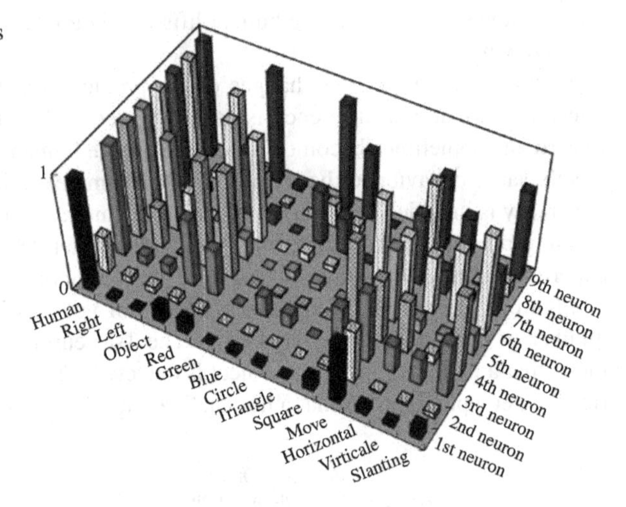

The 3rd, 4th, and 6th neurons memorize the perceptual modes that a human uses a red, square object (a cup). The 7th and 9th neurons memorize the perceptual modes that a human uses a blue, circle object (a ball). Figure 13.12 shows the connection weights for the transition of perceptual modes.

13.4.3 Additional Learning

Next, we conducted an experiment on the verification of the additional learning. In the experiment, a person showed three colored papers with three colored magnets (Fig. 13.13). Color combination is "green paper with a blue magnet", "blue paper with a red magnet", and "red paper with a green magnet". The person showed green and blue papers during first 23.5 minutes (Fig. 13.13(a)–(i)). Next, as a different behavior, the person showed a red paper for 5.5 minutes (Fig. 13.13(j)–(l)). And finally, the person showed the green and blue papers again. In this experiment, the training times is about 28,000. In this experiment, the number of neurons in the

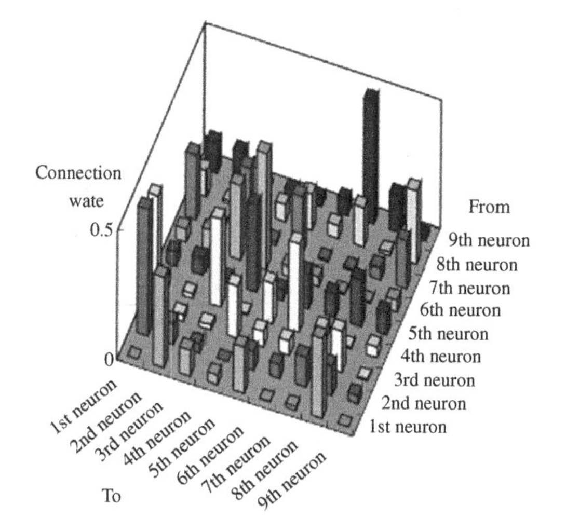

Fig. 13.12 Connection weights for perceptual mode transition

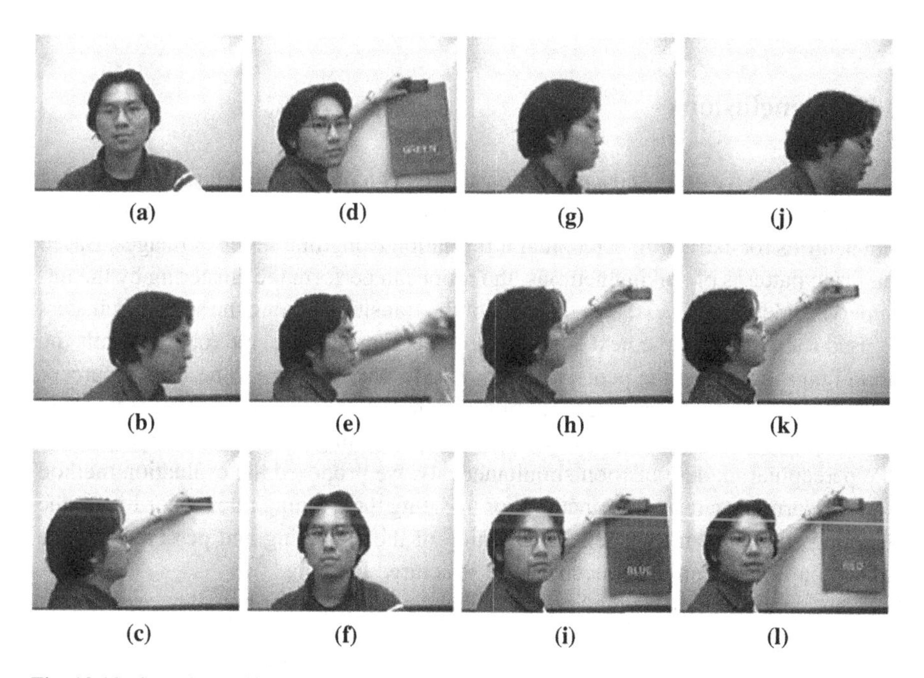

Fig. 13.13 Snapshots of human behaviors

C-layer is 25. Figure 13.14 shows the changes of the prediction difference. First, the robot learned two behaviors of the human showing green and blue papers and the prediction difference decreased gradually. After 23.5 minutes, when the person showed a red paper, the robot could not predict it, and the prediction difference increased. But the robot learned the new human behavior and the prediction difference started to decrease again. After 5.5 minutes, when the person showed green

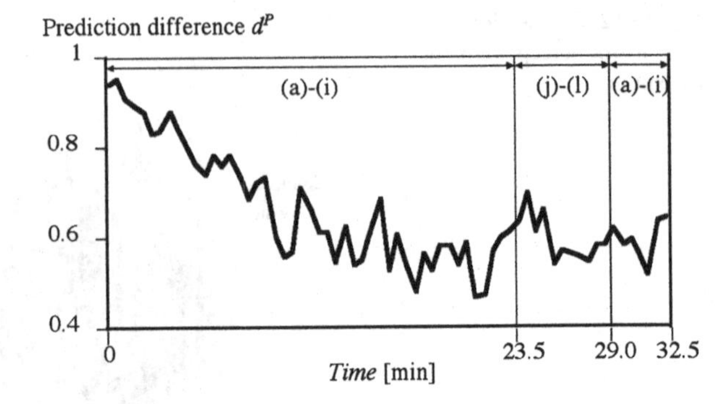

Fig. 13.14 Changes of the prediction difference

and blue papers again, the robot could predict it, and the prediction difference did not increase severely. The acquired clusters for the prediction of the green and blue papers are maintained.

13.5 Conclusions

In this chapter, we proposed a prediction-based perceptual system for a partner robot in order to realize the natural communication with a human. First, we proposed spiking neurons for extracting perceptual information from time series of images. Based on firing patterns of spiking neurons, the robot can perform the clustering by the unsupervised learning. Next, the sequence of the transition among the selected clusters is used for predicting the next perceptual mode. Furthermore, the robot controls the sampling intervals of the perceptual modules to pay attention to necessary perceptual information for the natural communication with the human. In order to perform the clustering of the perceptual modes and the learning of connection weights for the perceptual mode transition simultaneously, we proposed the evaluation method for the learning state and the method for updating the learning rates according to the learning state. The interdependent learning of the clustering and perceptual mode transition is based on the concept of the structured learning.

The proposed learning method did not evaluate whether or not all necessary perceptual information was extracted because it is one of unsupervised learning. Therefore, we intend to propose an interactive learning method between the robot and human by using supervised learning according to human evaluations and human reactions. We will use the human reaction to the action of the robot for evaluating how to extract perceptual information. As another future work, we will discuss how to construct the interrelation between a human and robot through long-term repeated interactions. As the prediction results become accurate, the robot can quickly behave suitably to the human. This indicates the constructed interrelation strongly restricts the coherent prediction of others. Therefore, we will propose a learning method for constructing the suitable interrelation with a human.

References

1. Anderson, J., Rosenfeld, E.: Neurocomputing: Foundations of Research. MIT Press, Cambridge (1988)
2. Bulthoff, H.H., Lee, S., Poggio, T., Wallraven, C.: Biologically Motivated Computer Vision. Springer, Berlin (2002)
3. Calinon, S., Billard, A.: Stochastic gesture production and recognition model for a humanoid robot. In: Proceedings of 2004 IEEE/RSJ International Conference on Intelligent Robots and Systems, pp. 2769–2774 (2004)
4. Carpenter, G., Grossberg, S.: Adaptive resonance theory. In: The Handbook of Brain Theory and Neural Networks, vol. 2, pp. 87–90 (2003)
5. Eysenck, M.: Psychology: an Integrated Approach. Longman, Harlow (1998)
6. Fukushima, K.: Neural network model for selective attention in visual pattern recognition and associative recall. Appl. Opt. **26**(23), 4985–4992 (1987)
7. Gibson, J.: The Ecological Approach to Visual Perception. Houghton Mifflin, Boston (1979)
8. Hawkins, J., Blakeslee, S.: On Intelligence. Owl Books (2005)
9. Jang, J., Sun, C., Mizutani, E.: Neuro-Fuzzy and Soft Computing. Prentice Hall, Upper Saddle River (1997)
10. Kartalopoulos, S., Kartakapoulos, S.: Understanding Neural Networks and Fuzzy Logic: Basic Concepts and Applications. Wiley-IEEE Press, New York (1997)
11. Kohonen, T.: Self-organized formation of topologically correct feature maps. Biol. Cybern. **43**(1), 59–69 (1982)
12. Kohonen, T.: The Self-Organizing Map, vol. 21, pp. 1–6. Elsevier, Amsterdam (1998)
13. Kubota, N.: Intelligent structured learning for a robot based on perceiving-acting cycle. In: Proceedings of the Twelfth Yale Workshop on Adaptive and Learning Systems, pp. 199–206 (2003)
14. Kubota, N., Nishida, K.: Fuzzy computing for communication of a partner robot based on imitation. In: Proceedings of 2005 IEEE International Conference on Robotics and Automation, pp. 4391–4396 (2005) (CD-ROM)
15. Kubota, N., Nishida, K.: Development of internal models for communication of a partner robot based on computational intelligence. In: Proceedings of 6th International Symposium on Advanced Intelligent Systems, pp. 577–582 (2005)
16. Kubota, N., Nishida, K.: The role of spiking neurons for visual perception of a partner robot. In: Proceedings of 2006 IEEE World Congress on Computational Intelligence, pp. 530–537 (2006) (DVD-ROM)
17. Kubota, N., Nishida, K.: Perceptual control based on prediction for natural communication of a partner robot. IEEE Trans. Ind. Electron. **54**(2), 866–877 (2007)
18. Kubota, N., Nishida, K.: Structural adaptation of prediction-based perceptual system. Int. J. Fact. Autom. Robot. Soft Comput. **1**(1), 19–26 (2008)
19. Kubota, N., Hisajima, D., Kojima, F., Fukuda, T.: Fuzzy and neural computing for communication of a partner robot. J. Mult.-Valued Log. Soft Comput. **9**(2), 221–239 (2003)
20. Kubota, N., Nojima, Y., Kojima, F., Fukuda, T.: Multiple fuzzy state-value functions for human evaluation through interactive trajectory planning of a partner robot. Soft Comput. A, Fusion Found. Methodol. Appl. **10**(10), 891–901 (2006)
21. Maass, W., Bishop, C.: Pulsed Neural Networks, pp. 3–53. MIT Press, Cambridge (2001)
22. Marek, A.J., Smart, W.D., Martin, M.C.: Learning visual feature detectors for obstacle avoidance using genetic programming. In: Proceedings of Genetic and Evolutionary Computation Conference, pp. 330–336 (2002)
23. Marr, D.: Vision: A Computational Investigation into the Human Representation and Processing of Visual Information. Henry Holt and Co., Inc., New York (1982)
24. Maurer, A., Billard, A.: Extended Hopfield network for sequence learning: application to gesture recognition. In: Proceedings of ICANN'05, pp. 493–498 (2005)
25. Miller, N., Miller, W., Werbos, P., Sutton, R.: Neural Networks for Control. MIT Press, Cambridge (1995)

26. Pfeifer, R., Scheier, C.: Understanding Intelligence. MIT Press, Cambridge (2001)
27. Rao, R.P.N., Meltzoff, A.N.: Imitation learning in infants and robots: towards probabilistic computational models. In: Proceedings of Artificial Intelligence and Simulation of Behaviors (2003)
28. Shaw, R.: The agent-environment interface: Simon's indirect or Gibson's direct coupling? Ecol. Psychol. **15**(1), 37–106 (2003)
29. Sperber, D., Wilson, D.: Relevance: Communication and Cognition. Blackwell, Oxford (1995)
30. Steels, L.: Evolving grounded communication for robots. Trends Cogn. Sci. **7**(7), 308–312 (2003). doi:10.1016/S1364-6613(03)00129-3
31. Syswerda, G.: A study of reproduction in generational and steady-state genetic algorithms. Found. Genet. Algorithms **94**, 101 (1991)
32. Trevor, H., Robert, T., Jerome, F.: The Elements of Statistical Learning: Data Mining, Inference, and Prediction, vol. 1, pp. 371–406. Springer, New York (2001)
33. Turvey, M., Shaw, R.: Ecological foundations of cognition, I: symmetry and specificity of animal-environment systems. J. Conscious. Stud. **11**(12), 95–110 (1999)
34. Wada, K., Shibata, T., Saito, T., Tanie, K.: Psychological and social effects of one year robot assisted activity on elderly people at a health service facility for the aged. In: Proceedings of 2005 IEEE International Conference on Robotics and Automation, pp. 2796–2801 (2005) (CD-ROM)
35. Yokoya, R., Ogata, T., Tani, J., Komatani, K., Okuno, H.G.: Experience based imitation using rnnpb. In: Proceedings of IEEE/RSJ International Conference on Intelligent Robots and Systems, pp. 3669–3674 (2006)

Index

H. Liu et al. (eds.), *Robot Intelligence*,
Advanced Information and Knowledge Processing,
DOI 10.1007/978-1-84996-329-9, © Springer-Verlag London Limited 2010